# LIFE'S SECOND HALF

Books by Jerome Ellison

*The Run For Your Money*
*The Prisoner Ate a Hearty Breakfast* (a novel)
*The Dam* (a novel)
*John Brown's Soul* (a novel)
*Report to the Creator*
*God On Broadway*
*A Serious Call to an American Evolution*
*The Life Beyond Death* (with Arthur Ford)
*The Light That Heals: A Modern Testimony*
*The Last Third of Life Club*
*Life's Second Half*

# LIFE'S SECOND HALF
## The Pleasures of Aging

Jerome Ellison

The Devin-Adair Company
Old Greenwich, Connecticut

Canadian agent: Beaverbooks Ltd., Pickering, Ont.

Library of Congress Catalogue Card No. 77-92703
ISBN: 0-8159-6116-2

Printed in the United States of America

If you're going to have a social life,
why have one based on small talk?
Why not a social life based on large talk?

—Phenix

# Contents

# 1 One Man's Meat

CALL ME JERRY. All you need to know about me is that I've been through the mill. I wouldn't be going into this business about myself at all, except for you. Together with some friends of mine I have made some discoveries about aging that our publisher thinks could benefit a lot of people. I have been nominated to do the writing part. When it was mostly written our publisher said, "But who is this Jerome Ellison who's doing all the talking about aging? People will want to know."

This seemed reasonable to me. How can you know about swimming, or about aging, or about anything else, unless you've done some of it? All you really need to know about me I have already told you. But the publisher thinks there should be more, and he is a reasonable man. "Something about your background, your life experience," he said. And so, for you, I oblige. Believe me, there is no egotism in it. What I am, I am.

My life has been wonderful and terrible. It began in Maywood, Illinois, seventy years ago. I can remember days from my childhood when the whole earth was on fire with delight, and days when I wished I would die.

At that time Maywood was a small town with its own identity and dignity, a remote suburb of Chicago. By the time I was eleven I had thoroughly cruised every foot of the place on my bicycle, usually riding no-hands, and took pride in the fact that I knew people on every street from First Avenue, where the town began, to Seventeenth, where it ended. To the east of town there was a great stretch of woods along both banks of the green, twisting, lazy Des Plaines River. Though the stream was already polluted, the lucky angler could still fetch up a catfish from it from time to time. The woods, still unspoiled, were full of marvels. I loved birds, and one time experienced a small ecstasy at spotting a scarlet tanager, a rare sight for those parts.

My father was an earnest, steadfast, faithful, plodding, violent, gentle man, who, though industrious and inventive and an adequate provider, could never make a big go of business because of his quixotic notion that business was for service. He used to beat me when he thought I needed or deserved it, and then be sorry for having done so. It took me a long time to understand that these often uncalled-for thrashings were for him a means of working out the frustrations of what we now know as "mid-life crisis." Except for these passing frenzies he was patient, generous, and wise.

Both my parents set considerable store by genteel respectability, or at least the appearance of it, and that may partly account for my own lifelong bent toward a mild bohemianism. Though my mother's education ended with high school, it was she who opened my eyes to the reality of the world of culture. She took me to see the great paintings hanging in the Art Institute of Chicago, arranged for piano lessons (so that one day I could play for an admiring audience in a gin mill!), and peopled the house with fascinating books by the great authors, making an avid reader of me while I was still in grade school. A meticulous housekeeper, she had nevertheless a fey side. To pull a wishbone with her was to experience the reality of wishing. She could impress you with the importance of always seeing the new moon over your *left* shoulder, because it was bad luck to see it over your right. I had a younger brother and sister I adored, cherished, cared for, hated, and fought with. Little five-feet-one mother bossed, fed, protected, and laundered us all, indefatigably, and got us to Sunday School regularly.

My childhood idyll had its rude shocks. By the time I was twelve two of my playmates had been killed violently—one in a fall from a building under construction where we'd all been playing, another by a train he didn't see coming. My grandfather suffered a stroke and died in our home, a lingering and pitiful death. My father's temper grew shorter as business grew worse, until the confrontations with his violence sometimes sent me into month-long glooms. My mother nearly died of hemorrhage giving birth to my brother in our home, and had a slow, desperate recovery. An aunt, an uncle, and three of my cousins were killed when their car was struck by a train while they were on their way to visit us. It was a kind of advance warning, I guess, that you can never know what to expect of life.

"I like coffee, I like tea," we used to chant when we were kids, "I like the girls, and the girls like me." This routine automatism of animal life called sex seems to strike us all as a marvelous and unique discovery each of us has made for the first time in human history, all on our own. In my own life it was so overwhelming as sometimes to be beyond my control, and I was well on in years before I learned to manage it.

During my first two years at the University of Michigan, when I studied engineering, I was under the remembered influence of my father, who held a degree in engineering and invented, installed, and serviced furnaces for large buildings. The last two years, when I switched to the study of literature, showed clearly the emergence of my mother's side. I graduated in the first year of the Great Depression—1930—and headed for New York City, where I earned my living for the next twenty-five years. It was often a chancy business. I have fed myself by serving as truck driver, construction worker, mechanical draftsman, free-lance cartoonist, free-lance writer, novelist, department-store clerk, publisher, essayist, short-story writer, editor, and teacher. At times in my life I have done well enough financially, but money slips easily through my fingers. My father noted this tendency and warned me about it. "You'll scratch a poor man's ass," he predicted, "all the days of your life."

His prediction, fortunately, has not come true. I have never been really rich, though there have been some affluent decades, and never really poor, though I have often glimpsed the bottom of the barrel. Once I was on my own, I waded into life with a rage to live-it-up so intense it sometimes scared me; I was afraid it would burn me out, and it almost did. During one thirty-year stretch I realized every single youthful ambition, both worthy and unworthy, I had ever entertained, even the most preposterous ones. My fulfilled professional opportunities exceeded my wildest dreams. I bought a fast car, a sailboat, and an airplane, and put them through all the paces. I liked to take my plane out, high over the open ocean but in gliding distance of shore, and do loop-the-loops, snap-rolls, wingovers, and spins until I was dizzy with a glorious skyborne abandon. I nearly flew the wings off that thing, finally crashed it, and walked away from the wreck. I spent a glorious year going to all the places I'd ever wanted to go, all the way around the world, including the joy of speaking the two languages I'd picked up (besides a smattering of English) in the cities and villages where they alone were spoken. I loved all the good things about marriage, and was married to one woman for fifteen years and another for twenty-five. I love children almost worshipfully—I have fathered three and stepfathered three more—but in recent years have avoided being too emotionally involved in their lives. They can wind powerful cords around your heart, and when those cords jerk, it hurts.

The life of Riley?

There was another side. World War II came. I was sent to

London to see what I could do for our government to help heal the psychological wounds of a splintered world, and lived under buzz-bombing for a while. There I learned the unspeakable tragedy of the fact that all the tribal vendettas of all the ages are caused by insane ego trips that are unnecessary, and that *even a little* understanding could heal all the major wounds of the world. I had a vicious, dangerous, long-drawn-out bout with booze, and barely extricated myself from it with life, limb, and sanity.

During one stretch I endured such sharp depressions, even when things were going well, that I thought I must be crazy. They now have a name for it, "mid-life crisis," and mine was a mean one. I spent seven expensive years on the psychoanalyst's couch trying to get over it—and learned only the Freudian futility of trying to comprehend the human spirit by fooling around with body-brain circuits. One terrible night the phone rang, and I picked it up to learn that my sixteen-year-old son had been seriously injured in a boyish adventure climbing an electric railroad trestle. I rushed to the hospital and was there when he died, and have been a torn man ever since.

In an age of rapid technological change it often happens that a whole industry is wiped out almost overnight. That is what happened to my chosen industry of mass-circulation family magazine publishing. The arrival of television deprived such magazines of their advertising revenue, and they dwindled. I shifted to the academic world, became a professor, wrote a widely read and much-praised article suggesting that Big Ten university education was not quite all it could be—and was fired for it. One of my wonderful sons-in-law, who had become a close and admired friend, was killed in a crash with a car driven by a drunken driver. My beautiful sister, beautiful in physique and in spirit, died wretchedly of cancer in her fifty-fifth year, after having raised four children. I, who enjoyed perfect health and had never been in a hospital except to visit friends until I was sixty, was suddenly hit with two serious sessions of major surgery, and learned what it is to be old and sick almost to death.

A life wonderful and terrible, as I told you in the beginning. Now you know. So let's get on with the business at hand, which is the introduction to this book.

About twenty-five years ago I, with a few friends, took a grim look at what was ahead of us: pre-retirement stress, retirement (which was only a polite word for compulsory unemployment), aging, old age, terminal illness, and death. We didn't like the looks of it, nor did we like what we saw happening to some of those a little way ahead of us on this route. So we began to read up. Was there anybody who knew anything about this business?

No, we discovered, there was not. The pre-retirement stress "experts" were peddling the same fall-winter-and-spring line of goods they are now: "Health, Housing, Income, Estate Planning, Activities-and-Recreation, Family-and-Friend Relationships, Budgeting-and-Consumerism."

We knew from firsthand observation that these things, all of them perfectly sensible routine measures, were not an adequate answer to the issues that arise in a person's life during this later stretch of years. We had seen too many of our friends, men and women well provided in all these departments and with even more going for them, collapse into a later-life existence so miserable that they either degenerated into grouchy misanthropy or, in a frightening number of cases, killed themselves. There must be *something else* that eases one through mid-life crisis and pre-retirement stress, we reasoned, that makes the difference between a wretched old age and a happy one. But what *is* this elusive "something else"? The experts obviously didn't know, so we bypassed them and began looking on our own—reading, discussing, meditating, asking questions, searching.

After about twenty years of this we found a clue. During this period we became acquainted with some of the ideas of that remarkable psychologist and man, Dr. Carl G. Jung. In one of his books Jung hinted at a possible plan of action.

We now had something we knew everybody needed. We also knew that you can't keep a good thing unless you give it away. But how to give it away? Here Jung dropped a timely clue. Noting that "nervous breakdown in the forties or after is a very common occurrence" in the commercial countries, he went on to say: "The worst of it all is that intelligent and cultivated people embark upon the second half of life wholly unprepared." He meant, of course, *mentally*—psychologically and philosophically—unprepared. "Are there perhaps colleges for forty-year-olds which prepare them for their coming life and its demands as the ordinary colleges introduce our young people to a knowledge of the world and of life? No, there are none. Thoroughly unprepared we take the step into the afternoon of life with the false presupposition that our truths and ideals will serve us as hitherto. But we cannot live the afternoon of life according to the program for life's morning."

This, by now, we had found out for ourselves. We also had found a "program" for life's second half that really worked. And now Jung had given us the clue we'd been waiting for: *start colleges for people now in mid-life, to prepare their minds for what's coming.*

We lacked the resources, of course, to start a regular college-with-walls, so we did the next best thing. Starting with a weekly reading-and-discussion group meeting, we created a college-without-walls in which we functioned as our own professors and found our own textbooks. To give our discussions form, aim, and purpose without doctrinaire rigidity, we mimeographed a discussion guide outlining major areas of life that, when studied and understood, opened the way to the fully realized later life we'd all been in search of. This was published without fanfare in 1973.* Since then my "retirement" has turned into the busiest time of my life.

Letters poured in from all over the country, asking for help

***The Last Third of Life Club*, Pilgrim Press, 1973.

in starting the kind of neighborhood reading-and-discussion groups ("Phenix Clubs" we call them) that brought about such startlingly happy life-changes. Since then I have made three cross-country motor trips and dozens of special flights to meet with people who wanted to know how to get going. Newspapers and magazines wanted data for articles; groups wanted study materials. Soon there were a dozen functioning Phenix Clubs, with a newsletter to keep them in touch, and more clubs forming. The "college-for-the-second-half-of-life" idea was taking hold. We got up a package called "A Philosophy for the Middle Years," to be used in adult, evening continuing education courses in colleges and universities. Everywhere these principles are applied they produce the same benefits they did in the original group.

But still we have not reached the point of greatest need, the home of Mr. and Mrs. America; that is, Mr., Mrs., Ms. or Miss You and Me. We needed something that would make available what we have learned in a conveniently accessible form. That "something" is this book. May it bring to you the same release, relief, and joy that we have known, and lead you to the goal we now know is attainable—victory over age.

# 2 Would You Settle for 137 Birthdays?

WHEN YOU TALK ABOUT "growing old," such questions arise as "What is 'old'?" and "How old do you want to grow?"

Carl Jung fixed "mid-life crisis" at around thirty-five, which suggests that he didn't expect to live much beyond seventy (he actually lived to eighty-six). His fellow psychiatrist Erik Erikson found that life contained "crises" not only at thirty-five, but also at seven other ages, with a particularly rough one at sixty-five. Erikson also apparently assumed a lifespan not much beyond seventy.

Actually, nobody knows how long it is possible for a human being to live with full use of all functions and powers. It is a great deal longer than the sixty-five point at which we Americans have, until recently, officially declared ourselves useless. (I managed to hang on until sixty-seven before they made me walk the "retirement" plank.)

The future *could* be somewhat better in this respect than the present. Dr. Alex Comfort, an eminent British biologist who has done considerable research on the subject of aging, agrees that great changes are in store for us elders—but only *if*, I might add, we get off our present dead center and start to move. Older people, Dr. Comfort points out, differ from other minorities in that they are a minority we shall *all* eventually join. "We are on the verge of a total revolution in aging," he says. "It will come through the increasing control of the physical process of age change, *and perhaps even more through a change in image of ourselves that we adopt or impose on others in later life*" (italics mine).

There are places in the world where aging is a much easier business than the hash we make of it in the United States. There are mountain farming villages in Eastern Europe, in southern USSR, in Turkey, and in South America where ages well over a hundred and even up to a hundred twenty-five are not uncommon. No factor of drinking or smoking, or of this diet or that, has been identified as relevant. Two factors, however, have emerged as significant: daily outdoor exercise and an occupation that is valued by the community. Since primitive rural societies can utilize a person as long as he or she can wield a hoe or milk a cow, both these factors are abundantly present in such surroundings.

But you don't have to dwell in a mountain fastness or be a farmer to live long. I know American businessmen who are still very much "with it" in their nineties—and without so much as planting a bean!

My favorite longevity case is a man who was neither farmer nor salesman, and who lived an incredibly strenuous life for 137 years. I refer to the East Indian teacher known as Govindananda, who died in 1963. His career has been documented by the well-known British physicist, chemist, author, and educator, J. G. Bennett.

Govindananda began his education at the age of five with his grandfather, a famous astrologer, as teacher. At eighteen he took religious vows of renunciation and withdrew alone into the almost impenetrable depths of one of India's wildest jungles. With only "wild" animals—tamed by his kindness—as friends and only roots, fruits, and wild grains as food, he remained in solitude until he was fifty years old. He taught astronomy briefly, then began a journey around the world.

This took him the forty years between 1875 and 1915. He made 80 per cent of the land part of the journey on foot, traversing some of the most strenuous and hazardous routes on earth. He began with the standard 3,000-mile round of India's holy places, then walked over the Khyber Pass to Afghanistan, where he had an audience with the Aga Khan. After walking to Persia, where he visited the Shah, he trudged over the fearsome desert route to Mecca. After learning the rituals of Islam, he went on to the Holy Land and thence to Rome by boat, after first covering Turkey and Greece on foot.

He next met the German Kaiser and the Dutch Queen, before leaving for England, where he had eighteen long visits with one of his favorites—Queen Victoria. He was in the United States for three years, and became a friend of President Theodore Roosevelt. Turning southward, he walked through Mexico and the Peruvian Andes, sailed for Japan in 1913, and finally, in 1915, returned to India.

By now eighty-nine, he settled in Nepal and actively carried on his work as teacher of philosophy for another forty-eight years. In his one-hundred-seventh year he sought medical treatment for cancer of the gums. The treatment failing, he cured himself by methods known to yogis. Thirty years later he contracted pneumonia and shortly died. Lucid to the end, he died what might be called an ideal death. It seemed almost planned. He gave a final blessing—"Live right life, worship God"—lay down, rested his head on his right palm in his usual sleeping position, and simply stopped breathing.

So some people can, and do, live a long time. For my own part, I always expected to die young, so find myself entering my seventy first year in mild surprise. The idea of life as a longevity contest has never appealed to me. Prolonging life can mean merely prolonging misery, unless the *inner* life is a joyous one. We recently had a startling illustration of this in our town. One of our citizens, in fairly good physical condition for an old man, was celebrating his hundredth birthday. The newspaper boys were on hand with pencils poised and cameras ready, all set for the usual friendly smile and birthday word about how wonderful it is to see a whole century roll by. He let them have it. He hated life, he said, and had been wishing for years he'd die.

How is it that some people find life a joyful excitement for far longer than a century, while others can't even make eighty without finding it an unendurable misery? There are several reasons for this. One is the almost universal American custom of shun-the-aging. There are only two ways to deal with this curse on our land. One is to eliminate it, the other is to get yourself into a state of mind where it doesn't bother you. Neither course is possible until we understand just what it is and just how it goes about its dismal work.

# 3 Our Strange Tribal Rites

TO GROW OLDER, and do a good job of it, is one of the most rewarding undertakings you'll ever know—and one of the most strenuous. By "older" I mean that half century or so that comes after thirty.

An unfailing way to invite disaster, however, is to imagine that the middle and later years will just take care of themselves, that you don't have to *do* anything about them, that you just let nature take its course, that there's nothing to growing older but letting time go by and finding a nice hobby.

Not so. It is true that the rewards of doing the job properly are glorious beyond description. This book records how a great many men and women have obtained these rewards. It also contains directions for how you can find similar fulfillment. But it's my duty to point out that the penalty for *not* wading into the job with a will and a positive program is as horrendous as the reward is splendid. You get with it, in other words, or you suffer the consequences.

When we imagine that we live in a rational, civilized, advanced society, we make life unnecessarily hard for ourselves. We invite disillusionment, and that, at any age, is an embittering experience. We should not let our glittering technological window dressing lead us to think that we are a culturally advanced nation. In many respects we are a primitive tribe. Some of our practices are downright savage. Among these is our treatment of our aging population.

In certain tribes of the southern hemisphere a ritual called "shunning" is regarded as so terrible a punishment that it is reserved for only the most serious offenses against the tribal code. In some of the voodoo tribes shunning is the death penalty. If the medicine man finds sufficient cause, the person found guilty is ostracized, soon sickens, and shortly dies.

By far the most savage of all the world's shunning practices is the one in vogue in that large North American tribe called the United States of America. This tribe has a very high crime rate and curious notions about what constitutes a "crime." The most widespread of all the common crimes, by the tribal definition, is "aging." In one stratum of the tribe the felony is to become more than thirty years of age. Another group sets the permissible age at forty. There is almost unanimous agreement on the maximum allowable age. Anyone who presumes to advance beyond that age is pronounced guilty ipso facto. Immediate and drastic social retaliation—shunning—is promptly invoked.

Whatever the presumed benefits to the tribe, the damage to guilty individuals is calamitous, particularly among males. The suicide toll alone is staggering. Suicides among males over fifty—I'm speaking now of known suicides, excluding the uncounted suicides masked as "illness" or "accidents"—are nearly twice those of males under that age.

During the past thirty years the life expectancy of a forty-year-old American woman has increased by nearly eight years. The same blessing does not redound to the male. The average forty-year-old male won't live much longer than his father. So heavy is the death toll that so far as males are concerned it almost cancels out the benefits of a third of a century of medical progress. In ancient times the mystical age of forty-nine, the seventh time for the seventh year, was hailed as a specially favored time, the year of the beginning of wisdom. Today it is all too often the Year of the Coronary.

The reason may be found in another curious practice of the same tribe, called "making a bundle"—which is not the same as an earlier practice of the same tribe called "bundling." A tribesman "makes his bundle" by engaging with other males in the strenuously competitive activity of "bundle-making." The practice is just as deadly for the winner as for the loser, but he doesn't mind because he hasn't got any sense. His medicine men tell him repeatedly that the coronary rate is directly related to the intensity of economic competition. But they are not listened to, and the game is made more and more strenuous year by year.

The tribesman begins early in his forties to think he'd better start making his bundle while there's time. Any year now, he realizes, he may be found guilty of aging and his bundle-making days cut short. He frets himself into a state of "pre-retirement stress," which is a common preliminary to the coronary. Heart ailments, chronic depressions, and premature deaths are as common among those who have made big bundles as among those who have made small ones or none.

I could go on in this vein at some length and produce some pretty grim statistics, but the point has been made. Shunning obviously is not working very well and needs to be reexamined.

So let's reexamine it.

What is there about this practice that is so devastating to so many people? Simply this: The victim is excluded from the principal tribal ritual—work. We are a work-obsessed tribe, having no comprehension of the meaning of culture or the uses of leisure. Things are so arranged that our total lives revolve around one focus, the job. On this depends our economic life, our social life, our married life, our family life, the welfare and opportunity of our children. Take away the job and, most Americans feel, you take away life.

This is quite literally true in an enormous number of cases. My own mother was widowed during her sixties and took a job that lasted into her seventies, when she was discharged for the crime of aging. It was the death sentence. Within a year she had a stroke, and a few months later died. In a suburb where I lived for some years there was an outstandingly successful and well-liked local merchant. He retired in good health at sixty-five and turned the business over to his sons. He could stand it for just about a year. Then one evening, while neighbors from across the road were visiting, he quietly excused himself, went to the bathroom, and put a bullet through his head.

Sometimes the shunning begins even before retirement. A banking executive of my acquaintance announced his retirement a year in advance. From that time on, though still on the job, he found his counsel completely ignored in policy meetings.

There are, of course, other solutions for the sixty-five-year-old condition than premature death. One is angry protest. When Frank Nelson, an English professor at the University of Hawaii, reached the mandatory retirement age of sixty-five, he sued for the right to stay on the job. A circuit court ordered him reinstated. Since the retirement age for other state employees was seventy, it decided that Nelson had been denied equal protection of the law under the Fourteenth Amendment and ordered him back on the job.

The U.S. Supreme Court has not yet rendered a decision based purely on age discrimination. However, early retirement suits have been brought against airlines, bus lines, and the State of Massachusetts. These suits are based on the U.S. Age Discrimination in Employment Act and apply only to persons between the ages of forty and sixty-five.

Some retirees in the "angry protest" category, instead of suing, take refuge in the chronic grouch, a protest movement, or petty spite.

An alternative to the Sorehead Syndrome might be called the Resignation Reaction. This takes two forms: one active, one passive. The passive form is a dull, defeated giving up. The victim takes the attitude that "you can't beat City Hall, so what's the use of trying?" He accepts the general verdict that he's no good any more and quietly obliges the tribe by *becoming* no good. The active form of this reaction is a great show of buzzing around doing trivial things and trying to pretend that they're really fun while, underneath, the personality erodes away.

There is a third solution to the aging problem, far more satisfactory than either the Sorehead or the Resignation routes. It is called the Philosophical Solution. The rest of this book will be concerned with this alternative.

We can agree, I think, that this national practice of shunning for the crime of aging is barbarous, brutal, and cruel. Along with some of our other tribal practices, it makes us unworthy to be thought of as an advanced nation. We have brought upon ourselves a national crisis in this matter of aging, and one that will deepen as time goes on, since by the year 2000 half our population will be over fifty.

How does the tribe defend its shunning policy?

By claiming that mental acuity falls off so markedly with age that older people don't have the wits to hold down jobs. This claim is based on research done during the 1930s, which seemed to indicate that intelligence reached a peak at about age twenty-five, held to a high plateau for about ten years, then slowly declined. For more than a generation this has been regarded as unalterable fact.

It turns out not to be so. A long series of recent studies, conducted by K. Warren Schaie, Paul B. Baltes, and associates in the United States and confirmed by sociologists in Germany, have demonstrated beyond any question that the most important factors of intelligence continue to increase well into the seventies, and, if certain simple conditions are met, throughout life.

The error in the earlier research lay in taking "cross-sectional" samples instead of "longitudinal" ones. In the first case a lot of people of different ages are measured at the same time. But intelligence rating of an individual can be affected by the quality and duration of the education received. Since educational opportunity is increasing year by year, the younger people in the sample have received more education of better quality than the older, hence they make better scores.

In "longitudinal" studies a lot of people of various ages are measured at one time, then the same people are measured again a number of years later and the scores compared. This gives a true measure of improvement or decline while aging. The results show a steady *increase* in the most important facets of intelligence as the years go by.

In other words, you are smarter right now than you have ever been in your life. And you'll be smarter still ten years from now.

Clearly, we must revise our national attitude toward aging. The notion, so assiduously instilled by our merchants, our youth cult, our society leaders, and our industrial managers, that one is "through" at thirty, forty, fifty, sixty, seventy, or at any other given age must be eradicated. One is never "through." This truth can be firmly established in our national consciousness in a single generation if we all put our minds to it.

Carl Jung estimated that the second half of life—as experience, if not as exact chronology—begins at thirty-five. At about this time of life a vast restlessness overtakes most people. This is a sign of great changes of inner dynamics soon to come, similar to the upheaval of adolescence, but more profound because freighted with a greater load of life experience. This is what is now called the approach to "mid-life crisis." It is nature's warning to begin to prepare for the entirely *new* dimensions of the *second* half of life.

And how do you do that? Here a philosopher named Henry Nelson Weiman comes to our aid. You do it, he says, by something he calls "creative interchange." This is accomplished by activities such as reading this book, thinking about it, discussing it, and acting on such portions of it as apply to your case. Once the process has begun it gains its own momentum. Imagined terrors of old age, and even of death, begin to fall away, and one finds oneself living the second half of life—whatever may be our outer circumstances—as zestfully as we lived the first half, and possibly even more so.

Having reviewed the three usual alternatives of aging in our tribe, let us evaluate them.

The first alternative—the Angry Protest—arises in people who really would like to keep on at the same old job as long as they live. Denial of this fixed desire results in the resentment and the chronic grouch, which do not make either for good health or for pleasant company. Besides, we must ask ourselves whether being in a rut for forty years is any justification for staying in the same rut for forty more. Retirement, both early and late, is here to stay. Being sore about it will not change matters.

We need not tarry long over the second alternative—either in its passive form of defeated giving-up or in its active form of buzzing around on petty errands that don't make any mature sense. A human life cannot be lived unless it has meaning. If the meaning is exalted, it's an exalted life. If the meaning is trivial, it's a trivial life. If there's no meaning at all the person dies, first emotionally, then mentally, then physically.

That leaves us only the third alternative—the Philosophical Point of View—as the only viable route to happy aging. "Nature would not have given mankind this great extension of years beyond the reproductive bloom," Jung said, "unless it had a special purpose for the life of the species." When this special purpose is being fulfilled, one has a very good chance of living out one's life-span energetically and pleasantly. And just what *is* the purpose of these later years? Jung is quite explicit. We are given these years to enlarge and expand our consciousness, to increase our awareness and understanding of the universe of which we are a part.

Humanity is something special on the earth. A salmon, to take another species as example, climbs the Columbia River fish ladder, spawns, and dies, its life purpose fulfilled. A human being climbs the ladder of ambition, spawns, then lives another whole lifetime as long as the first half. To what purpose? To acquire the perspective that the behavior of the tribe shows it so conspicuously lacks. To give the *whole* of life its *total* meaning.

# 4 Rich Man, Poor Man

ON A RECENT TRIP through the United States I was deeply impressed by the contrasts I encountered. Just before I left my home on the East Coast the aged mother of a friend of mine threw herself out the window of her high-rise apartment and plunged seven stories to her death. She had been in comfortable circumstances, in good health, and was visited regularly by her two devoted sons.

On the West Coast I visited one of the most expensive and luxurious retirement villages in the world. I found the population badly shaken. One of the most popular residents, a recently retired gentleman of health and wealth (you had to be wealthy to live there at all), of charm (he was the popular leader of the colony's "bridge set"), and of accomplishment (he had been an important figure in his profession) had just gone home from his usual afternoon of bridge, closed the door and shot himself.

Most of our larger cities have sizable populations of elderly persons trying to live on Social "Security" retirement payments alone. On such a stipend they cannot pay their rent and maintenance costs and still eat properly unless they are in subsidized housing, which is scarce. In one such city I talked with a social service worker. "Many of these people," he told me, "are slowly starving. We acknowledge that their stipend is inadequate to their needs, and that many have nothing else. To compensate, we have tried out a free hot-meals program. But we can't seem to get it going. They'd literally rather starve than come and eat our free food."

On the same trip I passed through New Mexico, where Robert Coles's fine book, *The Old Ones of New Mexico*, was attracting attention. These people, who have very little in a material sense, have achieved a clear-cut victory over age. They have developed a conscious philosophy of aging.

Consider this quote from Coles's book, the comment of a very old man about his aged wife: "I am talking about my silences, and I carry on as she does. She is not just an old woman, you know. She wears old age like a bunch of fresh-cut flowers. She is old, advanced in years, *vieja*, but in Spanish we have another word for her, a word that tells you she has grown with all those years. I think that is something one ought to hope for and pray for and work for all during life; to grow to become not only older but a bigger person. She is old, all right—*vieja*—but I will dare say this in front of her: she is *una anciana*; with that I declare my respect."

What is there about *our* culture that makes aging so difficult for all—rich and poor, ignorant and educated, male and female, sick and well? Since changing an entire culture is a slow, long-term business, what can we do *within the context of our own cultural climate* to attain a viable philosophy of aging we can apply to our own individual lives?

These are the kinds of questions one might expect to refer to our psychologists with some hope of an informed answer. But, as a class, the psychologists have no answers. Lest I be accused of maligning a respectable profession, I'll step aside and let a professional psychologist state the case.

Dr. Robert Ornstein is a research psychologist at the Langley Porter Neuropsychiatric Institute, a teacher at the

University of California Medical Center in San Francisco, and author of important books in his field. He has made this comment: "Those of us who are psychologists generally ignore the problems of old people. We think of them not as something worth our study, but as something to ignore. When we pay attention, we content ourselves with charting faculties lost or diminished with age."

Whatever it is that makes the difference between a joyous and a miserable advanced age is a matter of attitude, of philosophy. Just what "philosophy" works well, then, and what doesn't work at all?

Dr. Ornstein said, let us note, that psychologists "generally" ignore such questions. Fortunately, they do not universally ignore them. I shall cite a few such authorities, some of whom have themselves survived extremely severe later-life stresses.

I shall begin with Dr. Viktor Frankl, whose notable book, *Man's Search for Meaning*, has had a profound impact throughout the Western world. Dr. Frankl, professor of psychiatry and Director of the Neurological Department at Vienna Poliklinik Hospital, knows firsthand the grimmest of all possible stresses. He spent three years at Auschwitz and other Nazi prison camps. During this period he lived behind barbed wire and among the gas chambers and crematoriums. The odds for his survival were less than one in twenty; his death could have come at any hour or any minute of any day. The conditions of life—filth, starvation, disease, cold, exposure—were miserable beyond description. His wife, with whom he was deeply in love, was gassed, and her body burned in a nearby furnace. Here is a man, we can say, whose knowledge of his subject did not come exclusively from textbooks. What does *he* have to say about this matter?

In a nutshell, Dr. Frankl's message is this: *A human being cannot live a life without meaning; if the meaning is trivial, the life is trivial; if the meaning is exalted, the life is exalted—one is free to choose. Not to choose results in an "existential vacuum" that is life-destroying.*

"Let us think, for instance," he writes, "of 'Sunday neurosis,' the kind of depression that afflicts people who become aware of the lack of content in their lives when the rush of the busy week is over and the void within themselves becomes manifest. Not a few cases of suicide can be traced to this existential vacuum. Such a widespread phenomenon as alcoholism is not understandable unless we recognize the underlying existential vacuum. This is also true of the crises of pensioners *and aging people*" (italics mine).

Dr. Frankl underscores his points by citing case histories. Here is one of them:

In Vienna a high-ranking U.S. diplomat came to his office and asked Frankl to complete a psychoanalysis he had begun with another doctor in New York. Frankl asked why he thought he needed to be analyzed. He didn't like his job, the diplomat said. He disagreed with American foreign policy and found his round of duties boring. The New York analyst took five years trying to impress him with his own view, that the trouble started with the patient's boyhood friction with his father. After a few interviews with Frankl the diplomat began to see what was really troubling him. His diplomatic career was for him unfulfilling and constantly frustrated his "will-to-meaning." The solution, of course, was to find work he believed really mattered. This the man did. The result was a more purposeful and hence happier life.

Another of Frankl's patients was a rabbi in the last stages of senile depression. His wife and six children had been gassed and cremated at Auschwitz, he was in despair that no son would say the prayer for the dead for him at *his* death. After numerous unsuccessful attempts to rally his spirits, Dr. Frankl finally asked whether he expected to see his children after death. The rabbi burst into tears, and the real trouble came out. Since his children had been young innocents at death, they would be accorded the highest place in heaven—while he, a self-confessed "sinful old man," would have to be content with the lowest. Frankl explained that this might very well have been a reason for his surviving his children, that he might be purified. The rabbi brightened at once and went away completely reconciled to a new life of selfless service and self-purification.

"Suffering," Frankl explains, "ceases to be unendurable when the sufferer can see a worthy reason for it." He cites as illustration a case of an aging widow with a severely crippled son. The widow, pointing out their "utterly hopeless" condition, suggested a double suicide pact to her son. The son refused. Helplessly crippled though he was, he found life so full of meaning that, with the help of some therapy, he finally persuaded his mother to abandon her despair and start living in new and more significant dimensions.

C. G. Jung anticipated Frankl by a number of years in identifying *meaning* as the essential ingredient of the individual life. He took his cue, he writes, from his patients, who came to him "from many countries, because they could find no meaning in life. How often have I heard a patient exclaim, 'If only I knew that my life had some meaning and purpose, then there would be no silly story about my nerves!' "

This vital sense of purpose is something quite independent of economic status. It was graphically illustrated in the contrasting cases of the poor New Mexicans who are happy and the rich Californians who are not.

To be rich does not assure that one will be happy. There are authorities ready to go a step beyond that, suggesting that wealth can be a detriment to happiness. Duane S. Elgin of the Center for the Study of Social Policy at Stanford Research Institute is inclined in this direction. I quote him briefly:

> People are beginning to question that which we Americans do best, namely, creating unparalleled levels of material affluence for masses of people. We have extended a rational concern for material well-being into an obsessive concern for unconscionable levels of material consumption. However, . . . many who have experienced the affluent life have come to know an *inner* poverty and alienation. . . . Extravagant levels of material

> consumption inhibit the realization of our larger potentials. We are possessed by our possessions, consumed by that which we consume. We are now compelled to rethink what life means and where we want to go.

To a person brought up in a tradition of exaggerated rugged-independence it is such a disgrace to take alms that some would literally rather starve than do so. To a person whose society offers no values beyond the competitive commercial scramble of the first half of life, and whose life is plunged into utter meaninglessness at being suddenly excluded from such pursuits, suicide can seem logical and even attractive.

The tragedy is that the entrapped mind cannot see the truly beautiful alternatives, and there is no one to help it see them. But the beautiful alternatives are there. There is a potential in each of us for a joyous fulfillment of a *rounded* life, a life in which the second half is so full of inner richness that it reveals new meanings in even the recently lived first half, making the whole of life one of the greatest adventures the stellar cosmos can offer. Consider the tribute of the old New Mexican to his aged wife: "She wears old age like a bunch of fresh-cut flowers . . . I think that is something one ought to hope for and pray for and work for all during life; to grow to become not only an older but a bigger person."

# 5 Later-Life Fulfillment

IS IT POSSIBLE to emerge from the standard commercial-industrial scramble of our time and place and still enjoy the serene middle and later years known to the happy New Mexicans and their rural counterparts throughout the world? If so, what principles are involved in arriving at such a fulfillment?

Is it possible that the answer could be such a simple thing as removing the element of *competitiveness* from one's life and replacing it with the healthier principle of *cooperativeness?*

There is evidence that this may be the case. The happy New Mexicans, whose communities offered few rewards to aggressive competitiveness but many opportunities for friendly noncompetitive cooperation, are a case in point. The Phenix Clubs, which are found in every kind of standard U.S. milieu, from industrial Wallingford, Connecticut, to suburban Sunnyvale, California, are another.

These "clubs" are neighborhood "friendliness societies" featuring reading and discussion on the themes of good living and growing minds in weekly hour-and-a-half meetings. They are not restricted to low-tension rural surroundings, but bring their spirit of cooperative inquiry right into the heart of the competitive commercial strife. Except for small pass-the-hat donations to cover costs of a meeting room and reading materials, they cost little or nothing, and there is no membership requirement except the desire to participate. Members include men and women in the whole stress period from mid-life through retirement. They vary in age from thirty to ninety-five and have occupations ranging from stenographer to professor, from truck driver to poet, from housewife to psychiatrist, from business executive to retiree. A club can start with as few as half a dozen members. When the roster for any neighborhood group gets much larger than twenty, they usually split, because smaller groups make for better discussion, and good talk is of the essence.

The stated purpose of the membership is "to improve the quality of our lives," and, according to the participants, it works. When word got around that this book was in preparation and that this writer would like to know what Phenix Club members were getting out of their experience, a flood of heartfelt letters came pouring in, covering the whole span from mid-life to post-retirement. Quoting from a few of these is the best way, I think, to give you an idea of what ought to happen, and what *is* happening, in creative aging. I have permission to give on request the names and addresses of all who are quoted.

We may begin with a young (thirty-seven-year-old) man living in New York City. "I came across our group at a time when a very severe depression was developing. It would have been easy, at that point, to have fallen into despair or just waste my life in various forms of petty personal vice. Though I was actually ill with this thing, just thinking about getting to the meetings hastened my recovery and put me on my way. The members and the meetings opened up whole new worlds of ideas and opportunities that had not previously occurred to me. I see entirely new ways to use my life in the years to come."

A woman researcher of about the same age (thirty-eight),

also of New York City, was less desperate, but still feeling a strong need to reach out into some new life dimensions: "When I first heard of this group, it seemed that it might be part of the answer to my seeking. I was increasingly feeling a need to explore, with other concerned people, my own feelings about living, aging, and dying. After a year, has it worked? The transformation of a small band of diverse individuals into a caring, loving community is a difficult journey. As one member put it, 'to live by these precepts would require a revolutionary change for almost anyone. And which of us does not quake, at least a little, at facing our own foibles and giving up long-nurtured hang-ups?' Nevertheless, having my attention called to my 'outgrown value system' has helped me really outgrow it and get on with living. I feel especially strongly about this whole poisonous business, so common among us, of *possessiveness* in relationships. The group has made me much more aware of the possibilities of friendship in this nonpossessive way—just feeling enjoyment of people for what they are and for their potential in working together so that all of us may learn. It's the doing something worthwhile together that makes it exciting."

This comes from a fifty-year-old Connecticut management consultant: "There seems to be a point in our lives when we begin to question the meaning of our existence. Carl Jung said it was impossible to live successfully through the second half of life by the same standards that served during the first half. If we accept this, the question arises, 'How do we arrive at these new standards?' Each of us is unique, one of a kind, and each must build a philosophy of his own.

"I began to feel a vague discontent inside as I went through my mid-forties. As I approached fifty it grew to serious proportions. I couldn't precisely define it; things were very tense and getting worse, and for no apparent reason. All the things I'd once thought were 'fun' dulled. I felt myself to be in a vast swamp, with each step only carrying me deeper. I began to drink more heavily to dull the pain, and I needed heavier and heavier doses to maintain the level of numbness, and smoked continuously, pipe clutched desperately in my teeth. It was a pretty miserable existence.

"Then one day I read of a lecture being sponsored by the Phenix Club. As I walked into the meeting room a great feeling of openness, warmth, and friendliness enveloped me. This had never happened to me before. This 'cosmic' friendliness says to you, 'we accept you as you are and for what you are.' This allows you to open, and the more you open the more you receive. I like to think of us carrying this out, like a ripple, until we really achieve 'peace on earth.'

"I determined to find out more about this strange phenomenon and so began to attend meetings regularly. Gradually I began to grow in this supportive, encouraging atmosphere—began to realize that *I* must set the new life-goals and develop new standards if *I* was ever to emerge from the swamp and give meaning to my life. Week by week things began to fall into place. My fear and tension were replaced by an ever growing peace. The despair has faded. I no longer need liquor as a painkiller or a Linus-blanket pipe in my mouth. We (myself and I) are deeply engrossed in an exciting and rewarding search, and are too busy for time-killers. Each step further convinces me that the second half of life can be a more joyous and deeply rewarding adventure than the first half—*if* we are willing to open. That decision has to be made individually. We must each decide what's valid for us as individuals, and not according to what others say or think. To recognize an important new idea when it comes along, you have to be ready for it."

Next to comment is a mother and teacher, aged fifty-five. In this case all seemed to be going well on the surface—good job, fine husband, great family, and all that. But: "My husband and I seemed to spend too much time on the mere mechanics of living. When we talked it was about our jobs, children, social life, and immediate problems. I wanted to spend more time trying for a real philosophy of living. Our group gave me the chance.

"We soon found friends with whom we could discuss those aspects of life that ultimately matter. It was such a relief to be able to be honest and bring out one's fundamental doubts among understanding people. I shall never understand how I reached my fifties without ever having heard of such things as the development of consciousness, higher awareness, and the like. Now I was actually getting into some of the things I had long intended to look into 'some day.' We are experiencing absolutely no pre-retirement stress. There is no frenzy to compete for promotion (if you haven't made admiral by now, forget it!), the children are on their own, possessions can be enjoyed, but who wants to collect more, and why keep up with the Joneses? Feeling that the first half of life was just a preparation, with so much still to do in the way of deepening and broadening our awareness, we look forward to retirement and pity those still on the tribal treadmill. With a little effort you can win a deeper relation with spouse and friends, evolve your consciousness, and perhaps even draw closer to the 'Infinite.' "

A parallel account comes from a housewife in Kansas City: "While my 'nest' was not exactly empty—a husband and son at home and a daughter at college—I was, at fifty-three, certainly facing a crisis. I was at a job that included salting icy sidewalks, making coffee, helping with a meals-on-wheels program, editing, mimeographing, handling mail, telephone, and records. My pay was a little above minimum wage.

"I struggled and fretted. The world was going by, and it didn't know I was here. Grandma Moses had done it—why couldn't I? So I began the Great American Novel—the story of a person who within her had as many seeds as nature—varied, unique, fragile, strong. I would tell the world about that struggle. But it ended as so many novels do—unfinished.

"At about this time I read of this 'club' idea. There is a familiar hymn, 'Open My Eyes That I May See,' but for all the forty years I attended church I never really saw. Now I began to. Like a planned meeting at a crossroad, two friends expressed the same wonderings and wanderings. It was through the reaching out of Alice, my 'soul-sister,' that our group began to form. Dutch and I, Alice and Ed, Dorothy and Bob, then Fern and Jim, and now Joan and Bob.

"This is the most exciting time of my life. I woke early this morning and heard the birds saying hello to each other—and I was with them. The news of strikes, disasters, inflation, violence, and impending doom were little things. I was part of a bigger scene. I was part of Rachel Carson's sea; its beauty as she wrote lovingly of it became part of my experience. I thought of continuing life, which I had never grasped before—and it became clear. I feel I am on the threshold of something about to burst into being in me. I am part of all—the need to struggle and prove myself is gone.

"I even feel kindness toward people I once found abrasive. They no longer control me. I can let go of my children—I am too busy finding exciting days ahead to cling to them. The fellowship of our group deepens but does not tie. We are all free, or struggling to become free, and strolling through the world, knowing far beyond ourselves. The veil is ripping. At fifty-two my husband is taking retirement. He will find employment that will have less pressure and give him more freedom. We are breathing deeply. We are beginning to see."

This comes from a sixty-two-year-old man, a counselor in the correctional system of an eastern state: "Some two and a half years ago I'm afraid I was a typical individual going through what we call the 'mid-life crisis syndrome.' I had all the usual hang-ups: the fear of growing old, of losing my productivity, of not being wanted or needed, resulting in a loss of self-esteem and deflated ego—and, of course, I was going to DIE! Not exactly an attitude to make you feel comfortable, happy, and secure in your later years!

"Today the complete reverse is true. How did it all happen? The Phenix Club came along, and I discovered that my feelings and fears were not unique—others had them too! We could lay the blame to many things—our culture, our civilization, the mounting pressures of our times, the media, and the rest of it. This led to a very deep self-examination and at last to the basic question, 'What is the purpose of our life?' Answers began to come. I began to see that the purpose of life is that our energy-force, spirit—or whatever name you choose to give it—was created that we may come close to the universal consciousness, by whatever name one may choose to call *that.* How did I personally arrive at this? Through reading, discussion, and contemplation. And we had oh so many through the ages to confirm this. In our time the great psychiatrist Carl Jung probably expressed it most coherently.

"The effect of this, together with our weekly discussion meeting, has completely changed me. I no longer fear death. I know that the so-called 'last third' of life is every bit as important in my personal evolution as was the 'first third.' We are created that way. Therefore, as our discussion guide suggests, 'old age can be joyous.' Mine is!

"My feelings about *people* have changed. I do not necessarily *like* everyone, but I do not hate anyone either—I have learned to love my fellow human beings. I find myself widening my circle of friends. People seek me out; I am respected. Daily I discover that I am doing more and more for those who come into my life. I know that I am evolving to a higher state of consciousness—and that, after all, is the ultimate purpose of life. I cannot begin to tell you what a deep, rewarding inner feeling is engendered within me. As a consequence, at night I look forward eagerly and confidently to the next day. It is exciting—it is *good!*"

This comes from a retired couple: "The other day my sister asked what had happened to me, I used to get so mad. I have moderated, and there's a reason. When I saw the first article about the Phenix Club, I knew it was what I needed. I had survived the loneliness of retirement, a move to a new community, and a different mode of living, but I needed more. New vistas of thought, belief, and awareness have opened; every part of my life has been affected. Except for an occasional twinge, fear of death has been erased. I am able to communicate feelings I had long wanted and tried to share, but I did not always find a responsive audience. I am more relaxed. While I am still interested in maintaining our home, I haven't the same desire for material possessions. Of course the people we have met have added a full measure to our lives. I shall be eternally grateful for the discovery of the possibility of real development. Though I have barely touched the first rung, just the fact that I have begun is exhilarating. My retired husband is less vocal than I but readily admits to benefits received—and the fact that he resigned from the service club that had conflicted with our meeting night speaks volumes!"

A thirty-seven-year-old housewife and mother of school-age children expresses herself in this brief, lyrical declaration: "I've been going to these meetings for only a couple of months, but already I'm a changed person—for the better! I feel totally different from the way I ever have felt in my life, I have peace. These meetings are a source of great joy to me. I can be myself, without fear or tension, knowing that I can say what I want without being ridiculed or laughed at. 'Out there' my neighbors kind of tolerate me; 'in here' I have friends. I'm finding peace and joy, and that's what's important to me."

Let us now run up the age scale a few decades and hear from a retired gentleman: "I have found a companionable group of people united in simple friendliness who furnish a unique atmosphere and the right kind of encouragement to go on in my development. When I gave up the wear and tear of the competitive scramble, I was at a loss to find something that would replace the stimulation of competing. Here I find a refreshing emphasis on the responsibility we have to use whatever wisdom we may have acquired (by living this long!) to try to achieve a *planetary* view. In my new pursuit as an environmentalist (combating noise pollution and forced music) I've had sympathy as well as argument to help sharpen my approaches and ward off discouragement. But now I want to mention what to me is the purest joy in Phenix. It is to find kindred spirits to help me explore hidden subjects. I'm a small-time Dr. Faustus, and I want to delve deeper into the realms of speculation. How to develop our intuition, how to approach forbidden subjects like the right attitude toward what is called death—these are things that are truly worthwhile. We need a circle of congenial souls to meet with face to face (not just in books), to keep on in the pursuit of learning, or at least *peace of mind.*"

Finally, from St. Louis, comes news of a really *senior* senior. "Our group here at the Delcrest is the greatest! One of them, Nettie, who is in her ninety-first year, has been going with me to the Thursday night meetings at Kay's. Like Socrates, she is forever asking questions—and seeking truth as she might be able to accept it. She brings on the most intense discussions you ever heard!"

The point here is not just to extol the Phenix Clubs, which need no extolling, but to identify the principle that is producing such remarkable results in such an urgent category so that it may be more widely applied. Whether the group extending the principle is called Phenix or something else is secondary. Once identified it can be applied anywhere by anybody.

The principle is as old as man; it might be called Mutual Self-Help Power. MSHP bases its work on the proposition that all men and women have the potential for solving their own problems. Sometimes, especially during periods of challenge, the individual, immersed in an indifferent if not actually hostile society, requires a supportive social climate to activate this potential. No amount of superficial entertainment, development of manual skills, classroom instruction, physical or psychological therapy or financial aid can relieve the situation without first dealing effectively with the real need.

Continuing mutual self-education goes to the heart of the matter. The real need in all these crises is the same: to restore zest and interest to a life that has lost its meaning. Deprival of accustomed roles leaves the individual with a feeling of loss and worthlessness. Even the threat of such deprival can cause acute depression and personality disintegration during the crises of the second half of life. Life-meaning cannot be restored by lecturing or instruction not spontaneously sought by the individual. It *can* be restored by the friendly and intelligent interest of those around us.

The origin of the problem is societal. Called splinteredness, it is the inevitable by-product of the specialization that follows from industrialization. In our society neighbors no longer have a common base for neighborly human exchange. Each is involved in specialized work that is different from the work that occupies the others. This leaves only two bases for socialization—trivia and banality—and neither can fill the human need. When companionship is offered on the basis of "large talk" instead of small talk, growth is resumed and the problem is solved.

Though the Phenix Clubbers are the only ones who have *done* anything about it in the neighborhood context, a number of learned folk have talked about what needs to be done. One of these is the UCLA psychiatrist Roger Gould, who sees a *hunger for personal relationships* as a prime feature of aging. "Making sense of it all" arises as a stronger inner demand than it ever had before. Imbued with the idea that adults, once adult, are never supposed to change, people tend to be badly shaken at finding themselves different from what they were a decade or two earlier. Trouble arises when they try to deny these changes instead of accepting, comprehending, and learning from them. The best way to learn, of course, is through creative interchange with intelligent peers undergoing similar changes and coping with them.

The editors of the magazine *Synthesis* have stated the case adroitly from the global point of view. "One looks at the world," they write, "and sees immense need, immense conflict—troubled times. And simultaneously, one sees our own era as a moment of unique opportunity. As so many major thinkers have observed, the world is waking up. Everywhere beneath the surface disarray, one sees thousands, even millions, of individuals beginning to take conscious charge of their own evolution.

"People have seen that they can develop themselves, harmonize their inner conflicts, and align themselves with higher human purposes in a most deliberate way. In America, in Europe, and in Asia, among the young and among the old, among all peoples, there is this same stirring toward individual development in the interest of serving the evolution of mankind. This awakening is part of a broad process—the process of synthesis. One aspect of this process is a natural tendency within people to harmonize their inner lives at ever higher levels. It is a process that occurs in each one of us—and we can help it occur."

Our era must become one of *self*-help. With all our authority categories so seriously discredited—traditional religion, science, education, the media, and politics—our own inner sense of what fundamentally is becomes more important than ever. When you are in harmony with yourself, you are in harmony with the universe. And when you are in harmony with the universe, you are happy, friendly, creative. It's as simple as that!

# 6 Victory over Rage

ANGER IS THE greatest enemy of peace of mind. Since we can't win the victory without fighting the battles, we may as well have our bout with anger now and get it over with. My friends and I have been through it with this one and have a few tips that may be useful.

We are social beings. Being in a rage at another human being, or group of human beings, or circumstances created by human beings collectively is, therefore, the same as being in a rage at yourself. Being angry at yourself leads to depression. Sometimes depressions are suicidal. Suicides are people who have become so angry at themselves and their meaningless lives that they have killed themselves. Possibly even worse than the quick death, anger turned inward can make one deadly, cancerously sick. Anger is that dangerous.

The mechanism by which this happens is called "internalization," and is tricky. It usually begins with the model—perhaps a parent—who seemed to be able to appear effective and powerful by a display of anger. Imitation followed, and it seemed to work. Show a little temper, and presto, people would often let you have your way.

With essentially gentle persons, other and contrary learning soon overlaid it. It was discovered that, though you might have your way at the moment, the cost was more than it was worth. First there was the fear of retaliation that anger always brings with it; then there was the fact that you didn't want to lose a friend because of a silly temper tantrum; and besides, being an essentially gentle person, you didn't really *want* to be angry. So the anger was "controlled," in ordinary language, or, to use a more exact word, internalized. The model was not abolished, but only taken deep into the psyche to continue its dirty work there. It was neither safe nor socially advantageous to show anger publicly. It *seemed* safe to bury the anger inside.

But it is not. Because we are inextricably a part of the human confederation, being angry at anything human is being angry at ourselves, and this is misery. When anger breaks out of its confinement and is openly expressed, it may temporarily yield a false sense of power while people are scurrying for cover before the temper storm. When expressed repeatedly, it inevitably leads to being without friends and leaves you with no one to be angry at but yourself. This is intolerable, and produces various efforts to escape. Here are a few of the common ones, all of them unsuccessful:

Get a "noble cause." This seems to enable you to divide humanity into two classes, those who are with the cause (good guys) and those who are against it (bad guys). It restores you to a precarious but accepted social position, since you are no longer completely alone. But there is no peace in it, because the bad guys always seem about to win. As a former social reformer I speak from experience. I gave up social reform, replaced it with social education, and began to feel much better. The trouble with the noble cause route is that, though you do have a few like-minded friends, you are still sore all the time. Since the anger is still there, the whole thing can blow up at any time and the friends be dispersed in a gale of bad temper. Worse yet, the anger can turn against yourself and become depression.

Another dodge is booze, with which I have also done some experimenting. Being sore is a wretched state, from which alcohol in its initial effect gives momentary relief. But the first thing alcohol dissolves, as it works its way in, is social control, and there, raw and naked, stands the very thing you drank to get away from—anger, now unmanaged and sometimes turning into berserk rage. This brings the drunken quarrels, fights, crashes, accidents, and barroom killings that are the routine accompaniment of escape drinking. If the anger takes a decisively inward turn, it can become suicide. Self-murder in drunken rage is statistically one of the proved hazards of heavy drinking. Short of that, the special kind of post-spree remorse blithely dismissed as "hangover" is mostly incipient depression caused by anger turned against the self.

A third common device is the whipping boy. Direct the anger at a person of a class you think can't retaliate—a parent, a child, a spouse, an employee, a public servant, a race, a nation—and really cut loose. Again it doesn't work, for the good reason that we *are* social beings—that is a basic, unalterable reality. Anything antisocial is therefore anti-oneself, and can lead only to more misery. So let us not look around for a handy whipping boy; for whipping him would only be whipping ourselves.

Many people complain of "depressions" and "blues" without realizing that the root cause of these things is anger. There can be no depression without anger; be rid of the anger and you're rid of the depression. Temporizing with anger by working it off, running it off, bullying it off, quarreling it off, screwing it off, drinking it off, or talking it off won't work. The only cure for anger is to be rid of it altogether.

This is accomplished in four stages: acceptance, forgiveness, aggression, and reprogramming. We may begin by realizing that we cannot change anything by being angry except to make it worse. *Being angry will not get us the happiness we really want.* Therefore, we must temporarily accept things the way they are while considering more reasonable ways to change them (if we still think they need changing and that we can change them). There are things we cannot change. Such things call for permanent, unprotesting acceptance. Then we *forgive* ourselves for our former anger. This is vital. Only our own self-forgiveness can finally dissolve our anger.

The next principle is *aggression*. Some people are so afraid of their own anger that, in burying it, they try to bury along with it their quite legitimate aggression. This is a serious mistake. Some of the psychosomatic therapists are convinced that deeply buried angry aggression is one of the causes of cancer. Dr. O. Carl Simonton, a former Air Force radiologist now practicing privately in Fort Worth, has run up an impressive body of evidence along this line.

We must distinguish between anger and aggression. It is perfectly possible to be extremely aggressive in your own interest, and in the interest of society, without being the least bit angry and being, in fact, positively joyful. Aggression is a good and necessary factor in any creative action; you cannot even love without it. Aggression is necessary and can be constructive; anger is unnecessary and is always destructive. Aggression can and should be motivated only by a desire to help yourself, or others, or both, by contributing to your own or to the general welfare. We must strain out the anger from the aggression. There are calm, intelligent ways to stand up for what we believe, defend our rights and accomplish our objectives without being the least bit angry. This is aggression, and must by all means be preserved. Anger never produces anything but misery and must be abolished if we are to achieve peace-filled aging.

The final step is *reprogramming*. No deeply entrenched habit, whether that of being angry or something else, can be broken by just reading a book and saying you'll reform. We begin by recognizing that the thing is *nothing but a habit*—it is not part of our true nature. Like any other habit, it was caused by long programming of the inner computer. It is eliminated by reprogramming that same computer, and this may take a little time.

It calls for the kind of massive input achieved by repetition. It breaks the habit by the same means that formed it. I have, over the years, broken quite a string of unwanted habits this way—nicotine, caffeine, alcohol, and, the toughest of them all, anger. I know a man who even cured himself of cancer this way. Dr. Simonton uses the same autosuggestion treatment with his cancer patients.

The beginning step in reprogramming is to identify and acknowledge the anger. This may not be as simple as it sounds. Some people have buried their anger so deep they no longer know they have it, and may need professional help in getting it out. Others deny it, and may pound the table angrily shouting *"I am not angry!"* And it is always possible that there is an easily identified *ex*ternal cause. Possibly we have been somebody's whipping boy for too long. It may be something we fear and react to with defensive anger. It is best that irremediably negative relationships be given up, whatever the cost. But they must be broken in clear, calm necessity, not in anger. These are special cases. Most of us know perfectly well what we think we're angry at. Nonetheless, however we may project it, the true source is usually found to be *in*ternal. For us it's a simple matter of persistent reprogramming.

It's important to make clear that it's something worthless we're getting rid of rather than something valuable we're "giving up." This can be deceptive, since blowing your stack, like other undesirable habits, can *feel* good when it is in fact most definitely *not* good. We have to *want* to get rid of it or we *won't* get rid of it.

Reprogramming is repeated *affirmation.* This consists in finding short phrases and slogans that have struck a benevolent, healing chord in your higher mind, ideas you feel could make you a better, finer person if you could only lay hold of them and hang on to them long enough to make them really your own, part of yourself. You can do this by repeated affirmation.

Copy whatever sayings you have chosen onto placards to have in your room, and on small cards to carry with you. Look at them frequently, repeat them whenever you think of it, especially when you are tempted to revert into the old habit. If you really mean business, you might get yourself a cassette tape recorder and read the sayings into it, speaking slowly and in a calm voice. Add sayings to your collection as you run across helpful ones. Listen to your own fifteen-minute tape at least once a day, or spend an equal interval going over the written sayings. Memorize the most helpful ones and say them over to yourself in spare moments. These may do for a starter:

"Anger is the only bad guy. It is my enemy and I want to be rid of it. I choose not to let my anger make me into an old grouch."

"I cannot get the happiness I really want by being angry."

If you like biblical quotes, don't forget this one: "The wrath of man does not work the will of God."

"As I feel anger falling away, I feel a wonderful peace growing and expanding within me."

"Peace brings perfect happiness; anger only destroys it."

"I desire always to be on the side of inner harmony and inner peace. Outer peace and harmony will follow."

"Peace . . . peace . . . peace."

"I'm glad I'm alive and I thank God for life, just the way it is, right here and right now, with nothing added and nothing taken away. I am finished with complaining. I will complain no more, because I do not *want* to be a complainer."

Simple, but it works.

True happiness, as distinguished from just "fun," is more than a simpleminded absence of anger and depression. The joyous fulfillment of a rounded life has more to it than merely getting rid of these most common killjoys. We mention them merely to clear the way so we can go on with our victory march.

# 7 Some Case Histories

THERE ARE ESSENTIALLY two kinds of fears that beset the average human being, fear of life and fear of death. We'll start with the first category. The fears of mid-life and of the pre-retirement stress years stem from the fear of life. Let us look at the following situations.

*Case 1.* At age forty-eight Edgar found himself in a pre-retirement stress situation that was unbearable. Everything he owned or could borrow was staked on a business venture that would neither succeed nor fail but hovered year after year on the verge of bankruptcy. Meanwhile his weight and blood pressure were increasing. He was tempted to follow his equally pressured business friends into the local bars for occasional temporary relief. What would you have done had you been Edgar?

*Case 2.* When Ruth was forty-eight, her fifty-two-year-old husband became despondent over his health; he closed himself in the garage and sat in the car with the engine running until he was dead from carbon monoxide. Ruth was left with two dependent children and no income. She had no marketable skills in the usual business sense, but was considered a good cook. Ruth had never worked and was accustomed to a comfortable life in an upper middle-class suburb. What could she do?

*Case 3.* At age forty-three Paul found himself plagued by deep and uncontrollable periodic depressions that were all the more exasperating because they had no apparent cause. He was brilliantly successful in his profession, and this success had brought him both wealth and prestige. He loved his wife and children, and the whole family was in good health. Any kind of setback, however, even a relatively minor one, in his unbroken string of successes could trigger another of his long and miserable depressions. What to do?

*Case 4.* Myrtle had been an exceptionally attractive and winsome young woman. She married a handsome and accomplished man with the expectation that he would take care of her for life. They had three healthy children, lived in an exclusive suburb, and traveled with an expensive crowd. Then, quite suddenly, when her husband was approaching fifty, he found himself out of work and without any job prospect. Due to unforeseeable technological developments, the skills he had spent his life perfecting were rendered obsolete. In two years their savings were gone. Myrtle had never worked, had no marketable skills, and no income except the chancy charity of relatives. What might Myrtle do? What about her husband?

*Case 5.* At age sixty, Harold experienced an uncontrollable desire to pursue women—a recurrence of a pattern manifested before his mid-life span. His wife Winifred disapproved so strongly that a divorce resulted. What kind of life did Harold finally work out for himself? What happened to Winifred?

*Case 6.* Robert, a man of some intellectual resources, knew well in advance that retirement was coming and that some preparation for it would be necessary. Nevertheless, when at sixty-five the mandatory retirement day arrived, he found himself at loose ends. He could not abide the childish pastimes of the Senior Citizens centers in his area. Though he mildly enjoyed fishing, golf, and table games, he did not feel that such pursuits ultimately added up to a meaningful life by themselves. What did Robert do?

*Case 7.* At mid-life (late thirties) Allan and Marie were a couple who had everthing settled: successful marriage, solid professional status as teachers, healthily growing children. Only one fly in this otherwise pleasant ointment—they were deadly bored with the rut they were in. What did they do?

*Case 8.* Sometimes special circumstances can bring on the crisis earlier than usual. When Anne was twenty-five, a car driven at high speed by a drunken driver suddenly leaped the divider and crashed into the car in which she and her husband were returning from a vacation. Her husband was killed instantly, their seven-month-old baby was thrown out of the car with a broken leg, and Anne was taken to the hospital with multiple injuries. Anne's father and mother were divorced; her mother, a domineering person, moved into the crisis to such an extent that, as her thirties approached, Anne found herself virtually a psychological prisoner of her mother. Did she give in to Mamma—or just what she do?

Eight cases. The successes and failures of these people in dealing with their own life cases will presently be revealed. First, however, a few general remarks on life crisis are in order. Words like "crisis," "depression," and "stress" cover a wide spectrum of life disturbances. Some are so severe as to be classed as true "insanity" and to require professional treatment or even commitment to an institution.

Recent experience, however, indicates that a much broader range of difficulties can be straightened out simply by rallying a person's own inner resources to cope with the emergency. The prolonged psychoanalysis-on-the-couch, so popular at mid-century (among those who could afford it!), is now generally regarded as obsolete. The psychiatrist who knows his business now tries to set his patient on the road to recovery in a much shorter time. William James touched off his recovery from a serious and debilitating depression just by reading an essay—*and acting on it.* Many thousands of "loners" have rescued themselves from seemingly hopeless addictive and neurotic complications by reading the book *Alcoholics Anonymous.* I myself was carried through a rough stretch on John Cowper Powys's little book *A Philosophy of Solitude.* Many will be helped by reading *this* book. The decision rests within the individual. When something or somebody touches off an inner chord of harmony and well-being, and one *acts* on that something, new life and new hope begin to open up.

One must never confuse simple, normal, cyclic ups-and-downs of mood and morale with a true life crisis. The biorhythmic cycles, ranging from three to seven weeks, that cause us to be a little above or a little below par on any given day, are to be expected, and accepted, without alarm. I cite the case of a salesman who sometimes found himself bereft of the power of speech and simply unable to present his product. This plagued him for years until he discovered that his "bad days" followed a certain rhythm. When he decided just to stay home on these days, the difficulty vanished and he became outstandingly successful. George S. Thommen's little book *Is This Your Day?* has some interesting things to say about cycles.

The "stages-of-life" disturbances, however, unless resolved into higher levels of awareness, can become genuine life crises and make very rough going. The eight cases cited above are not matters of mere transient moods but are genuine life crises. Here are the actual box scores on the successes and failures among the eight cases:

*Case 1.* Edgar looked at himself in the mirror one day and decided that the flabby, paunchy man he saw there was not what he wanted to be. He took up jogging, and gradually increased his range until he now runs two miles every morning. Then he added Hatha Yoga to his routine and began reading mind-expanding materials not connected with his business. His vocational situation is as chancy as ever, but he's safely on top of it. "Stress?" he says. "Sure I have stress. I run it off every morning. If the business went broke—which I'm sure will not happen—I'd just find something else." Crisis solved.

*Case 1.* Ruth immediately applied for Aid to Dependent Children, which provided enough to support her children but not enough to give them the education she wanted them to have. To earn enough to send them to college she took a part-time job, at the allowable ADC $2,500-a-year maximum of the time, as cook in the local high-school cafeteria. After one child had finished college and the other had been entered and was doing work at the honors level, Ruth remarried.

*Case 3.* Paul couldn't think of anything to do about his depressions except go into psychoanalysis. The analysis has been under way now for several years, and the periodic depressions still continue. Paul thinks talking it out kind of helps, but is not enthusiastic about the results. Obviously, Paul has not yet solved his problem.

*Case 4.* Myrtle could think of nothing to do about her situation but go into hysterics. Her cultured bearing and excellent sense of fashion would have made her an asset to any department store or dress shop. Her response to the suggestion that she seek such a job was to increase the pitch of her hysterical screaming—no one was going to make a slave of her; it was all her husband's fault for not meeting his obligations as a provider. Her hysterics grew worse and turned into threats of suicide. Finally, she had to be placed under the care of a psychiatrist. Meanwhile, in a fit of petulant rage, she had divorced the offending husband, who had landed a part-time job paying enough to support him. As of now Myrtle's case is a failure.

*Case 5.* Harold, it turned out, was one of those individuals who could not endure marriage, no matter how hard he might try or over how many years. Once divorced, hence free to chase all the women he wanted, he lost interest in the game. To everyone's surprise, his thinking took a serious turn. He joined a reading-and-discussion group, found his companionship among the single persons and couples that made up its membership, and settled into a contented and studious celibacy. Harold believes that now, at age sixty-four, he has won his victory over age. Winifred is also doing well. An assertive woman with proved administrative ability and experience, she had no trouble finding a responsible and interesting job where her talents were appreciated. Each partner in this marriage was clearly better off for the separation.

*Case 6.* Allan and Marie constitute an almost classic example of the onset of mid-life crisis, boredom with genteel sufficiency. They are also a classic illustration of the creative resolution, arrived at spontaneously and without outside help. Here's how it came about: One day Allan remarked to Marie, "You interested in Africa?" Marie replied, "Fascinated. I've always been enchanted by the mystery of Africa and wanted to go there." Allan said, "Me too. Let's go." Over the wails of their well-meaning friends, they sold their house, resigned their comfortable jobs with their accumulated seniority, found overseas assignments through a government world-service agency, took their children out of school, packed up and went to Africa. They stayed for seven exciting years and returned—as professors of African history at a U.S. college.

*Case 7.* Robert recalled that in 1905 a Chicago businessman formed a luncheon club just so he'd have some friends to banish his loneliness. They called it the Rotary Club, and it was the grandfather of many similar organizations. Rotary was not the answer for Robert. He longed for companionship, but not at the business level; he wanted friends with whom he could discuss a general philosophy of living—the meaning of life and of the universe, the relation of human life to the totality of things. He heard of such a group, joined it, and became an active member. This led to part-time jobs in areas of social service and community betterment, jobs that yielded little financial return but great inward satisfaction. His discontent gone, he now feels he has conquered the major hazards of aging.

*Case 8.* Anne, after the tragic and life-shattering accident, desperately needed exactly the kind of immediate and personal support her domineering mother provided. Slowly her spirit roused itself from the depths of her depression. She shouldered the responsibility of being a good mother to her growing son. She had basic teaching qualifications from her college years and began attending graduate classes until she was fully qualified and found a teaching position. When realization dawned that her mother's solicitude was being parlayed into complete domination, Anne broke the tie by resigning her teaching job, finding another in a different part of the country, and moving. Her son is developing into a fine, robust boy. Though Anne has male friends, she has never found a romantic replacement for her brilliant young husband and has not remarried.

We have examined a wide variety of cases. Those that have been solved, despite differences in circumstances, have had one thing in common—a resolute and fearless summoning of the individual's own inner resources to deal with his and her own life situation. This, and this alone, is the saving principle at *any* stage of life, and it is available to everyone.

Some people find this solution spontaneously and intuitively, and thus escape altogether what to others becomes a dangerous transition. Rut-jumping is the key. Allan and Marie, the couple who packed up and went to Africa, had no crisis. But we need not change job or residence to find this saving release. Any mind-expanding, spirit-expanding life adventure—the development of a latent but neglected talent, the exploration of new avenues of learning, the cultivation of a new and more elevated circle of friends—all these and other personality expanders have also obviated the severity of crisis, *which is essentially nature's evolutionary demand that we continue growing.*

# 8 With or Without Anesthesia?

THE SCENE IS A posh and expensive retirement village in the West. I am in conversation with the social worker who has been hired to help with some of the emotional crises that arise. I ask him what are the most pressing problems around the place, and he doesn't even have to stop to think before responding, "Loneliness, depression, and booze." Then he smiles and adds, "And not necessarily in that order."

It doesn't take much heavy thinking to reach the conclusion that alcohol is a nothing that leads to nothing, though a lot of people, caught up in the mystique of the social institution, have had to endure a lot of hell to find it out. Retirees who imagine that now at last they can "enjoy themselves" with this chemical are particularly vulnerable. Since this is a book on victory over age, it must ask whether alcohol, which is so universally present in our national society, is friend or foe to the aging individual.

Alcohol is to mental health what heart disease is to physical health—the number one killer. Cirrhosis of the liver, a frequent concomitant of habitual drinking, by itself ranks high among the statistical causes of death. But that only begins its ravages. Alcohol is a depressant. Severe depressions sometimes turn suicidal, and the roster of alcoholic suicides is long. Drinking-induced accidents add to the death toll. In these cases, death at least ends the problem. Possibly an even greater tragedy is the year-in year-out wretchedness people bring on themselves and others in the name of "gracious living," "instant happiness," "release from tension," and "fun." Drugs, including alcohol, are mind-muddlers. You can't have a clear head and a muddled mind at the same time—and the victory we seek calls for a clear head. Much painful experiment could be avoided if we could just manage to bear in mind one simple truth: alcohol does not solve problems, it makes them.

Until very recently, though it was known that problem-drinking was extensive among people of retirement age, there were no dependable statistics. Concerned about the growing seriousness of the situation, the Administration on Aging, U.S. Department of Health, Education, and Welfare, commissioned a study on the extent of the problem.

The research was carried out by Bruce Carruth, University of Arkansas, and E. P. Williams, Rutgers University, Center of Alcohol Studies. "The last decade," these researchers say, "has produced an increasing awareness and an increasing commitment to act on two significantly related social problems of our time, aging and alcohol and drug abuse." Using statistics from alcoholism information and referral centers, it was found that fully one-third of all problem-drinking callers, or those called about—an estimated three million persons—are over fifty-five.

Further study revealed that there are distinct types of drinking problems developed among the late-middle and older-age persons. Some were those whose involvement with alcohol was already established, but who entered the "acute problem" category only with advancing age. Some were people with no alcohol history at all before retirement, but who developed serious drinking problems after retirement or bereavement.

Unless the life-crises of mid-life are successfully weathered, the later transitions, including the one the psychiatrist Erik Erikson called "the sixty-five-year-old crisis," also known as "retirement" or "empty-nest syndrome," are likely to be stormy. Erikson maintained that each of life's eight crises must be solved; delaying the solution only prolongs the crisis, so that a "mid-life" crisis can still be with a person in the sixties—and will be if not solved before then. Carl Jung warned about entering this stretch of life "wholly unprepared." Preparation means cultivating those habits of thinking that lead to a continued, clearheaded, and ever expanding interest in life and its meaning.

A couple we will call Joe and Irma provide one instance of inadequate preparation—and there are millions of Joes and Irmas. They would be known in any American suburb as "nice people"—honest, conscientious, industrious, and friendly. But their retirement years were almost a classic illustration of what might be called "Retirement, American Plan."

Joe had been employed by a large interstate merchandising firm that had a liberal retirement plan. At Joe's executive level every employee who had served the company for more than twenty-five years received a lump-sum cash payment of $150,000 at the retirement age of sixty-five.

Though inflation eroded away some of its glamor as the years went by, that huge sum had glittered in Joe's and Irma's imaginations all those years as the American Dream come true. How glorious it would be when such wealth came to them, unencumbered by any necessity to earn it! They could live in a milder climate. Joe could go deep-sea fishing whenever he liked or spend the day on the golf course. Irma could go to concerts, lectures, and the theater, play a little bridge, have a reasonably gay social life among congenial friends.

"All that time," Joe said ruefully a couple of years before he died, "I was the donkey, dragging the cart along to get the carrot that was dangled in front of me from a fishing pole."

By the time the great day finally came Joe and Irma had already selected their apartment in a Florida condominium and made a down payment on it. Now they set about furnishing it and moving in.

For the first year it was pretty nice. They assessed their new neighbors in the condominium as a fair-to-middling mixture of good, fair, and awful, but they never had to be without *some* kind of company when they wanted it. Joe had no trouble finding somebody to go around the golf course with him, and quickly struck up acquaintances among his fellow fishermen aboard the party fishing boats.

Irma, at first, was like a child in the toy department. Exploring the fashionable shops, dining in the famous restaurants, looking in on the jai alai and the racing, joining one of the condominium bridge clubs, rubbernecking at the fantastic variety of tourist exhibits filled her days for many months.

Beginning about the middle of the second year, though, the magic began to wear thin. When the novelty of having a winter when you didn't have to shovel snow was gone, they found themselves thinking furtively that it would be nice to *see* a little snow. They found themselves wondering more often how their old friends up north were getting along, how things were going with the kids.

Joe felt it first. In the old days he had grumbled about the nine-to-five "grind." Now he was beginning to feel that maybe a little of the old nine-to-five wouldn't be so bad. He didn't mention it to Irma because he didn't want to "burden" her idyll with any flaw of discontent. But he did let it out to Ted, one of his new golfing friends.

"You know," he said one day when they were having a couple of drinks at the nineteenth hole, "this isn't making any sense, this kind of a life."

"Come on, get with it, man. It isn't *supposed* to make any sense. You've spent your whole *life* making sense. This is the time when you have *fun.*"

"But it *isn't* fun. I guess I'm just not that much hooked on golf."

Ted laughed and slapped him on the back. "Have another

drink and cheer up. You're only having a bout of guilt at not having to drag your butt down to the treadmill every day. Let go. It passes. Pretty soon you learn it's all *right* to loaf."

After the next round, Ted confided, "I don't know if you've noticed, but there's *stuff* around here, man. Wouldn't advise fooling around in your own condominium, but if you had a look in the dovecote next door, you might be surprised."

It was not the first time that Ted, who had not been getting on well with his aging playgirl wife, had hinted at his experiments with extramarital sex. Joe shrugged off this part. For himself, he had always thought of forty-eight as about the outside limit of when a woman could reasonably be thought of as a sexual temptation. Besides, he had Irma.

But in the matter of relaxing and having fun without feeling guilty about it, he was ready to concede that Ted had a point. Work, work, work. Worry, worry, worry. He'd had enough of it. He'd change.

He tried out his version of The New Heartiness on Irma. Around five o'clock, for a while, he'd say, "Well, doll, the sun's over the yardarm, how about a little drinky?" Irma went along with the five-to-seven drinkies for a while. But she had always been a little suspicious of alcohol. She thought Joe was making it a habit. He never used to take a noonday cocktail. Now he'd have two or three regularly, and begin ever earlier in the forenoon. At the evening end of his drinking day he no longer waited for the sun to be over the yardarm. The day came when Irma, asked whether she wanted a drinky, said, "No. And if you've got any sense you won't have one either. You're drinking too much."

Joe was stunned. Had he ever failed to hold his liquor like a gentleman? Had he ever been arrested for drunken driving, or smashed a car, or been sent to a drying-out place, or failed to be on the job when expected, or to pay his bills? Can such a man be "drinking too much"? How outrageously unfair! Irma was becoming a nag, a common scold.

This happened in the fourth year of their retirement. Until

that time, no matter what the ups and downs, Joe and Irma had always been able to say, when asked how things were going, "Well, we're still speaking to each other," with a twinkle and a smile and a glance at one another that carried real affection. Now they couldn't do that any more. They'd go out to "a nice restaurant" and sometimes sit through a whole meal without speaking except to the waitress. There were weeks when they spoke very little, and days they didn't speak at all.

Joe's fun program began to be punctuated by fits of depression. Try as he would not to worry, there were times when he couldn't help it. His once fabulous fortune of $150,000 was eroding away. Expenses had been higher than he thought they would be. There were always unplanned and unexpected extras—club dues, greens fees, little trips here and there, car maintenance, eating out. Income from investments had not covered their expenses. He was "going into capital" at an alarming rate.

One time, when the sun was over the yardarm after an afternoon of golf, he went out with Ted "for a couple of drinks" and ended up at midnight in bed with a sixty-year-old widow in a condominium a few miles up the shore. He didn't really enjoyed it much, and the rest of that week didn't like himself for it. He felt no need to "confess" to Irma, and even had a funny feeling she somehow suspected. Once the downward spiral of depression began, over this or something else, it would deepen until it became intolerable.

Then he'd strive to pull himself out of it by main force. Dammit, Ted was right, these retirement years were for *fun.* Or *were* they? What else was there to do—but die? What else could you do when the "fun" stopped? Get a job? Doing what? Toll-taker? Cab driver? Parking lot attendant? An executive of *his* background? Ted was right. This was the time to have fun. So *have* fun.

Irma, meanwhile, was coping with her own version of the retirement syndrome. She didn't enjoy having Joe around the house—"under foot all the time"—any more than he did, and was just as bored with his constant company as he was with hers. Besides that, she had found the community resources for a cultured leisure unexpectedly wanting. There were no concerts, theatrical productions, or mind-expanding lectures to be had in several hours' drive in any direction. Even at this distance such attractions were rare, and hedged about with difficulties of traffic, parking, and availability of tickets. Her bridge club friends had turned out to be disappointing. Some were so fiercely competitive in the game that they weren't much fun socially. Some were petty gossips. One was so open about her sexual hopes, expectations, and connivings that Irma found herself thinking, one time, "Why, the woman is nothing more than an ordinary old *bawd*!"

She was deeply concerned for Joe, not only because of his drinking, but because of his general health and spirits. His depressions were deepening, and the heartiness of his "comebacks" was an obvious pretense. Never, even in their happier days, had she been able to get Joe to submit to the annual physical examinations she had herself. This was "dragging him to the doctor" when there was no need—and, indeed, he had always enjoyed remarkable physical health. But she knew the statistical medical facts about business executives. She suggested several times that he at least go and have his blood pressure checked. But no, not him.

As for his drinking, she found unexpected help, not for Joe, who was convinced he needed no help, but for herself. She had read about the Alanon Family Groups, made up, not of alcoholics, but of their nonalcoholic spouses. She began attending meetings and here she found orientation that served her well. Don't nag, she was told. Don't plead, implore, beg, threaten, or weep. No one can accept help for the problem drinker but himself. Develop your own inner strength to a point where you are less dependent on the wayward drinker.

With this steer, Irma undertook an assessment of what, if any, her "own inner resources" might be. She attended the discussion meetings, and met a few people who, having found a sustaining interest they could share, had been able to make a go of retirement as a couple and were genuinely happy. She wondered again whether she and Joe ever would. It seemed so strange, so many years of happiness, then *this*, this bitter standoff.

The problem was suddenly, shockingly solved. Joe died. It happened one night after a routine evening of over-the-yardarm rituals and a heavy meal at a popular restaurant. Joe came home and lay down on the sofa. A few minutes later he staggered to his feet, gasped convulsively, and pitched forward on the floor. The emergency siren was heard—a familiar sound in retirement communities. There was no need for oxygen; all they had to do was take the body to the morgue. It had been a stroke, sudden, massive, fatal. Joe was seventy-two and had "made" seven years of retirement.

It took a few months for Irma to find her bearings again. How strange, she thought, is the life of this world! She had a new perspective now on the life she had lived and the life she saw in progress around her. You are born, you grow up, you go through the family years, you retire—or rather, are retired—you live more or less aimlessly for a few years, and then, for this reason or that, in this way or that, at one time or another, in this place or that, you die. It happens to everybody. It had happened to several residents of their condominium just since she had been there. It had happened to Joe. It would happen to her. Was there anything else? Could this possibly be all?

Irma lived for four more years. She avoided the partying set at the condominium and seldom went out. She continued to enjoy the weekly reading circle and sometimes visited with one or two of her Alanon friends. Her reading, like her life, took a more reflective turn. She learned, about three years after Joe's death, that she had an inoperable cancer and would probably die of it. She was not greatly alarmed. She had had enough of the life that had been offered her, and that she had made for herself, and was ready to die. She only hoped it would not be too painful. It wasn't. Irma went through the first four stages of dying quickly and without great stress. Her denial was only momentary, her despair was brief, her anger negligible, and her hope was not fevered. By the time it became necessary to move her to a hospital she had achieved acceptance. She was there only a week when she drifted peacefully away.

Those who are thinking of retiring in the American style may consider the case of Joe and Irma. We are a work-obsessed, job-obsessed, money-obsessed, "success"-obsessed tribe. "Success," in middle-class America, is amassing dollars in six figures or more. Its symbols are luxurious residential surroundings, a prestige car, and time entirely free from the work-and-job necessities that are the lot of commoner folk.

Joe and Irma had all this. They entered their retirement years in full expectation that the reward of their labors would be a happy, or at least a satisfactory, life experience in their later years.

It did not pan out that way. With our standard cultural expectations, it seldom pans out that way. When leisure is thrust upon us, we don't know what to do with it. One of the saddest sights in the United States is the line of old people at the cut-rate liquor counters in our retirement centers on the day the Social Security checks arrive. This is a national confession of the failure.

As distress cases go, Joe and Irma were not so badly off. They were, in fact, about average. They knew the misery of meaninglessness, yes, but there are far worse cases.

There are far better cases, too, and that brings us to the central message of this book: you do not need to be aimless and unhappy just because you are old and jobless. It is statistically true that retirement in the United States is, more often than not, a disaster to the retiree. But, though the odds at this writing are against you, no single individual *has* to be a *negative* statistic. You can be a *positive* statistic. The victors over age are still a small minority, but their victories *are* being won, and regularly. They can be increased until the trend is reversed. Radiant happiness, right to the end, can be achieved. This book contains some suggestions as to how. And as we will presently see, the so-called "end" may really be only the beginning.

# 9 Sex, Sublimation, Synthesis, and Satisfaction

PEOPLE ARE ALWAYS ASKING, "What do you do about sex in the later years?" The only satisfactory answer I have been able to find is, "Do whatever seems to work best for you, with only two provisos: don't exploit or hurt anybody, and steer clear of dogmatic 'expert' pronouncements and pat formulas on what has to be done about it." The truth is that nothing has to be done about it. If one chooses to do something about it, he will come to no grief if he follows the axioms given above. All avenues of discussion, research, speculation, and debate seem, in the end, to wind up in that simple statement. No "right way" has yet been found that applies to everybody.

However, since sex seems to be much on people's minds these days, we may as well "review the literature," as the academics say. The one outstanding fact about sex is that it seems to be the most satisfactory way yet found to continue the species. This is about the only positive, definite statement that most people have been able to agree on. Beyond that, all is speculation, debate, turmoil, confusion, and dealer's choice.

Human beings have manifested every imaginable quantity, quality, and variety of sexual activity at every imaginable age—and have lived splendid, noble, productive, creative lives with no overt sexual activity at all. There was once a Turkish sultan who had four thousand concubines and had been to bed with them all. He accomplished little else; his regime was a blight on his country. At the other extreme have been celibate monks, nuns, renunciates, and eunuchs who have made spectacular, even immortal, contributions to the evolutionary development of humankind.

Simone de Beauvoir, in her overlong book, *The Coming of Age*, makes a good deal of the fact that sexual appetite and ability continue into the later years. It should not surprise anybody when a woman well past the half-century mark gives birth to a child or when a man in his nineties demonstrates sexual potency. Alfred Kinsey made such a big thing out of tabulating sexual "outlets" that he had a whole generation running around trying to make sure they were having the standard number of "outlets." This corresponds to Wilhelm Reich's insistence on almost continuous "organ satisfaction." It is, of course, ridiculous, and belongs in the same category of human activity as pie-eating contests. If you've been reading Masters and Johnson you already know that sex can appear in freak, illegal, and grotesque forms as well as in the standard garden variety, so that takes care of that.

My own observation suggests that the thing that inhibits sex in the later years is not sheer animal capacity at all. Many men sixty-eight and over with whom I have discussed the subject declare they are as competent in physical relations now as they were at fifty-eight, forty-eight, or even in their thirties. Perhaps more so, because over the years they have gained a better understanding of how women respond to sexual stimuli and a keener appreciation of intimate sharing and caring. Older women with whom I have talked express similar feelings about sexual life in the later years.

There are, of course, natural and societal inhibitions to biological sex in the later years. One of these is the discovery, made with the years, that sex involves one with a great deal more than the mere organ responses of another human being. One gets involved with another person's complete constellation of crotchets, neuroses, preferences, aversions, and idiosyncracies. These may or may not harmonize with one's own set of reaction patterns, and one is not so sure, for all the pleasure of sex, that one wants to "get into all that." Then, too, it must be said that even sex, for all its delight, can begin to get a little repetitious after two-thirds of a century or so, and one tends to be on the lookout for adventures in other dimensions. One is more inclined to look at a sex possibility with a slightly jaded eye, "all this fuss and folderol—for *that*?"

Perhaps the greatest inhibition is a societal one—one does not like to make a fool of oneself. Nature has rigged the game so that young persons have greater sex appeal than older persons. This factor of sex attraction is quite independent of sex capacity and applies to both sexes. Both young and old women are sexually aroused more by young men than by old ones; both young and old men are sexually aroused more by young women than by old ones. Even in our "sexually emancipated" era we are all supposed to play the game, and if we don't there are penalties. Old men who are too obvious about their sexual appetites are "dirty old men," and there are even harsher names for old women with the same proclivities. There is indeed no fool, in society's eyes, like an old one who can't keep his or her sexuality well under control. One naturally does not like to be made a fool of, or to make a fool of oneself, so one watches one's step. Taken all together, the emotional and social odds against sex are generally sufficient to slow down its urgency in the later years. Not that that alone necessarily extinguishes it, but it does help make whatever solution one has decided on acceptable.

Other factors tend to put an entirely new light on sex as we age. As usual, on this subject Carl Jung has been and remains far ahead of his psychiatric colleagues in observing the special psychology of aging. First, he pointed out that, at any age, no human being is, psychologically speaking, entirely of one sex.

Psychologically, we are androgynes. Each man, Jung postulated, carries in his psyche a female counterpart called the anima; each woman carries in her psyche a male counterpart called the animus. This is represented in oriental philosophy by the yin-yang symbol:

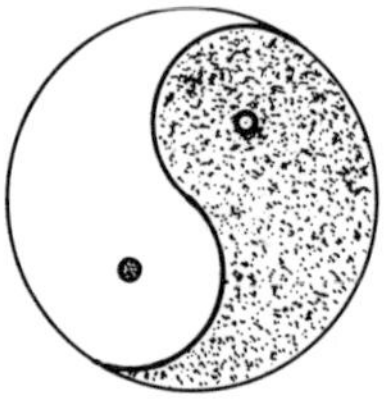

The whole—the complete circle—is made up of the intertwining yang (male) and yin (female) principles. Each of these carries within it an element of the other, represented in the symbol by the small dot circles.

In both Eastern and Western philosophies the male and female principles embrace far more than raw animal organ sexuality. The yang principle represents all that is idea-sparking, penetrating, illuminating, and inseminating, in intellectual as well as physical dimensions. In these same realms the female represents all that is nurturing, receptive, caring, and compassionate. And each principle, to make it completely functional, has within it an element of the other. In these perspectives the sun has often been spoken of as a great male archangel continually inseminating, with his rays, the great female archangel earth—who takes his rays into her dark womb and there, with tender nurture, sees to their development, finally yielding them up, in her great fertility, as grass, trees, flowers, fruits, grains, birds, animals, fish, people, and all the incredible abundance of this planet.

Thus, theoretically, there can be such a thing as a rich and rounded sex life that is entirely philosophical and has no organ expression at all. In the Western tradition this has been known as sublimation; in the East it is called the raising of the kundalini force. Both postulate a basic *evolutionary* thrust that can express itself in infinite ways without ever surrendering its purpose of raising consciousness to ever higher levels. If it is channeled into biological sexuality, it seeks to fulfill its purpose through many generations of offspring. If channeled into the upper mind, it seeks to accomplish the same evolutionary purpose in a much shorter time, possibly bringing a millennial evolutionary leap in the lifetime of a single individual. The yogic kundalini force is conceived as a current of cosmic energy running strongly up the spine from the lower sex-energizing centers to nourish and vitalize the brain and higher mind. Similarly, Western sublimation assumes a conscious direction of thought-energy from the lower and narrower toward the higher and broader reaches of intellect. This, it is claimed, opens the practical possibility of a nonorganic yet entirely fulfilling sex life—the philosophical one. Those who have achieved it insist that there is a joy so far exceeding the thrill of orgasm that the latter must be considered a pale and remote reflection of the prime, the original, the pristine source-ecstasy. Tantric philosophy maintains that this prime joy is approached through full realization, within one's individual psyche, of both male and female principles. There are, of course, many other versions of this same very ancient and fundamental human experience.

Apropos such things, a story is told about the Hindu Swami Muktananda. During a question session at Muktananda's New York headquarters a young man asked, "Swami, why are you celibate?"

"You know," the swami replied, "the sense of joy a rich sex experience gives you?"

"Oh, *yes*," replied the young man eagerly.

"The feeling of power, fulfillment, release?"

"Yes! Yes!"

"The experience of competence, of satisfaction?"

"Yes."

"That," said Muktananda, "is what *I* find in celibacy."

Still another aspect of the universal male-female duality

shows itself spontaneously, without any conscious direction, in the later years. Jung pointed out that we do not need to theorize about our yin-yang natures, since we see them expressed all around us. Once basic animal sexuality has been satisfactorily demonstrated, the *other* side of each sex seeks to express itself. Men who have been active in the world's affairs tend, in their later years, to be more content with pottering about the house, digging in the garden, and perhaps even doing a little cooking. Women who had expressed themselves in the creativity of the home often develop a strongly masculine tendency to branch out and take charge of affairs on a broader scale.

So we find the latent "feminine side" of men allowed more opportunity of expression in the later years, and vice versa. A man known in his early years as a go-getter may in his retirement become a contented stay-at-home. (World War II Admiral Chester Nimitz declared on his retirement that he intended to become a baby-sitter!) On the other side of the sex equation, many women who had given their lives to domestic pursuits have, in their later years, burst out into "masculine" business activities ranging from gift shops and delicatessens to trucking firms and steam-engine factories.

An enormous and sometimes spectacular creativity, such as that of the paintings by the famous Grandma Moses, may arise from these long-denied potentials—from these "I've-always-wanted-to" drives. These leftover urges can also take a destructive turn. One successful businessman who had been a teetotaler into his fifties abruptly decided that his life had not contained enough carousing. In four flaming years he drank himself out of his fortune and into his coffin.

We who may be a little ahead of you on the age track have found much satisfaction in some of these later-life fulfillments of early-life wishes. By all means, we would say, take the trip, start the business, bake the cake if the urge is strong. Even burn a few late candles if you must, while bearing in mind the hazards of carrying the thing past the point of no return. The trick seems to be to stay on the upward spiral and not get caught in a downdraft. Just as the global weather sequence causes equinoctial gales with the change of the seasons, so does a change in life's seasons bring its inner storms—and sometimes these can blow up quite a breeze!

# 10 What Needs to Be Done?

OUR NATURAL CONDITION is to be happy. If we are not, something is blocking it. It behooves us to find out what the block is and get rid of it.

As an aid in our self-diagnosis we will list some of the things that bother a great many people. None of these things are in any way "abnormal," if we take as a norm behavior patterns experienced by a large number of persons who, though burdened, are nonetheless making continuous and relatively successful contributions to our society. We are not talking about individual situations that require specialized professional psychiatric care. Rather, we have in mind problems that are quite commonly overcome by ordinary means by ordinary men and women who have learned to mobilize their own resources on their own behalf.

Here's the list. Go through it and check any that apply to you. If you are troubled by something not listed, jot it down in the blank space provided at the end. Be ruthlessly candid. Remember, you're looking down on yourself, and on all the rest of us, from a higher plane.

My life is marred by:

**Booze.** I suspect that I'm probably drinking more than I should. There have been some unpleasant episodes, and I'm feeling foggy and jittery too much of the time. I try to kid myself that I "need" it to "relieve the tension," but sometimes I think that in the end it causes more tension than it relieves. Anyway, I'm bothered about it. I'm beginning to admit the thought that maybe my life would be better if I didn't drink.

**Financial insecurity.** I worry about money all the time. When I'm not worrying about how to get enough to keep body and soul together for basic survival, I'm at my wits' end from scheming how to get more, or consumed with envy of people who are rolling in it while I have to slave for nickels. I wish I had a million dollars. Money, or rather the lack of it, is the curse of my life.

**Sex (not enough).** I'm just not getting enough sex, darn it, and what I get, if any, is not the right quality or in sufficient quantity. It bugs me to the point where I could bite my nails, which I sometimes do. I think about it all the time. If I could only somehow get enough, or just get it off my mind, I'd be a lot happier. I guess I'm just a sex nut, and not a very happy one.

**Sex (too much).** Honest to Pete, I'm satiated. It's repetitious and humdrum. It's got me up to my neck in kids and made a slave of me trying to provide for them, and still it goes on. Sex is for the birds, not for people. I want out, but can't see any way to get out. There it is, everywhere and always, sex, sex, sex. Blah.

**Loneliness.** I'm just plain lonely. I wish desperately that something or somebody would come along and fill that great, gaping, lonely emptiness in me. There are enough people around, God knows, but I feel empty and lonely even when I'm with them. We can't seem to communicate in any real way, or in a way that gives me a satisfying sense of deep relationship. Empty loneliness is the blight of my life.

**Depression.** I get so low sometimes I almost believe I could shoot myself. I can't think of any real reason for it, but that only makes it worse. Actually, things aren't really so bad with me, except for these unreasoning fits of the blues. If I could only get rid of them, I think I could be fairly happy. But I can't.

**Anger.** I seem to be sore as hell all the time and can't figure out why. It's got so bad some people are beginning to think of me as a grouch, and that makes me mad all over again. Sometimes I'm mad without any reason, but what makes me really mad is not getting my way. If I have my heart set on something and don't get it, or if I have a failure when I wanted a success, or if I get turned down when I wanted to be accepted, boy, then I get *really* mad! The worst part of all is when I get mad at myself for a failure of some kind. That gets me so *down* that I find some reason to fix the anger on somebody else so I'll feel a little (though not much) better. I'm tired of being sore. Anger is spoiling my life.

**Fear.** What the dickens am I afraid of? Of course, the papers, the radio, and television are always carrying on about the murders, rapes, mayhem, wars, bloody accidents, assaults, robberies, beatings, danger of explosions, earthquakes, and poisoning everywhere and always, and I must say that doesn't add much to my peace of mind. But I have a funny feeling I'd be just as scared if there weren't any papers, radio, or television. It's the darndest thing—just plain raw, unexplained *fear.* Maybe I'm just plain bughouse. But it certainly pulls down my enjoyment of life to be unreasoningly afraid all the time.

**Bereavement.** I can't get over the loss of someone I loved very dearly. Just can't get over it. The sadness.

**Marital.** I'm one of those mentioned earlier. Spouse. Used to think he (she) was pretty nice, now I can't see how I ever got myself involved with *this.* Ruining my life.

**Greed.** What gets in the way of my happiness is that I can never get enough of whatever it is. Eat too much. Smoke too much. Then feel guilty and worry about it because both are bad for my health. Might kill me, but can't help it—*got* to go on. And it's not just the oral, either, but everything: money, power, position, prestige, property, grabgrabgrab, never any rest, never any peace. How can you be happy with *that* going on in you all the time?

**Children.** Sometimes I wonder why I ever was a party to bringing children into the world. No respect. No obedience. No appreciation for what you do for them. No sense of the right thing. Ruining themselves. I used to take pride in them and think children would mean happiness. Not any more. They're nothing but sorrow and disappointment. They've taken most of my life and ruined it, and now I don't have anything left but sorrow and disappointment. Happiness? Huh!

**Boredom.** I tell you, mister, I get so sick of this rut I could . . . well, I'll use a *nice* term, throw up. Same routine, day in, day out, plod, plod, plod. If that's life, mister, there's no sense to it. You can have it. You can take it home with you, and you know what you can do with it when you get there. There's no such *thing* as happiness, far as I know. Just boredom, and I'm sick of it. Sick almost to death.

**Impatience.** I'll tell you right away what's ruining *my* happiness, I know myself *that* well already. It's *impatience.* I want what I want when I want it, and that's *right now.* Fellow ahead of me is a little slow getting away at the light, I give him a blast of the horn—*roll it*, mister! Long red light waiting for a "walk" sign runs up my blood pressure a hundred points. I'm overdue for a promotion and a raise, and don't think the boss isn't going to hear about it. Lost my last two jobs that same way. Wrecking my life, but can't help it. Impatience.

**World.** Happiness? In *this* world? Come off it. War. Hate. Strife, conflict, wrangling, disagreement, and disagreeableness. Greed and ruthlessness and violence. Pollution—earth so totally poisoned that even the snow at the North Pole is full of lead from auto exhausts. Inflation. Unemployment. Tax exemptions for the rich, poverty for the poor. All nations bankrupt from buying atomic bombs and means to deliver them. And don't think I'm *imagining* this—it's all true. So don't preach to me about "happiness," Mister Pollyanna. Anybody who goofs off on a happiness kick under these conditions is a blind fool or an idiot. And I already "know myself" this well—I am neither.

**Self-Pity.** Nobody loves me. Nobody even *cares* for me. Not even enough to send me a postcard. Wouldn't you think they'd care enough about their old mother (father, uncle, friend, benefactor, cousin) to do at least *that?* But no. No gratitude. No human feeling at all. Take everything, give nothing. And now that I've given everything I have to give, I'm cast off like an old dishrag. That's it, exactly like an old dishrag, and that's all they care. Well, it's my own fault. I was foolish to ever think anybody could *have* any consideration. Certainly my unhappiness, if that's what you want to call it, is not *my* fault. It's just the general, total lack of consideration.

We could go on at greater length, but the above is a fair sample of the kind of happiness wreckers that plague the middle and later years of millions.

But remember, we are not talking about the "millions," we're talking about *you.* If nothing in the above list struck home, write in the space below, from your vantage point of high objectivity, what you consider to be *your* primary life wrecker.

By now you should have made your preliminary diagnosis. Now—and this is very important—*remain in a diagnostic frame of mind about yourself until your problem is solved, even if this turns out to be a considerable time.*

Success hangs on the trick of maintaining the same calm, concerned objectivity about yourself and your predicament that a kindly doctor would about a favorite patient. Never, *never* get angry or impatient at yourself for any reason. The intelligent doctor is fully aware that getting mad at the patient will not gain him his objective of making the patient well and happy.

Once you have made your diagnosis, write it down in a place where you can refer to it frequently. *This* is what is blocking the fullness of your experience of life; *this* is what must be dealt with. Never let it get out of mind long enough to get the best of you again. If you have checked more than one "roadblock to happiness," jot them all down. But work on them only one at a time, beginning with the most bothersome.

Remember at all times that it is only yourself you are going to reform, not anybody else, or the world. Seek "the serenity to accept the things you cannot change, the courage to change the things you can, and the wisdom to know the difference." Once you embark on changing somebody else, the wrangling begins and you have lost the ball game. At this point it is yourself, and no one else, that you want transformed into a better, happier person.

The sequence, then, is this:

1. Diagnose your case.
2. Decide whether you want to do something about it.
3. If the decision is negative, forget it.
4. If the decision is affirmative, go into action.

The critical point is the decision about whether you want to hang on to the quirk that is marring your happiness or get rid of it.

So decide . . . and act on your decision.

At this point some new considerations will arise. Suppose you have decided to do something about it, whatever "it" may be, but don't know what to do. Of the many things you might do, your desire, of course, is to select the best. But "best" is a value judgment of a kind that calls for the maximum use of your whole brain.

We have said—and it is true—that you can bring yourself to a fuller, happier life by mobilizing your own inner resources. Some of your greatest inner resources come to you through your brain. And now prepare yourself for a shock: Most of the "best scientific minds of our time" have been trying to solve their problems *by using only half the brain* while suppressing the input of the vital "other half." It would seem appropriate, in our extremity, to consider the appropriate use of the entire brain.

# 11 That "Other" Brain

DR. WILDER PENFIELD, world-acknowledged Canadian leader in the field of brain surgery, clarified the distinction between *mind*, which is an immaterial force, and *brain*, which is a material structure. He showed that the brain is not an organ that produces thought, as was once supposed, but an instrument *used* by thought for its own ends. He demonstrated that mind can use even a damaged brain to accomplish its purposes, even though the brain circuitry normally used for these ends had been destroyed. Sir John Eccles, the Nobel Prize-winning British neurologist, has also endorsed this theory.

This profoundly shifts the ground of any philosophical discussion of thought. The question "How does the brain work?" becomes secondary if not obsolete in the face of the more urgent question, "What are the objectives of mind?"

At this point it might reasonably be asked what this has to do with winning our personal victory over age. The answer is that we have to think clearly in order to achieve this victory. Science, as we are now aware, has badly fuddled the thinking of the past three centuries by its insistence on using only half the brain circuitry in reaching its conclusions. Modern science is quite literally half-witted. If one is to think straight, correctives must be applied to one's "normal," "logical," "scientific" mode of thought.

A brain is not a mind, but it can be a great convenience to the mind that knows how to use it. Recent discoveries in the structure and function of the brain are so fundamental that anyone who is seriously trying to think clearly should know about them. Winning the victory over age calls for the clearest possible thinking.

It has long been known that the cerebrum, a vital center of consciousness, is divided into right and left cerebral hemispheres, each governing the side of the body opposite it. It is now known that the difference in function between the two lobes goes a great deal further than mere "handedness." They govern two entirely separate "modes of apprehending reality," distinct in quality and divided in function.

One of the outstanding research teams in this area was headed by Brenda Milner of McGill University in Montreal. This team devised special tests so that specific areas of brain damage could be related to specific tasks. They demonstrated that the left lobe specializes in speech, memory of words, arithmetic, seeing things in distinct parts, and being locally and immediately "practical." The right lobe has jurisdiction over judgment of movement and position in space, seeing things whole and entire, musical discernment and "spirit," in the sense intended by the Indian musician who said, "True music is not for the mind or the body but for the spirit, which exists independently of each."

Another approach to studying the specialized right- and left-lobe functions was employed at the California Institute of Technology, by Professor Roger Sperry and his associates. These researchers, working at first with animals, separated the two lobes by direct surgery, then studied the resulting behavior of the creature operated on. The two cerebral hemispheres are joined by a bundle of connecting nerves, the *corpus cal-*

*losum.* When this cord is cut, the two hemispheres, normally utilizing this connection to act in cooperation, are forced to function independently. This allowed more precise study of their special jurisdictions. Surprisingly, this drastic brain surgery did not seriously affect the ability of the animals to continue satisfactorily with their usual activities.

The next step was to try the same operation on humans. Such a drastic step could not be justified, of course, except in cases of extreme suffering that had been declared hopeless of cure by all less extreme measures. This operation was performed on humans by Drs. Bogen and Vogel of the California College of Medicine. The theory was that if a "hopeless" epileptic had a bad seizure in one cerebral hemisphere, the other would be functioning well enough to enable him to take the proper medicine or summon help. It worked, to an extent that allowed many of the most extreme cases to leave the hospital.

Rather as a side effect, these operations provided a supply of "split-brain" human beings that allowed further research into the functions of the two halves of the brain. The left hemisphere, it was found, is the verbal one, governing things having to do with words and symbols—writing, arithmetic, and the like. The right hemisphere is dimensional, presiding over the location and relationship of objects in space—drawing, vibrations (music), mental structures, and so on. In the special case of left-handed people, who comprise about 5 per cent of the population, there is somewhat less specialization and somewhat more overlapping of function, but the same general principles apply.

One experiment involved literary and artistic ability. Split-brain people were asked to test out each hand in turn, first at writing, then at drawing. The capacity to write longhand was still in the right hand, but this hand could no longer *draw* in a way that made any sense. When asked to copy a square, for example, the result was a jumble of four disconnected corners. The left lobe simply could not see things *whole*! The left hand, however, could draw a perfectly satisfactory square, but could not write. It was an impressive lesson in the necessity for utilizing both lobes, the one that can "see the whole picture" and the one that can interpret it verbally.

Let us now leave the operating room and psychology lab and consider some evidence from the laboratory of our own lives. I will prime the pump by citing an example from my own experience, then ask you to comb your memory for similar cases:

One time I was walking down one of the main avenues of New York City when I ran into an acquaintance from the Midwest whom I hadn't seen for ten years. He greeted me warmly, called me by name, and shook my hand in a wave of glowing friendliness. Normally I am verbally active enough, perhaps to a fault. But on this occasion words literally failed me. While my mind (the mind I now realize is my left lobe) desperately scanned its data for some kind of verbal clue, I stood there foolish, blank, stupidly tongue-tied and embarrassed. This seemed to go on for an eternity of seconds, when suddenly, out of "nowhere" (though I now realize it was my right lobe seeing things in total perspective), the *whole picture* of my total relationship with this man swam into view: where I had known him, when, and all the circumstances of our relationship, including such details as his street address and telephone number! All I had to do was read it off from the visual record presented to my mind. I take this as a classic example of "seeing the whole picture" when the analytical, fragmented, word-by-word process fails to solve the problem immediately presented.

The intent of this chapter is to invite you to use your whole brain instead of following our national custom of using only half of it. This involves temporarily silencing the endless flow of words proceeding from the left lobe while we open ourselves to the mind-expanding input of our nonverbal but far-seeing right lobe. Then we begin to see things whole. We are then capable of action that takes into account our total cosmic situation. This is vital to any permanently satisfactory solution to the challenges of aging.

People of genius have always combined the functions of the two lobes, naturally and intuitively. Mozart, as an example, conceived his symphonies all at once, in a flash ("holistically")—a characteristically right-lobe thinking function. Then, turning the project over to his logical, workaday left lobe, he undertook the systematic effort of writing them out so they could be played by an orchestra. Both activities were necessary before anything could be produced that might be called a symphony. Without both, the intuitive conception would have amounted to no more than a passing fancy. Newton, in the famous "falling apple" formulation of the gravitational pull as a function of inverse squares, and Einstein, conceiving his limited theory of relativity, followed, by their own accounts of the process, the same pattern. Madame Curie had a clear hunch that whatever caused photographic plates to be "exposed" in the dark was to be found in a rock called pitchblende. She then worked through tons of the stuff to separate a tiny speck of a new element, radium.

The reason for this review is, of course, to enable us more wholeheartedly to bring our whole brain to bear on the problem of winning our victory. During the past decade there has been, in the West, a great rash of techniques, variously called yoga, relaxation, meditation, or mind control, all aimed at restoring our mental balance by strengthening our right lobe "whole seeing" vision so that our left lobe "practical" action may function more practically. It is a clear necessity of our situation as persons who want to think our situation through, then act appropriately and creatively, to make the very best of it.

# 12 The "Death" Sequence

WE NOTED A WHILE AGO that fears tend to divide themselves into those of living and those of dying. Having examined the first of these, let us now have a look at the second.

Historically, the question of whether there is such a thing as death has always rested in the realms of philosophy and religion, and the consensus has always been that there is no death, but only continuing life. Science has, until lately, taken the opposing view, that there *is* such a thing as death and that it ends the experience of human life. Since the thinking of our century has been dominated by linear-thinking, left-hemisphere scientists, we have all been painstakingly taught to believe in death, no matter what the philosophers say.

Quite recently, however, science has produced unshakable evidence, amenable to repeatable investigation, that human personality survives death with memory, purpose, vigor, and meaning. In other words, science is now ready to concede that the philosophers and religionists were right, on this point at least, all along. This makes a profound difference in one's personal outlook as we move into life's second half. We must attend carefully as the new evidence unfolds.

The most consistently influential voice in this historic overturn has been the quiet one of a professor of psychiatry at the University of Virginia School of Medicine. His name is Ian Stevenson. As a behavioral scientist Dr. Stevenson was impressed with the frequency with which children, particularly in countries where reincarnation is accepted as doctrine, speak of remembering a previous life in another home with another family.

Stevenson knew that reincarnation had never been scientifically investigated. As a religious doctrine it was part of Christian belief until the sixth century after Christ. Then, in a close vote that was dictated more by politics than by factual investigation (if a few bishops had stayed home that day, all Christians would now be doctrinaire reincarnationists), an adverse decision was made and reincarnation was declared an anathema. Nevertheless, some of the best of our thinkers, both ancient and modern, have endorsed the doctrine. These include Pythagoras, Plato, Plotinus, and Origen, among the ancients, and Hume, Kant, Fichte, Schopenhauer, and Broad, among the moderns.

Stevenson decided the time had come to go after the issue scientifically. He went to the Middle East, India, and Ceylon early in the 1960s and began gathering data. His method was simply to check out the details of the stories the "reincarnate" children told. Such tales, of course, are also common among American children, but U.S. parents "shush" such talk so sternly that it disappears into the rubbish heap of nonsense to forget. In Eastern countries the information circulates more freely and is more readily available for careful investigation.

Stevenson uses two methods of research. First, he takes down with great care what the children say when they tell about their supposed previous lives, then checks out any possibilities of their gaining the information by means other than their own memories. After that he arranges to have the children taken to their alleged former home environment and have them identify streets, houses, locations, schools, teachers, friends, and relatives. His second method is to trace possible reincarnation through birthmarks. The Eastern legend is that when people die violent deaths with hatred in their hearts, they may reincarnate with a birthmark at the site of the death wound.

He gathered hundreds of cases, so the University of Virginia collection is large and of high quality. The cases were astoundingly veridical. Small girls would recognize grown men on sight as their children in a former incarnation, and remind them of verifiable—and verified—details of the grown-up's childhood. In one case (not Stevenson's, but investigated by another reliable American professor) a child was taken to visit her former home in a city at some distance from her present home. She recognized family members by name and personal history, and, as final proof, declared that she had hidden a considerable sum of money in a certain wall. When the wall was torn open in the presence of witnesses the money was found. In a birthmark case the subject remembered that in a previous life he had been a gangster in Istanbul and was killed in a fight with the gendarmes. At the spot where he claimed the bullet had entered his head there was a round birthmark of bullet-hole size. Puzzling over the case when back home in Virginia, Stevenson reasoned that a high-velocity rifle bullet, fired at the close range claimed, probably would not lodge in the victim's skull, but would pass right on through. He flew back to Turkey and asked the subject if he'd let him have a look at the *back* of his head. Parting the man's bushy black hair, he found, just where it should be according to the remembered angle of fire, another birthmark of bullet-hole size the Turk didn't even know he had.

Stevenson's work, of course, is not the only evidence, or even the best evidence, of reincarnation—nor does he represent it to be. But he did the best job so far done in presenting the material in a form suited to the close scrutiny of his fellow scientists so that any who cared to could dip into the ceaseless flow of available cases and repeat the investigation. In 1966 he published his now famous book, *Twenty Cases Suggestive of Reincarnation*, a 400-page careful and detailed documentation. His book might be compared with a tremendous underground explosion, slowly working its effect deeper into the substructure of comprehension and higher toward surface awareness. There has been no attempt to debunk it, for the excellent reason that it is not debunkable. One scientist tentatively ventured the observation that these phenomena *might* not be reincarnation at all but cases of involuntary mediumship in which a discarnate spirit had taken possession of a flesh-and-blood body. But in such case, if reincarnation has not been proved, survival has, and we are back to the inescapable fact that human personality survives bodily death. Qualified scientists stand in two camps on the issue. The survivalists hail Stevenson's work as further evidence to support what they knew all along. The materialists solve the problem of embarrassing evidence by simply not looking at it.

The immediate effect of Stevenson's work was to throw into new light and fresh perspective some of the better mediumistic communications that have emerged during the past half-century. If at least some of us are reincarnates, a host of questions demand answers, around the general theme, "What is the mechanism of reincarnation? Why do some people reincarnate while others do not?" Mediumistic communication takes place in a continuous and boundless flow. It has been discounted because so much of it, like the communication of flesh-and-blood life, is trivial and inconsequential. But some of it is sustained, relevant, high-level discourse—clearly the work of superior intelligences. The best of it has been available in book form and yields at least tentative answers to many of our questions. Among these we would have to list *Raymond*, the account, by the outstanding British physicist (father of the neon light) Sir Oliver Lodge, of the communications from his son Raymond, who was killed as a soldier in World War I. Certainly *The Betty Book* and *The Unobstructed Universe*, accounts by the American naturalist Stewart Edward

White of the mediumship of his wife Elizabeth, would need to be included. Arthur Ford's *Unknown But Known* and *The Life Beyond Death* shed much light. The classic communications of F. W. H. Myers, through the Irish medium Geraldine Cummins, rank with the very finest. They appear in the form of two books, *Beyond Human Personality* and *The Road to Immortality.* Myers, a Cambridge professor in life, had been president of the British Society for Psychical Research, which was the pioneer sponsor of modern survival research. The Myers communications confirm the general content of all such messages, asserting that the evolutionary impulse toward ever expanding awareness is cosmic and eternal rather than local and temporary, and does not stop at death. The main thrust of the Creation is not physical forms but mental ones, easily capable of casting off one physical form to take on another (as we give up the child's body to take on the adult's) or existing in abundant, energetic life with no physical form at all.

Bodies can destroy bodies, true. But since bodies are only temporary conveniences, say our discarnate informers, the loss or destruction of one does not involve losing anything of permanent value. The wise human being on earth, these communications emphasize, progresses toward wisdom through ever widening and ever deepening understanding. Though the exact experience of a life beyond death is still hidden from our earthbound senses, there is no longer any question that there *is* such a life. The evidence is in. It is multiform and abundant.

Belief, however, hinges not on "proof" but on choice. We believe what we *choose* to believe, not what has been demonstrated to be true. Many societies still do not believe that any human being ever set foot on the moon. They "know" that such a thing is impossible. The television tapes recording the event are, to them, just another instance of Hollywood trick photography.

In a time of nearly total breakdown of trusted authority, what we choose to believe is personally decisive. We live in such a time. Business and commercial authority has lost its hold because the messages of advertising's hired liars finally became unrelated to the clear truth. Political authority has shown itself untrustworthy by the repeated scandals throughout recent political history. Religious authority no longer holds because its deviations from its own doctrine became so obvious that all could see them. Science, our reigning authority for the first half of this century, has lately turned out to be so frequently and so catastrophically mistaken, in a half-century-long series of disastrous blunders ranging from the miscalculation of the effects of chemical fertilizers and pesticides to foolhardy errors in estimating radioactive fallout, that we must sadly abandon this last stronghold of belief.

This leaves us no authority but our own inner instinct for the truth. The feeling within us, the deep something that says "this has the ring of authenticity," the *instinct* for truth, is the only authority remaining to us—and, of course, it is, has always been, and will always be the best. It is this instinct that we must invoke when presented, as we so often are, with two or more disparate alternatives that have been "proved" with equal certainty. It was this instinct that Jung referred to when he wrote:

> Here my physician's conscience awakens and urges me to speak. I have observed that a directed life is in general better, richer, and healthier than an aimless one, and that it is better to go forward with the stream of time than backwards against it. To the psychotherapist, an old man who cannot bid farewell to [earth] life appears as feeble and sickly as the young man who cannot embrace it. In many cases it is the same childish covetousness, the same fear, the same obstinacy and wilfulness, in the one as in the other. As a physician I am convinced that it is hygiene—if I may use the word—to discover in death a goal toward which one can strive; and that shrinking away from it is something unhealthy and abnormal which robs the second half of life of its purpose. I therefore consider the religious teaching of a life hereafter consonant with the standpoint of psychic hygiene. When I live in a house which I know will fall about my

> head within the next two weeks, all my vital functions will be impaired by this thought; but if on the contrary I feel myself to be safe, I can dwell there in a normal and comfortable way. From the standpoint of psychotherapy it would therefore be desirable to think of death as only a transition—one part of a life process whose extent and duration escape our knowledge. Despite the fact that most of mankind does not know why the body needs salt, everyone demands it because of an instinctive need. It is the same in the things of the psyche. A large majority of people from time immemorial have felt the need of believing in a continuance of life. The demands of therapy, therefore, do not lead us into any bypaths, but down the middle of the roadway trod by humankind. And therefore we are thinking correctly with respect to the meaning of life.

A startling confirmation of this professional view has recently been provided by the psychiatrist Elisabeth Kubler-Ross, whose 1969 book *On Death and Dying* attracted such widespread attention. Adhering to a viewpoint of strictly clinical observation, with no philosophical overtones, Dr. Kubler-Ross identified five "stages" of mental reaction commonly observed among dying patients in hospitals: Denial, Anger, Bargaining, Depression, and Acceptance. Staying safely within the bounds of accepted clinical jargon, she in no way jarred the standard medical-materialist stance, and so her book was received with great interest and caused no controversy. Continued research, however, led her to the "discovery" of an astonishing "sixth stage": After the acceptance of death comes, sometimes right at the last moment, the discovery of life! Dying patients consistently evidenced glimpses of a continuing life beyond the physical-material plane. This called for another book, published in 1975 with the title *Death, The Final Stage of Growth*. In this work she concludes: "Only the body dies. The self or spirit, or whatever you may wish to label it, is eternal. You may interpret this in any way that makes you comfortable."

The latest, and still unreported, work of Dr. Kubler-Ross

involves her observations of people who have undergone death-and-return experiences. These are patients who have "died" in the sense that all vital signs—brainwaves, heartbeat, breathing—had stopped for an extended interval. In many cases they had out-of-body experiences while the "dead" physical body lay on the operating table, discovering that life continues with peace, ease, harmony, and a sense of general cosmic "all-rightness." On returning to their bodies and physically reviving, they have astounded doctors by giving totally *accurate* accounts of what was said and done by those present during the "death" episode. From their position above and apart from their dead bodies, they were able to see and hear all that went on. Dr. Raymond A. Moody, Jr., has studied 150 similar cases, and reported on many of them in his recent book, *Life After Life.*

The work of Drs. Kubler-Ross and Moody gives independent confirmation of an important study by Dr. Karlis Osis, Director of Research for the American Society for Psychical Research. In his excellent 1961 monograph, *Deathbed Observations by Doctors and Nurses*, he reports the findings of 1,600 American doctors and nurses who observed thousands of patients who were conscious during the last hours of their physical phase. Many had experiences that careful analysis showed could not have been caused by medication, shortage of oxygen, or the illness itself. As with the Kubler-Ross patients, they knew happiness and joy, serenity and peace. Some produced strongly evidential material pointing to factual knowledge gained through contact with the world beyond. In one case a woman we'll call Mrs. B. lay dying in a British hospital. During her stay in the hospital, her sister Vida had died, but this fact was kept from her because of her own serious illness. Shortly before her death Mrs. B.'s face brightened, and she said, "Oh, it is so lovely and bright. I can see father" (her father was long "dead"). Then, with a puzzled expression, she said, "He has Vida with him." In a few moments she died. Mrs. B. had had no "normal" way of knowing her sister had passed on. These experiences of visions and bliss at the end and the meeting of relatives and friends who had gone before are not due to childhood teaching, or to religious or cultural influences. Dr. Osis demonstrated this by repeating his investigations in India, where the culture and religions are totally different, though the essentials of the experience of passage were the same.

Some people have no fear of death, but do fear a crippling illness that might reduce them to a vegetable state far worse, to them, than death. As insurance against such an eventuality many members of the Phenix Clubs carry cards that read: "I am not afraid to die. I therefore request my family, and medical attendants acting under their sanction, to withhold heroic life-saving measures when these could not restore effective functioning. Should my body be rendered permanently incapable of the useful transactions of living, I wish to be allowed to die without unnatural delay."

To a fully death-adjusted person the death of a loved one can be a greater concern than one's own death. Even when survival is intellectually accepted, the emotional vacuum left by the removal of a beloved life from the immediately apparent scene can produce a temporary but wracking crisis. There is help to be found in such crises from people who have lived through similar ones and have written some good books about coping with them. One such book is *On Becoming a Widow*, by Clarissa Start, with its strong message that sorrow is bearable and that the world is so bountiful it can fill even the greatest vacuum. *A Manual of Death Education and Simple Burial*, by Ernest Morgan, is full of practical information concerning costs and methods of body disposal, and includes a valuable checklist of things to be done by the family when a death occurs.

The greatest of all fears is the fear of death. When that is overcome—by the knowledge that there is no such thing—the other fears become manageable.

Now let us return to our original undertaking, the self-diagnosis of you. *You* are the real doctor in everything that affects you, and *you* are the patient. If you are in the grip of fear, and it is Big Fear—the Fear of Death—you can remain ill or you can heal yourself. You choose what you believe. It is all a matter of belief.

Belief, however, can be rational or irrational, depending on whether the believer has a reasonable basis for his belief. Rational belief grows out of experience that has stood the test of scientific scrutiny. The survival evidence reviewed in this chapter makes possible a fact-based, national belief in a thrilling and meaningful life beyond death.

# 13 Moments of Seeing All

A FAMOUS STORY about the ancient Greek philosopher Thales of Miletus points up the fact that we probably generate more anxiety than is necessary over the time and manner of our eventual "death." Thales, convinced that death was no more than a transition to an altered state of consciousness, was perfectly calm about this matter. He used to teach that, since physical existence is only temporary, it doesn't make any difference whether you stay in it or "die" and move to another mode of being.

"Well then, Thales," said a heckler, "if it doesn't make any difference whether you live or die, why don't you die?"

"Because," replied Thales, "it doesn't make any difference."

The great seers of history, including current history, have lived at all times in complete equanimity concerning their personal death. They "see all," to use Browning's phrase, and, seeing all, have no fear. This same insight, normally lacking in less favored people, often comes instantaneously in moments of stark "this-is-it" confrontation with the immediate likelihood of one's own death. These insights are so numerous and so consistent as to suggest strongly that they point to an important truth.

The late Alan Watts told of a visit with the former German diplomat Karlfried von Durkheim, who had studied Zen in Japan and later had a school for meditation in Germany. Durkheim was working at the time with people who had great spiritual crises during the war. "Sometimes people have a natural spiritual awakening when it appears that all is finished for them," he told Watts. "This happened often in the war. There were three principal ways in which these crises came about. You heard the whistle of a bomb coming at you and knew that this was quite certainly the end. You accepted it, and quite suddenly the whole universe made sense. All problems, all questions vanished, and you understood that there is no 'you' other than the eternal. But the bomb was a dud, and you lived to remember the experience. . . . You were in a concentration camp, and had been there so long that you were fully convinced that you would stay there for the rest of your life. Finally you had to accept it, and in that moment you understood everything. . . . You were a displaced refugee far from home. You had utterly lost your friends and relatives, your possessions, your job, your very identity, and saw no hope of regaining them. You accepted it, and suddenly you were as light as a feather and as free as the air."

The explorer David Livingstone (famous for H. M. Stanley's "I presume?") had one of these experiences. While walking on a jungle trail in Africa he was attacked and badly mauled by a lion. His gunbearers came up barely in time to save his life by killing the animal. Livingstone reported that, though he was quite certain that his death was at hand, a total serenity overtook him, so complete that he felt quite detached from the proceedings.

I have known something like this. One time during my aviation days I got lost in an extensive cloud that seemed to reach from the ground to the moon and cover the whole earth. I was flying a plane that was intended for aerobatics and had no instruments for blind flying. I knew my life was in immediate danger. Many a pilot, in exactly this situation, had ended splattered against a mountain, a tall chimney, or a power line. I prayed that God would bring me out safely, then put the plane's nose down into the murk, in a glide that could have ended in death. Suddenly I saw a farmhouse dead ahead, only that, through a hole in the cloud. I pinioned on it until I saw a pasture, and put the plane into a glide for a crash landing. When about twenty feet from the ground I could see under the cloud ceiling. The farmhouse was on the summit of a 500-foot hill. I could see the open valley below. I opened the throttle, flew on into the clear air of the valley, and was soon safely landed at my home airport. On solid ground and safe, I spent some time realizing, with a strange otherworldly calm, that I had been living on the border of death for a full half hour.

"I would gladly give my life if . . ." had always seemed to me a rather hollow and rhetorical phrase until I actually experienced it. While in the hospital at my son's death I would gladly have given my life if his could have been saved. There was absolutely nothing noble, selfless, or sacrificial about the feeling. It was simply that at that time my earth-life did not matter to me and his did. Such moments far transcend reason, logic, intellectual weighing of values, or examination of motives.

Sometimes these experiences occur in a dream or coma, but are remembered so vividly they are nonetheless real. My friend, the late trance medium Arthur Ford, had such an adventure. "One of the most vividly real experiences of my life," he told me, "lifted the question of survival, for me, out of the realm of faith and brought it clearly down to the plane of realism. I was critically ill of a serious heart malfunction. The doctors said I could not live, but as good doctors will, they kept fighting. I was in the hospital at Coral Gables, not expected to last the night. I heard a doctor speak to a nurse: 'He might as well be comfortable, give him the needle.' I wondered how long it would take to die. Then I was floating in the air over my bed. I saw my body, but took no more interest in it than in a cast-off coat. Then total, timeless unconsciousness. I awoke from this floating through space with no sense of having a body. Yet it was I, *myself*. I was in a green valley with mountains on every side, and everywhere an indescribable brilliance of light and color. Coming toward me from all sides were people I had known and thought of as dead. Those who had passed on in old age were young, those who had passed over as children were in mature spirit bodies. Then a gentle, radiant person stood by me. It was made clear that I would have to go back. At the doorway of return I balked like a spoiled child in a tantrum, braced my feet, and fought against going. There was a sudden sense of hurtling through space, and I found myself looking into the face of a nurse—in Coral Gables. I had been in coma, almost lifeless, for two weeks."

The Swiss psychiatrist Carl Jung had a comparable experience, which he described in his autobiography. Jung had a severe heart attack at sixty-nine and hung in coma, on the edge of death, for three weeks. His nurse told him afterward that he seemed "surrounded by a bright glow" of the kind she had often observed around dying patients. His images were so real and "so tremendous" that he himself believed, behind his coma visions, that he was close to death. He writes of it:

> It seemed to me that I was high up in space. Far below I saw the globe of the earth, bathed in a gloriously blue light. I saw the deep blue sea and the continents. . . . A short distance away I saw in space a tremendous dark block of stone, like a meteorite. It was the size of my house or bigger. . . . An entrance led into a small antechamber. A black Hindu sat silently in lotus posture on a stone bench. He wore a white gown, and I knew he was expecting me. On the left was the gate to the temple. . . . I felt sure I would receive answers to all my questions as soon as I entered the rock temple. . . . Then an image came floating up from below. . . . It was my doctor. . . . He had been delegated . . . to tell me [that I] must return. The moment I heard that, the vision vanished. I was profoundly disappointed . . . the world struck me as a prison . . . I felt violent resistance to my doctor because he had brought me back to life!

Long after his recovery from his illness, at a time when he was in excellent health, Jung had an experience of a kind that parapsychologists engaged in survival research call "highly veridical"—that is, producing information by paranormal means that can be checked out in waking-state observation. This account also appears in his autobiography:

> One night I lay awake thinking of the sudden death of a friend whose funeral had taken place the day before. Suddenly I felt that he was in the room. It seemed to me that he was at the foot of my bed and was asking me to go with him. I explained it to myself as fantasy. But in all honesty I had to ask myself, "Do I have any proof that this is fantasy? Suppose it is not a fantasy, suppose my friend is really here and I decided he was only a fantasy—would not that be abominable of me? . . . I might just as well give him the benefit of the doubt." The moment I had that thought, he went to the door and beckoned me to follow. . . . I followed him in my imagination. He led me out of the house, into the garden, out to the road, and finally to his house, which was several hundred yards from mine. I went in, and he conducted me to his study. He climbed on a stool and showed me the second of five books with red bindings which stood on the second shelf from the top. Then the vision broke off. I was not acquainted with his library and did not know what books he owned.
>
> Next morning I went to his widow and asked whether I could look up something in my friend's library. Sure enough there was the stool standing under the bookcase I had seen in my vision, and even before I came closer I could see the five books with red bindings. I stepped up on the stool so as to be able to read the titles. They were translations of the works of Emile Zola. The title of the second volume read: "The Legacy of the Dead." The contents seemed to me of no interest, but the title was extremely significant.

The theme of all these experiences, and of countless similar ones, is—and let us note this well, for it is important to us—acceptance. It was *acceptance* of an apparently imminent death that opened the way to realization of life in eternity. It was *acceptance* of the possibility of a personality surviving death that led to its demonstration. It was *acceptance* of mortal danger that led to its transcendence.

The principle of acceptance is an essential in winning the final victory over age. It applies not only in situations of desperate, life-and-death emergency, but also—and perhaps more especially—to our everyday living. Only when we calmly accept the "worst" can we know the *best* that is "yet to be"—this is the principle of acceptance.

It applies with special force to the matter of age. *No matter what your age is, accept it calmly, with all its implications.* Strange though it may seem, people have been actually suicidal over being thirty years old. Now, at seventy, it seems laughable that I had a brief but nasty depression at becoming forty. But it was no laughing matter at the time.

*Not* to acknowledge something that is obviously true is a mechanism psychologists call "denial." It is a strenuous activity and consumes enormous amounts of energy that could be put to better uses. We who have consciously and intentionally applied full and calm acceptance to such obvious truths as our chronological age and everything that goes with it, including societal shunning and our closer approach to the transition called death, can testify to the benefits of the practice. If it is true that life goes on, we might reasonably begin to consider the possibilities of further realms. Obviously one cannot have peace of inner mind if one is constantly fighting truth. If we want peace, we have to stop fighting, since fighting is the opposite of peace. John Cowper Powys stated it beautifully when he wrote, "People find to their astonishment that when they drop their eternal striving and clutching, real happiness flows in upon them in a brimming flood."

Those who compiled a Phenix Society handbook (*The Last Third of Life Club* to be reissued with the title, *The Life Club*) put the matter this way:

> We found that if we stuck with it, and utilized all available

supports, we would win through to a point where we could calmly (or with only brief moments of panic!) consider our approaching death as one of the ordinary facts of life. Our fears diminished to a point where they stimulated rather than inhibited constructive action. And then some amazing things began to happen. As the energy we had been pouring into denial mechanisms was released, new resources of mental, physical, and emotional vigor came pouring into us. As we directed our imaginative powers away from morbidity, they took hold of the problems of the second half of life with surprising skill.

Answers to formerly unanswerable problems began to appear, sometimes with unexpected ease. Were our physical and mental powers diminished? Very well, we'd put to the best possible use whatever was left of them. Were we going to die before long? Then we'd better pull ourselves together to do a good job of it. As we began to look around the universe with open eyes, we found worlds of resources opening to us we had never known existed. Why, the last of life might be a really splendid thing! With a new will, we began to explore life's new possibilities.

# 14 The Bottom Line: From Time to Eternity

ONCE HAVING MADE the intellectual leap, we have taken a long step in the trip from time to eternity. But we have by no means made the whole journey. The nimble intellect sprints ahead of the lagging emotions, which are likely, I have found, to remain timebound and reluctant. They must be coaxed to keep up.

The thing was well said by a seventeenth-century Zen poet named Angelus Silesius:

*This body is my enemy and friend*
*It ties me down*
*Is disobedient*
*I hate it and I love it,*
*And when at last we part*
*I shall be much relieved.*
*But—it will break my heart.*

We're all in pretty much the same predicament as friend Angelus. The victory over age is a victory over time. This implies a cheerful acceptance of the trip from time to eternity.

We won't make the journey cheerfully unless we're convinced that eternity is better for our purposes than time.

The only clear-cut evidence that it is so is the experiential kind. The only way we have to experience eternity, short of death or spontaneous mystical revelation, is to enter the meditative state. The vibrant, meaningful, vividly alive peace that is found in this altered state of consciousness is clearly better than anything offered in the turbulence of time. The point doesn't even have to be argued. It has, in fact, been demonstrated in behavioral science. Research recently completed by Dr. Charles Garfield at the University of California has shown that those with meditative experience have significantly less anxiety about their own death. That's why it's so important for us aging people to allow time each day for quiet inner contemplation.

There is other evidence that eternity is not only *ultimately* better but immediately real and present, that one doesn't have to die to begin experiencing it, and that it is not just for older people. It is nothing but the unutterable charm of eternity that keeps so many young people interested in what they call meditation. We begin feeling it the instant we let go of time and release ourself from its tyranny. We need only to step outside on a clear night when the moon is down to perceive that the universe is not upset. It sustains the turbulence of the stars in perfect peace and serenity. It sustains our personal turbulence with the same calm. We need only to identify with the universal peace, instead of with our individual turmoil, to hear its vast and still, small voice and let it share its calm with us. We will recognize this peacefulness as eternity, and find it far superior to the hubbub of time.

Much has been printed lately about people who have died and come back to life. Many of these have berated their doctors for bringing them back, finding the serene joy of eternity so much better than the conflicts of time.

In certain important details the evidence of these modern clinical note-takers squares precisely with ancient testimony.

Many centuries ago in certain Eastern monasteries it was the custom to stay by the bedside of a dying elder brother and record his whispered reports of his voyage to eternity. Over the centuries quite a bit of death-transition information accumulated. The much-discussed "white light at the end of a tunnel," so frequently reported by people who have known the first stage of legal and clinical death in modern hospitals, then returned to body-life, corresponds exactly with the "clear light" seen in the first stages of bodily death according to the ancient *Tibetan Book of the Dead.*

The Tibetans were quite explicit about what they meant by "light." Light corresponds to what modern psychologists call the positive or constructive emotions—love, compassion, generosity, selfless devotion to the welfare of humanity. Darkness is what we call the negative or destructive emotions—hate, greed, anger, envy, violence. The things the ancients said were good during the life *after* death are the same things modern authorities are saying are good for the life *before* death. If we were to infer from this that what is good is eternally good we would not, I think, be too wide of the mark.

There is, of course, no reason for not starting now to acquire that which is endurably good. During the stresses of mid-life and pre-retirement such acquisition is certainly advisable, if for no other reason than the relief of desperation. After retirement it is downright urgent. And indeed, what is there to do, in these priceless years of release from competitive striving, that is more important than acquiring the lastingly good?

Here one must beware of the trap of "stamping out evil." The realization comes that conflict and destructiveness are artifacts of time. In eternity there is no evil to stamp out. To make war on it is to fight something that has no existence in eternity, to set oneself against oneself, and to take the one certain road to bigotry. Forgiveness of self and others is the path. The destructive emotions are in every case caused by an unrealistic clinging to a physical body and the experiences and possessions that circle around it. They are nothing but body-fortress panic. The troubled mind has come to think that a body is its fortress. Any sign of this stronghold crumbling brings on panicked reactions of defense, shoring up the cracked walls with work, money, medicine, prestige-symbols—or bluff.

We need not scale lofty philosophical heights to take the first long step toward a more workable formula. It is contained in one of the oldest, commonest, and simplest of our folk sayings: "You can't take it with you." Eminently practical, conspicuously sane, it gives us the key to our further progress toward the *ultimate* victory over age.

Clearly we cannot take our physical body with us to where we are going in eternity. Nor can we pack along any of the material clutter with which we have surrounded that body, or any of its appetites and pleasures, or any of the prestige or other importance we may have imagined as attached to it. These things are of no ultimate importance. So, says the ancient voice of the folk wisdom of our species, begin breaking our attachment to them. Give them up. You can't take them with you.

"Then," we ask, "what *can* we take with us?" Our inner peace, love, and joy, no more. These are ours for eternity.

This simple statement must not be taken as an invitation to smugness. Time is crafty. It snatches us back, after we've had a spell of release from it, as fast as it can—and that, if my own experience is any indication, very often is instantly. We're no sooner out of the hubbub than we're back in it again. That is our reason for placing such emphasis on *daily* periods of quiet inner reflection. The peace found in these intervals is the peace of eternity. It is better than time, and is a good thing to become accustomed to. After one has known it a few times, not only does the management of daily affairs go more smoothly, but we know there need be no terrors when one gets to the Bottom Line. The peace that comes with the Big Trip will be that same peace.

# 15 Do You Run Your Body or Does It Run You?

SO FAR WE HAVE talked about what our basic situation is. Now we begin to get into what we are going to do about it.

The first thing we need to consider is the thousand and one ways people have of saying, "Doctor, please help me, but don't under any circumstances ask me to stop doing what's making me sick." This comes from patients of *all* kinds of doctors—doctors of the body, of the mind, of the soul, of the spirit. It comes in unnumbered forms, but, as the French say, the more it changes, the more it's the same thing: "Heal me, but don't change me."

In the joint adventure we have here undertaken, you are *your own* doctor. You make your own diagnosis. We are about to take steps toward dealing with whatever you may have found to be the trouble, and this will involve a review of the relationship between you and your body.

"But," some will protest, "there's nothing wrong with my *body*. I'm in fine physical health." If this is so, congratulate yourself and read on. If you can get all the way through this chapter without finding something that, if taken to heart, would make you a better, stronger person, it will be unusual.

Many people, when confronted with a habit that's damaging to them, coupled with a suggestion for change, protest at once: "But I *can't*."

This is rubbish. What such people are really saying is not "I can't," but "I won't." Resolute human beings can accomplish far more difficult things than those we are about to suggest.

If you run into something here that you know you should do and hear yourself responding "But I can't," bring yourself up sharp with a stern "Hogwash," or, if you have one handy, some stronger word. We have assured you that victory over age is possible, have offered proof, and will offer more. But to win the victory you have to fight the battle. And it *is* a battle, a challenge of an order you have never met before. If you lose it, the most likely reason will be that you said, "I can't," meaning "I won't." Some people, alas, *choose* to lose. Don't be one of them.

To win you must arouse that often maligned, usually sleeping, but potent power within you, your willpower. William James, one of the dozen greatest Americans in my opinion who ever walked this continent, the most influential philosopher our country has yet produced, is both our authority for the efficacy of bare willpower and one of our most exciting demonstrations of it.

Our undervaluation of willpower is due mostly to the spin-off from a half century during which orthodox Freudian psychology reigned supreme. The Freudian rule is ended, the Freudian fallacy has been exposed, and his system, after granting several basic contributions, stands as an interesting intellectual antique, a brilliant neurotic reaction to the neurotic excesses of the brilliant Victorian Age. We can no more use it for today's living than we can use another brilliant offshoot of the same era, Edison's original carbon-filament lamp, for lighting our homes. Willpower has been restored to its classic dignity, partly through the efforts of such straight-thinking contemporary psychiatrists as Dr. Rollo May.

"Will," writes Dr. May, "is the capacity to organize oneself so that movement toward a certain goal may take place." He continues with some appropriate remarks about the general situation of members of our species: "Man can and must, unless he abdicates his consciousness, *choose how he will relate* to necessity, such as death, old age, limitations of intelligence, and the conditioning inescapable in his own background" (italics mine). Those who wish to go into exhaustive psychoanalytic detail about this matter are advised to read Dr. May's excellent book, a late twentieth-century landmark in its field, called *Love and Will.* For our present purpose it will suffice to accept as authoritative the simple statement: Your will is the strongest ally of your intelligence in winning your victory over age.

We will mention briefly the highlights of the James case. William James was plagued throughout his life by poor health. His mental health was no better, during the first half of his life, than his physical health. He was subject to long depressions during which he frequently considered suicide. This affliction threw him into a five-year period in his late twenties and early thirties during which he was almost completely helpless.

How, you ask, did he pull out of it? One day he ran across a short piece by a French essayist on the subject of free will. It "hit" him. "My first act of free will," he declared, "will be to *believe* in free will." From that moment he began becoming THE William James—psychologist, philosopher, one of the world-acknowledged founders of modern psychology, a pioneer who built so soundly that research psychologists still consult his writings for inspiration. The rocket of free will, once ignited, carried him swiftly, during the second half of his life, to one of our country's most brilliant victories over age.

This digression on willpower has been a necessary prelude to establishing the fact that *you can win* if you recognize your resources and mobilize them, as Dr. May says, "so that movement toward a certain goal may take place." Now let's get down to cases.

In a book such as this one, intended to be useful to a great many people, we have to bear in mind the factor of statistical averages. In the United States the Number One opponent of victory over age, the prime premature killer and inflictor of terrible wounds, is cardiovascular disease. Most such malfunction of heart and circulatory system is caused by lack of cooperation between higher mind—that part of our mind that has our welfare at heart and knows what is good for us, and which includes our intelligence—and body.

The body has usurped part of mind as its own. It is a lower mind, reaching upward to invade the brain from lower centers of consciousness (such as the stomach). It is wily, crafty, in most Americans overindulged and hence badly spoiled, and can give one a hard time until one lets it know who's boss. Speaking of the stomach and its will, let us remind ourselves that death from cardiovascular disease is frequently brought on by long years of injudicious eating—the wrong foods, eaten in the wrong combinations and in the wrong quantities.

Let me begin with an instance from close to home. As I write this I am practicing what I am preaching. Because I have so far been winning this confrontation with age and would like to continue to, I keep watch on my waistline. There is a notch on my belt that I call the "this-far-and-no-farther" notch. When my waist expands to this notch I stop eating for a day, then eat lightly until I am *two* notches *slimmer* than the no-farther notch, after which I can, with only my customary precautions, eat more or less as I please for quite a while.

Yesterday I hit the no-farther notch, so today I am on a water fast (nothing but water all day and all night). I wish you could hear the fuss my stomach is making. In these dialogues I always treat my stomach with tender, loving care, and never get mad at it. It has been a good friend to me through thick and thin for a long time, and I want it to go on that way. My stomach is not mannerly, but I don't let that ruffle me. It is, after all, a poor, stupid organ that can't even talk. It needs my avuncular guidance as much as I need its faithful service. On a fasting day my stomach doesn't actually talk, of course. But if the signals it makes were translated into words, the dialogue would go something like this:

"I'm hungry. Gee, am I *hung*-gree!"

"There, there, it can't be that bad. I'm giving you a nice rest. A whole day off. Relax and enjoy it."

"Enjoy *starving*? You're a masochist, that's what you are. You *enjoy* torturing yourself. You're sick, sick."

"But I'm not being tortured. Fasting gives me a kind of nice, light feeling."

"Of course, always thinking only of yourself. *You* don't suffer, no. *I* do the suffering. You *like* to make me suffer. You're a sadist. You take pleasure in causing pain."

"Now, now, you and I are old friends, and I don't like to hear you talking that way. Besides, you can't have it both ways. If I'm a masochist, I'm not a sadist. The truth is I'm neither."

"Right. You're neither. You're a fiend. A plain, simple, garden variety fiend."

"You're overwrought, my friend. I'm just a person trying to plan the best possible life for both of us. Here, have this nice tall glass of cool water and you'll feel better."

After the water the fuss subsides for a time. But it eventually starts up again, perhaps adopting a plaintive note this time.

"Nobody gives a damn about me. Nobody cares whether I live or die."

"Come off the self-pity kick. Can't you just rest and enjoy a nice vacation? I'm doing this for you as much as for myself, you know."

Then stomach may turn nasty. "I have allies, you know. I'm not alone in this. I can get your nerves on my side. I can make you so jittery you'll want to jump out of your skin."

"It will pass. You're getting overwrought again. Here. Try this nice hot cup of herb tea."

That's the principle involved. Never, *never* get mad at your stomach, or any other part of your body. Jolly them along. Treat them as you treat a fretful baby—a gentle rocking of the crib, a few soothing words, a little water, and things are quiet again. But hold firm against their tantrums. In the case of fasting, the fuss gradually subsides until, by the end of the second day, it's gone altogether and you feel neither hunger nor restlessness.

The benefits, besides simple weight control, are numerous. Plato and Socrates went in for ten-day water fasts from time to time because they had reason to think it improved mental acuity. There is a considerable fasting cult in the United States, whose members, for reasons mainly of health, fast under medical supervision for as long as forty days. One of the important principles in any fast, short or long, is to bring the body under the control of the higher mind, the intelligence.

In this context our aim is victory over age. Part of this victory is to achieve sensible, healthful eating. The connection between diet and the cardiovascular killers—high blood pressure, stroke, and heart disease—is so well known that every schoolgirl can run off the no-no list without thinking. Alcohol, nicotine, caffeine, sugar, saturated fats, the heavy cholesterol foods (eggs, chocolate, dairy products), and heavily salted foods are bad for you and should be got rid of. Fresh fruits and vegetables, nuts, whole grains, lean animal proteins (fish, fowl, meat), dried fruits are good for you and may be eaten in generous quantities. If you want more details on healthful eating you'll find them on the book counter of any good health food store. Our thrust here is more toward general principles than toward detailed practice. And the general principle involved is that if you want to win the ultimate victory over age you have to bring your body under the control of your intelligence.

This will mean a sequence of showdowns in which you make it clear who's boss, you or your bodily organs. The saying "too many of us dig our graves with our teeth" is quite literally true. Once the sovereignty of the higher mind is established, however, things go along with a harmony that is more and more pleasurable and sometimes even beautiful.

The mouth, which gets itself involved with addictive

drugs—nicotine, alcohol, caffeine, and the like—is a much tougher customer to deal with than the stomach. Usually this is because of its tie-in with an addictive drug that immediately involves the whole nervous system. Oral discipline can bring on some nasty systemic temper tantrums until things are brought under control.

The same principles apply, however, and they are these: Never lose your temper with one of your bodily organs. Treat them in the light of the true situation, as inferior intelligences that have been overindulged and misled and that need to be brought under the command of enlightened intent.

Don't take on more than one project at a time. If it's eating you're working on this year, let smoking (or whatever) go until next year, and something else the next, and so on by priority rating. That way, if you have put your higher will and your higher intelligence into the effort, and if you are resolute, you can win. And you *will* win.

You will get very little support for your efforts from the mercantile society you live in. The media are single-mindedly concerned with selling you junk foods and poison drinks and inhalants at high prices simply to make a profit for purveyors who would have no compunction at all about your eating or drinking yourself to death on their products if it will make them a few extra dollars. It's a case of buyer beware—and it's a life-and-death game.

Two of the most dangerous words in this ploy are "need" and "enjoy." "Boy I *needed* that!" is the standard remark of the competitive businessman, already registering an elevated blood pressure, on knocking off a couple of ounces of raw whisky.

The statement, of course, is not true. Neither he, in the sense of his higher being and ultimate development, nor his body, as his vehicle for his life journey, could derive anything but damage from his infusion. The higher mental functions are the first to go out of kilter in alcohol narcosis, and the temporary mild bodily anesthesia is bought at a systemic cost that far outweighs its benefit. He did *not* "need" the drink. The false sense of need was created by a conditioned habit, supported by a systemic addiction that demanded, like a spoiled and undisciplined child, to continue the same indulgence it had become accustomed to. Such indulgence is not "needed." On the contrary, it shortens the man's life and dulls his wits and thus reduces his chances of emerging as a victor over age.

So also with the word "enjoy," a word commonly employed by waiters in fashionable restaurants just after they have loaded the table with everything that's bad for you. It goes even deeper, to the philosophical level. We'll deal with that at length in another chapter. Here we touch on it only long enough to say this: "Enjoy," as commonly used, means tweaking our sensory nerve ends in ways that produce a transient pleasurable input. If this kind of "enjoyment" is left ungoverned by a higher intent, we can literally "enjoy" ourselves to death, and prematurely. Worse than that, we miss the higher, deeper, broader, and more joyous fulfillment of the later years.

One of the inexorable laws of the universe is that we get what we deeply *will* ourselves to get. If we *will* to waste our years playing kid games of titillating our nerve ends, that's what we'll get, and it's an empty category. It's important, therefore, in this business of mastering the *second* half of life to find out what's *ultimately* good for us, and will *that.*

In body control the principle next in importance to diet is exercise. Here again, anybody who's been in contact with the media has heard of the importance of regular daily exercise to the maintenance of health and sanity, so we needn't labor the point. Most Americans are overfed, underoxygenated, and underexercised. It's simply a matter of applying decision and will to knowledge and intelligence and getting on with actually *doing* it.

To deliver the basic benefits it need not be an elaborate routine or take much time. My own minimum routine on a busy day doesn't take quite half an hour for the whole day's program. On getting up I take four very deep breaths, do four push-ups, four knee bends, four sit-ups, four standing forward toe-touching bends, four trunk twists, and, to reach the upper arm muscles (biceps), eight arm lifts with five-pound dumbbells. (A trunk twist, by the way, is standing with arms outstetched to the side, then swiveling the whole upper body slowly and gently to the left to look as far toward what's back of you as you can without forcing, then gently swinging back and doing the same on the other side.) This is the morning minimum. On days when I feel extra vim I may double or triple the count. Then, twice during each day, I either jog a hundred paces or walk briskly for half a mile, depending on my mood. All this is not very much, but it does the job. If you want more guidance in this area, many good exercise and Yoga manuals are available at your local library or paperback shop, including Suza Norton's excellent *Yoga for People Over 50*.

Don't overdo; let it be training without straining. And be regular; don't under any controllable circumstance miss a day. The principle in both diet and exercise is the same as the principle of good motoring—take proper care of the vehicle you're using to carry you through this world.

There are, of course, other ways than overindulgence and underexercise to misuse a body. One can simply abuse it, and this is sometimes done even by ethically elevated persons. One of the great saints of the Western tradition, Francis of Assisi, did *not* win a victory over age (he died at forty-four). On his deathbed he apologized to his "brother-ass body" for the way he had mistreated it. He saw belatedly that the ascetic route was a blind alley. So is its opposite, the overindulgence route. Getting the most out of a body, particularly through life's second half, is a simple matter of feeding it wisely and exercising it daily.

# 16 Reawakening the Higher Mind

"SECOND CHILDHOOD" is a phrase that has traditionally been used to indicate senility. There is another sense, however, in which it means a surpassingly alert and expanding mind—the kind of mind that was implied in the biblical injunction to "become as little children." A normal child is alert, full of curiosity about the world he finds himself a part of, full of questions about it, and full of confidence that satisfactory answers to his questions would be found.

Senility is caused, not by the reemergence of the healthy child mind, but by prolonged rebuffing and putting it down. Classic signs of senility have been produced experimentally in persons between the ages of twenty and thirty after they were consistently ignored, belittled, and treated as if they were a burden to society.

While we're on that subject, a recent report of a professional task force in one of the populous California counties should be noted. "Senility," says the report, "can be induced largely by extreme personal frustration. The lack of a productive role for the elderly in society today produces such frustration. It adds to the risk of personal deterioration and premature dependency."

Recommending "preventive/supportive" measures, the report goes on to say that the tendency to "treat only crisis demands" compounds the difficulty. "Inappropriate services to the elderly, such as early institutionalization in place of needed outreach programs, add to the problem. Some of the resulting effects are isolation, confusion, depression, suicide, malnutrition, acute lack of self-esteem, alcoholism, fearfulness, and premature death. All these are costly problems, which accelerate with neglect. Long-term hospitalization and convalescent care are increasing with frightening speed and to prohibitive levels."

You and I are aging, and this sort of thing concerns us. It has been made quite obvious that the needed "preventive/supportive measures" are not soon forthcoming from our society. Society offers us nothing but a putdown. Where, then, can we look for the needed support?

Within ourselves, of course. There we will rediscover our own inner resources, the means of reawakening our higher minds, and of providing each other with the reinforcement and encouragement our society has withdrawn from us.

Not long ago a popular magazine reported the doings in retirement of a famous football player of the 1920s. He had been a successful insurance man and had retired to Florida. He told the reporter that he did not plan his days and lived by whim. One day he might decide to go fishing, another day he might play golf, another he might just decide to do nothing. That's the way his life added up—fishing, golf, and nothing.

It has a sporty sound, but there's a serious flaw in such a program—it doesn't work. After a few months the thrill of the possibility of landing a big one wears thin, and going around in one or two fewer strokes doesn't matter very much. That leaves the third alternative—nothing. And "nothing" has ever and ever been proved inadequate nourishment for the human mind and spirit. We soon come up hard against the fact that we are social beings. Many a "successful" person has discovered that success, once gained, means nothing. What made it fun was the camaraderie that went along with its attainment. When that is withdrawn there is nothing, and the nothing has to be replaced with a something.

That something has to be real. Artificial things such as games and manufactured activities in which one is not really interested don't fill the bill. Only a reawakening of the higher mind can fill the vacuum created by society's withdrawal of its approval and support. When you are drifting downstream there is nothing to do but take the oars in hand and start rowing again.

*The alertness, curiosity, and confidence of the higher mind never dies*, but sometimes can be put to sleep by years of specialized activity in a specialized rut, even though that rut may have led to what is loosely called "success." But self-confidence is not based on "success." Personality breakdowns are as frequent among overachievers as among underachievers. Self-confidence is not having everything hunky-dory. It is not a house where you establish yourself and stay. It is a road you travel.

The purpose of our personal universe is the unceasing expansion of our personal awareness. When we are on this road we are in harmony with ourselves, our companions, and the universe. By simply continuing to learn—and learn about things that naturally interest us—we are forever experiencing fulfillment rather than frustration. Boredom vanishes, depression evaporates, and a new, ever expanding, noncompetitive universe opens up. We have found our own "preventive/supportive measures." All that is required is to reactivate the same alert, curious mind that has been ours from the beginning and reopen avenues of inquiry we had long ago felt obliged to close because of the pressure of circumstances.

The more these avenues are related to ultimate meanings of our personal life, of all human life, and of the universe itself, the more satisfaction they bring. This may mean enrolling in a continuing education course, joining a special interest group (during my astronomy phase I joined an amateur astronomer's club), or attending meetings of a personal development group.

The Phenix Clubs, a recent development now become nationwide, are pioneers along this route. Defining themselves as "friendship societies of men and women who seek, through reading, discussion, and reflection to improve the quality of their lives," they are easily started, low-expense neighborhood groups. They are finding great pleasure and reward in pursuing expanded awareness together, comparing notes in neighborhood meetings once a week. Such groups supply the vital elements of social welcome, warm friendliness, cooperation, exchange of experience, and mutual support that society withholds.

Getting back into the study-and-learning habit is a little hard for some people. This is partly because these brain circuits have been long unused, partly out of fear of appearing stupid, and partly out of fear of ridicule ("My gawd, lookit what Pop's doing, reading a *book*!").

Each of us resumes our personal growth from wherever we are and go on from there. We have reliable assurance that no one on earth knows everything, and that all are equally entitled to peer into the fascinations of the all-surrounding mystery. Orientation for beginning travelers along this route is provided by such organizations as the American Library Association and the Phenix Society. Obviously, anybody who is reading this book can read, and that is all the special preparation required. All you need is the appetite, the personal resolution to make certain that your later years be a feast and not a fiasco.

# 17 Relaxation Does More Than Relax

RELAXATION OPENS THE WAY for the ideal—the cosmic, the transpersonal, the long view, the nurturing—to manifest its benign and balancing influence. This is separate from, and in addition to, the purely physical benefits—better body tone, sense of physical well-being, and so on. The holistic mind is not competitive. It will not buck the bombast of the superpractical runaway left-hemisphere mind. It is the "still, small voice." It can make itself fully known only in deep relaxation, and this is the real reason why such intervals are so essential during life's second half. The victory over age cannot be won without these pauses.

Next, logically, we ask, "How do you relax? How long should I give to a relaxation interval?"

An expert was once asked how long a relaxation period should last. He answered that about twenty minutes was a good interval. "But," he added, "five minutes that you *do* is better than twenty minutes that you should do but don't." It was good advice.

There are three relaxation techniques that can be immediately useful. The first is the kind of thing you might do, say, once a month, setting aside in advance a sufficient block of time for it so that you can really let yourself get into it and become relaxed as you've never been relaxed before. The second is for daily use, no matter how busy you think you are, and requires only fifteen to twenty minutes. The third is something you can do in momentary snatches, while waiting for a stoplight to change, for a traffic snarl to clear up, or in whatever spare moment may come your way.

The longest one is placed first for a reason. It is to plant deep in your mind some notion of what really deep relaxation can be like, so that when you come back to it (as you will if you really intend to *do* this) your deeper awareness will have had some preparation. The best way to use this chapter, for one who seriously plans to learn relaxation, is to read the whole thing through at a sitting, and then, as time allows, come back to it and actually do it. You will be mildly surprised and greatly pleased at the results.

Now for the long (once a month or so) relaxation method. When you get around to actually doing this, you'll want to fold your book back so it stays open, and crease these instruction pages so they turn easily. When you first read the instructions through to get the general idea, you'll be wide awake and alert. When you get around to using them, there will be times when you'll be in a lazy drowse that's close to the dream-state. It will not be necessary to memorize this sequence. In this state you'll be glancing at the book just long enough to pick up the clue on what you do next. You will keep minimal attention on the book, maximum attention on relaxing your body. The brief interruption to glance at the next instruction will not hamper your relaxation. You will return to your procedures and gain their full benefit. If you can imagine yourself, instead of reading the next step, just *dreaming* of reading it, you've hit the right note.

These are the directions for occasional long periods of deep relaxation:

1. Arrange to be in a place where you can be entirely by

yourself for several hours, a place that is quiet.

2. Free your body from any binding. If your shoes bind, kick them off. If belt, collar, or bra feel tight, loosen them.
3. Get some circulation going. If you have your own regular setting-up exercises, do them, but for not more than five minutes. If you don't have your own exercises try these (also for not more than five minutes): Bend forward and reach toward your toes several times (don't strain to touch toes if it doesn't come easily). Hold your arms straight out to the sides from your shoulders and turn head and trunk to the left as far as they will *comfortably* go, looking as far to the rear as you can without straining. Swing and do the same thing to the right. Repeat four times. Do four deep knee bends and four push-ups.
4. Seat yourself comfortably in a chair that provides full support for back and head. Legs crossed at the *ankles*.
5. Take four of the kind of deep breaths called yogic breathing, inhaling until the lower chest is filled with air, the downward thrust of the diaphragm forcing the abdomen out, then filling the upper chest, full as you can, with air, then exhaling.
6. Close your eyes and relax into your chair. Imagine a small bottle of a warm, fragrant lotion marked "relaxation" being unstoppered and poured over the top of your head. A pleasant, warm, relaxing sensation now begins to spread down over the scalp, over the forehead, back of the ears, down neck and shoulders, over the face. Muscles of face, neck, and tongue relax as the feeling descends.
7. Close your eyes as tightly as you can. (Except for glancing at these directions, your eyes will remain closed throughout, but right now you're holding them tightly closed.) After a few seconds, release the eyelid strain and let your closed eyelids completely relax.
8. Pull in all your face muscles toward the tip of your nose, scrunching them in as tight as you can (never mind how

silly you look, no one is watching). Now relax the whole facial muscular system completely. Do one round of breathing as in step 5.

9. Pull up the corners of your mouth into a great big silly circus-clown grin, dragging each corner toward the cheekbone as far as it will go. Hold it for a few seconds. Let it go. Take a deep breath as before.
10. Now tense the whole hand-arm-shoulder-chest muscular system, one by one, with all your might. Starting from each fist clenched into a hard ball, tighten the muscles of the forearm, then upper arm—biceps against triceps—then tighten the shoulder muscles (deltoids), then the same for the chest muscles (pectorals), all very tight, the whole system right across, fist to fist, tense and tight. Hold this for a few seconds. Now relax the whole system completely, all at once. Take a deep breath.
11. Rest for a few seconds.
12. Pull your abdominal muscles as far back toward your spine as possible, straining as hard as you can. Hold a few seconds, then let go, totally and all at once. Deep breath.
13. Now tense the whole muscular system from the toes through the buttocks, tightening the muscles of the calves, of the thighs, drawing the toes into a tight curl, firming up the buttocks good and hard, each muscle straining against its opposite throughout the whole system, very, very taut. Now relax the whole system entirely, letting it go all at once. Deep breath.
14. Turn the feet outward, splayfoot, as far as they'll go, forcing them, holding them splayfoot for a few seconds. Now turn them inward, pigeon-toed, as far as they'll go, and hold a few seconds. Center the feet and relax them totally. Deep breath.
15. You are now in a deep state of relaxation. Remain there for a few seconds, deeply, totally relaxed.
16. Now you will go still deeper. Imagine yourself descend-

ing on an escalator to the fifth level below ground. Count one floor at a time, for five floors, at each floor going into a deeper state of relaxation. From ground floor where you start, board the escalator and let yourself drift gently down to level one. Turn, taking your time, board the down escalator and flow down to level two, then three, then four. Then, ever so easily, down to the deepest, level five.

17. You are now in a very deep state of relaxation. Tell yourself that you will remain there as long as you like. Just as long as you like. When you choose to return to normal awareness, tell yourself you will do so gradually. Not suddenly, but gradually, at the slow count of five. But there's no hurry about this. Remain relaxed as long as you like, until you feel entirely rested.
18. At a long count of five, imagine yourself ascending the escalators to ground level one floor for each count, slowly rising to the surface—one going up, two going up, three going up. Pause at four and tell yourself that at five you will be in full normal waking consciousness, feeling immeasurably refreshed and buoyed by a beautiful lightness. Before going on to five, open your eyes half way, blink them once or twice, close them again. Taking your time, take the up escalator for the count of five. You are now wide awake, fully refreshed, feeling finer than you've felt in years. *Remember this feeling.* If you fix it well in memory, you will be able to return to this depth of relaxation in the future in a matter of seconds, when your available time does not allow all the preliminaries. If you're really after relaxation, however, this "long-form" should be repeated about once a month, for maintenance. Also, you can use it for occasions when you're *really* uptight, that is to say, *beside* yourself, emergency occasions when you need relaxation more than you need anything else and can therefore afford to give it first priority.

Now let us consider normal, daily, brief relaxation. This is tailored for the hectic hurly-burly of daily urban living. It goes like this:

The first step is to bring to mind something that to you (remember, the person being relaxed is *you*; nobody else's imagery will do) symbolizes, or stands as something closely connected with, the Infinite. If you are religious, you may choose a religious symbol of some kind. Francisco Geddes, who is religious, describes his imagery as follows:

> I visualize myself sitting in the center of an inner sanctuary, circular in design, whose ceiling is a round glass dome above me to admit the light. I then seek to become completely silent and still. My surface consciousness, however, has a lot of *thoughts* that it keeps popping into my mind. I look at the thoughts and acknowledge them. Some I agree to come back and work on later, but *all* thoughts are briefly acknowledged and then placed outside the sanctuary. It may take five or six minutes, with a parade of many thoughts, before my sanctuary becomes quiet enough for the stillness to settle around me. This stillness takes the place of the many intruding thoughts.

If you are not religious, you may choose to draw your imagery from nature. In creating your image only two things matter: it must be an image of peace and quiet, and it must have some connection with the infinite. For many the image of a quiet beautiful lake nestled in a remote, deserted mountain valley serves both purposes very well. It is quiet and peaceful. It reflects the brilliant, distant sun by day and the immeasurably distant stars by night, and is therefore a reflection of the tangible infinite.

Next, have a relax-word to repeat to bring your mind to peaceful relaxation. In the Eastern tradition this is called a "mantra." My favorite is the traditional mantra "OM" (pronounced "ohm"), and I recommend it. This word, or whatever you choose in its place, is silently repeated over and over.

Now you are ready to begin. All the following steps can be taken, and enjoyed with benefit, in five to twenty minutes. If you are a beginner, it may be well to start out with ten minutes and gradually stretch it to twenty.

One preliminary admonition: This sequence works best when you do it the first thing in the morning, and that means the *first* thing, *before* you turn on the news and *before* you get into the morning hassle over making the coffee. Remember, coffee is a stimulant, not a relaxer, and the morning "news" is deliberately compiled by well-paid merchants of panic whose sole job is to rile you, alarm you, scare you, and generally unsettle you with whatever scare stuff they may have on hand. Don't fall for it. The news can wait until *after* your relaxation, and so can your coffee, once you get used to this slight change in routine (it may take a little effort at first).

The procedure:

1. Find a comfortable chair in a quiet place.
2. Sit in the chair. Back should be reasonably straight, though not ramrod stiff. Hands may be in the lap or on the arms of the chair, legs crossed at the ankles. Clothing should be arranged so that nothing binds.
3. Take four deep breaths of the yogic breathing type described earlier.
4. Close your eyes and visualize as clearly as you can whatever symbol or scene of peace-and-quiet-associated-with-the-infinite you have chosen.
5. Start repeating your relax-word at your own speed, consciously relaxing your body as you do so. By this I mean calling to mind the deep state of relaxation you were in the last time you did the "long-form" relaxation.
6. Tell yourself how many minutes you intend to meditate. It will surprise you to discover that you have a dependable internal alarm clock that will let you know when your allotted time is up. If you don't yet trust this, by all means set your alarm. Leave it in another room, though, so it doesn't jar you when it goes off. And get rid of it as soon

as you have won confidence in your internal timekeeper.

7. Continue repeating your relax-word. When thoughts interrupt, set them aside gently and resume your relaxation. This gentleness in putting thoughts aside is important. If you let yourself get sore—either at the thoughts for interrupting or at yourself for allowing them to interrupt—you have defeated your purpose, since an angry person cannot be a relaxed one, and our purpose here is relaxation. Flora Davis has put the case well: "When I closed my eyes, the first thing I noticed was a kind of rosy-orange color. Then I became conscious of what it means to concentrate on the repetition of a single word. My mind kept leaping off on tangents. Even when I managed to focus on the word briefly, thoughts kept trickling in around it. The mind is apparently perfectly capable of chanting a word while simultaneously worrying away at some problem. After ten minutes the phone rang. I dealt with the phone call, then went back to finish my twenty-minute stint—gladly. The peripheral thoughts began to slow down. At the end, I felt unusually relaxed."

You too will feel unusually relaxed if you follow this method, and persevere. Perseverance is vital. Every reasoned, sincere effort to gain inner calm, by following the directions of those who have experienced it, has an accumulating beneficial effect. Though surface evidence of this effect may not be instantly apparent, it reveals itself beautifully when given a little time.

Distinct benefits can be had from relaxations as brief as one or two minutes. This brings us to technique number three of the triad announced at the beginning of this chapter as "something you can do in momentary snatches." Instead of fussing and fuming at life's petty delays—waiting for the dentist to see you, for the "walk" sign to let you cross a street, for a package to be wrapped, for your lunch to be served—you can gain both inner and outer serenity by using these brief intervals to relax. You might even, as Dr. Herbert Benson has suggested, take a relaxation-break instead of a coffee-break. Caffeine, a nerve and heart stimulant, tends to make you tense and jittery. Relaxation relaxes, which is what you have implied you wanted. So relax.

The technique here requires a little experience (though surprisingly little) in the first two methods, and is based on this experience. With a trial run of techniques one and two behind you, you're ready for a go at the quickie method, number three.

The procedure:

1. Call instantly to mind your chosen scene that is sublimely peaceful and related to the infinite. Dwell on this for a few seconds, long enough to fix it firmly in your mind. Then, without letting go of this image entirely, let it fade into the background while you proceed to—
2. Remember as vividly as you can how you felt the last time you got yourself deeply relaxed by following "long-form" relaxation as in Procedure One. Give yourself over entirely to these two things—relaxing the mind into the image of peace and infinity, relaxing the body into the recollection of its last deep relaxation. If your snippet of time is not yet used up, you can go on to—
3. Start repeating your relax-word, keeping intruding thoughts out of the scene by one of the methods given you in this chapter.

I'm giving our closing passage to a quote from the meditation instructor, Flora Davis, on a theme whose importance cannot be overemphasized in all undertakings bearing on relaxation. That theme is EASY DOES IT. The moment you begin to get anxious, tense, or impatient, or to strain for results, or to cuss yourself out because you failed to keep all stray thoughts out of your adventure into peace, you have lost the ball game, because, of course, these feelings are the opposites of relaxation.

EASY DOES IT.

"Effort," Flora Davis writes, "is what keeps you from reaching the restful depths of the mind. . . . I'd become annoyed with myself for drifting—even though I knew that annoyance made me tense and counterproductive." Perseverance, yes. Effort, beyond a calm assertion of your own will, no. *Easy does it.* In this gentle way does our *whole* mind lead us toward the infinite.

# 18 The Heaven That Is on Earth: The Love Response

AT THE END OF World War II an aging American woman returned from Europe in the last stages of nervous exhaustion from her experiences in a prison camp. She compounded her difficulties by trying to drown them in alcohol, and finally wound up in deep depression and in the care of the famous psychiatrist Dr. Erich Fromm. After some weeks of treatment Dr. Fromm dismissed her with these words: "I have done all I can for you. You will have to do the rest for yourself. Go to some place where you can be quiet and alone. Say these words over and over to yourself until you feel better, no matter how long it may take: 'THE TIME IS NOW.' "

"I took myself off to a small house in New Hampshire," she reported, "and did as instructed. For two months nothing happened at all. Life remained dreary, dull, and despondent. Then one day things brightened. This was *now.* It was not those terrible remembered years in prison camp, but *now.* I began to see the sun shining and the flowers growing, and hear the birds singing and the children laughing. I was myself again. I still use this saying whenever the inner clouds threaten to close in. The true world is beautiful and eternal. Our contact point with eternity is not somewhere in the past or future, but now."

Our habitual concept of time as a flying arrow or flowing river is mistaken. We are not vehicles on a road called time, with a past beginning at a point of departure and a future ending at a destination. We are caterpillar tractors, carrying our own road with us as we go. In reality there is no past and no future. What we call "past" is the memory of the now that has just passed under our wheels; the "future" is the part of the track descending to pass under them, to be experienced as a "now." Any possible future time will be experienced as a now. Now is eternal. It is the only time we have ever had or ever will have. Our own "now" is the only road under our wheels—this is the reality, and it is beautiful. When we abandon the dim shapes of the unreal past and future, and enter the shining real world of now, it is no wonder that things get brighter.

If one has adopted what Jung called a "religious point of view," a similar and perhaps even greater "brightening" can be realized. It does not involve dogmatic acceptance of any kind of doctrine, and it is no longer "unscientific." Three-quarters of a century ago the "unscientific" charge could be made because so many natural scientists tended to be atheistic. This is not so today: to believe that is out-of-date. "The religious feeling of the scientist," Albert Einstein wrote, "takes the form of rapturous amazement at the harmony of the natural law, which reveals an intelligence of such superiority that, compared with it, all the systematic thinking and acting of human beings is an utterly insignificant reflection . . . What is impenetrable to us really exists, revealing itself to us as the highest truth and the most radiant beauty."

The intangible force of thought is the most powerful force in the universe. Negative, hating, or fearful thought cannot yield harmony, joy, and creativity. The universal life thought is loving-nurturing, since it has produced life so abundantly. To deny this thought is to deny yourself the inner peace it offers. To accept it means that you do not have to wait for pearly gates in a sweet bye-and-bye. Since the inner peace is "within you," its pleasures can begin to be enjoyed at any time, such as now.

But what about the body? Isn't that important too, as well as the mind? Don Johnson, author of *The Protean Body* and an acknowledged expert on bodily well-being, has this to say:

> The body itself is of no ultimate importance. What *is* of ultimate importance is love among us all. There are lifelong cripples and people with twisted bodies who are filled with love and whose lives have served humanity. There are people with balanced, flexible, and energetic bodies who are selfish and cruel. Work with the body is of ultimate significance only within the context of the larger energy fields of life, when it served the more comprehensive goals of human existence.

If your biocomputer is heavily programmed with money-thinking and other material worries, you will have to reprogram some of the more troublesome circuits. Worry, incidentally, is simply a runaway computer circuit in which a thought-package chases itself round and round in circular fashion thoughtlessly, like a dog chasing its tail. Such circuits can be damped by reprogramming so that even thought can lead to serenity, contentment, and joy. Daily nonpossessive love outlets are essential vitamins. Always ask, "What is my nonpossessive love outlet for today?"

Here is an assortment of thoughts you may, if you find them appealing, add to Dr. Fromm's "The time is now." If you wish, you may copy any that seem to speak to your need, inscribe them on placards or small cards, or record them on a cassette tape and make them inwardly your own by repetition.

"I am finished with complaining. Instead of complaining about what is not, I resolve to enjoy what is."

"I have been given everything I need to make my own contribution to the general welfare, day by day, including this day."

"Humanity, past, present, and future, is in reality one great friendship society, of which I am a member."

"I am not my body. I am only a temporary resident in it."

"I understand my fellow man well enough to be patient with him during his temporary insanities caused by misunderstanding and hurt feelings."

"The old triggers-to-panic that used to plague me are losing their force. I am released from competitive alarms to take my place in my new, noncompetitive world of voluntary helpfulness."

"Panic is neither sane, nor useful, nor helpful, and if it is none of those, it is nothing. I can safely ignore it."

Any truly existing heaven is just as real on earth here and now as it is in any other place and time. All our hells are of our own manufacture through our own mental states. The inner struggle is between an ego that does not love us, but only drives us mercilessly to maintain itself, and a true Self that does love us, and hence evokes the love-response in others. The only real response to life is the love-response. All other responses are illusion and lead to misery. We have a choice. It is just as easy to help yourself to a bit of heaven as to a chunk of hell, and a good deal more intelligent.

# 19 Our Connection with the Infinite

THE WORD "INFINITE" IS awe-inspiring and incomprehensible. Even the most advanced astrophysicists can't make up their minds about the basic mechanics of the stellar universe. Because it is incomprehensible, the bafflement of the average citizen in confronting it is usually resolved by simply pretending it isn't there, or taking it casually, as part of the earth's weather ("The stars are pretty tonight, aren't they?").

But it *is* there. Your address as "Earth, Sun System, Milky Way Nebula" is just as real as your street address and zip code. And, like your local environment, it affects you. More important still, it *supports* you.

This is the aspect that concerns us here. We need to utilize consciously this cosmic support if we are to win the victory we have set our hearts on. To discover fully and realistically that one is actually part of a cosmos is a consciousness-expanding experience. It is a long step toward cosmic consciousness, which is our later-life goal and final evidence of our ultimate triumph over age.

There is an historic reason, closely bound in with one of the disasters of our intellectual history, why this reconnection with the cosmos is so difficult for the twentieth-century mind.

Early in the nineteenth century a school of philosophy called "positivism" took root and spread through Western thinking. Further reinforced in the 1920s and renamed "logical positivism," it denied that anything having an *innate* quality of mystery or metaphysics was proper material for the attention of the human intelligence. This exactly suited the materialistic-industrial mood of the past two centuries. It threw our thinking and teaching into exclusively left-lobe fragment channels, excluded the whole-thinking, nonverbal right-lobe channels, and resulted in the lopsided minds we have already discussed.

In recent years science has been forced by its own discoveries to acknowledge the presence of powerful but irreducibly mysterious forces. Psychology, as we have observed, has been obliged to accommodate parapsychology, physics has had to make room for metaphysics, and psychiatry for metapsychiatry. We are being reunited with a cosmic totality that was temporarily lost but has always been an historic source of strength and inspiration to the members of our species.

The work of Professor F. A. Brown, of Northwestern University and the U.S. Marine Biology Laboratory at Woods Hole, is an interesting development. Biological scientists had long supposed that all living creatures follow seasonal, solar, and lunar rhythms and take their cues only from the usual sensual sources—weather, water movements, sunshine. What would happen, Brown wondered, if all such cues were shut off from the creatures? He set up sealed quarters where environmental factors such as temperature, atmospheric pressure, humidity, and illumination were always the same. Then he spent years watching them under these unnatural circumstances.

To his surprise the confined creatures maintained their habitual rhythms even when moved far from their native surroundings. Oysters, when moved a thousand miles from their home beaches, timed their opening and closing to what the tides would have been had the sea extended that far inland. In some mysterious way they were in continuous contact with the forces that generate and control the tides. Potato buds followed the seasons as precisely as potatoes living outdoors.

How they came by this knowledge is one of the countless perennial mysteries of organic life. For several centuries natural scientists have enjoyed ridiculing astrology and expressing annoyance when the preponderance of "hits" of the better astrologers is pointed out. It is ironic that it should be a natural scientist who has provided one of the strongest proofs of the basic assumption of astrology, that residents of earth are affected by extraterrestrial forces! Professor Brown established that living creatures react to forces from outside the earth. Some of these forces operate, he writes, "at energy levels so low that we have hitherto considered the organisms completely oblivious to them. There are still unknown temporal and spatial, subtle and pervasive forces influencing the behavior of living things." Not all these forces operate at low energy levels; some, such as cosmic rays, arrive at extremely *high* energy levels. It has long been known that cosmic rays affect human genetic material, but where cosmic rays come from is still a mystery.

It is not without reason that doctors prescribe a bit of quiet outdoor life for the businessman overwrought to the verge of nervous and emotional collapse. Our instinctive sense that an outdoor vacation once a year is good for us springs from the same deep wisdom-root. All these insights, together with the similar intuitions that can lead us to final victory over age, arise from the fact that man is not an exclusively earthbound creature. His essential nature is cosmic, and he never feels completely serene until he has brought himself into harmony with the cosmos of which he is a functioning part.

Carl Jung felt that the "decisive question" for human beings is to decide whether or not they are "related to something infinite." He regarded it as the really "telling" question of human life, and the only true orientation for life's full meaning. "Only if we know that the thing which truly matters is the infinite can we avoid fixing our interest on futilities, and on all kinds of goals which are not of real importance."

Granted that a "relationship to the cosmos" is vital, how does one go about creating such a relationship? It begins with contemplation. And that could begin with contemplating so commonplace a thing as a day.

Yes, a *day*, a simple, ordinary day. But was there ever a truly "ordinary" day? Is not every day extraordinary, different from all other days, having its own magic and its own wonder? To a hard-driven, jaded, workaday mind this may sound a little Pollyannaish, I know. But like it or not, it happens to be true. Each day *is* unique and full of its own special excitements. And part of winning the victory over age is breaking the grip of the hard-driven, jaded workaday mind and adhering to that which is true.

To "contemplate" means "to consider thoughtfully." Walt Whitman, possibly the wisest of our native poets, suggested that we take time now and then to "loaf," and "invite our souls," and this is just the right mood for the most successful contemplation of a day. One of the finest things that can happen to a person is really to come to know a day.

"Day" is formally defined as "the period of the earth's revolution on its axis." To undertake the thoughtful consideration of one you have to pick your time and be in the mood for attempting a contact with the infinite. The best time, for me at least, is after a period when I have had a lot to do with people and am ready for a little solitude. This complete break from people, this getting by one's self and trying to know a day, is high in dividends. It throws you into consideration of one complete revolution of the earth, which obliges you also to consider the sun, which puts you in contact with the astronomical universe, which is infinite in dimension and therefore is, by definition, a particular kind of infinity—*outer* infinity.

I like to get up when it's still dark on the day I have chosen and watch for the dawn, the beginning. The way dawn comes is always a special and fine excitement, too often missed. Things seem to know it's coming; the quiet of the night seems to deepen a little just before first light. Then the sky changes and the predawn twilight is there, a filtered presence. I like to think of the great arc of light advancing over the earth and of all the things greeting it—a forest full of birds in song, a rooster crowing somewhere. Presently the sun breaks the horizon, quickly clears the earth, and begins its climb.

All day I keep track of the sun. If I can manage it, I walk on a wooded road or in a park, or along a river or lake or the sea, or I drive to a distant place to have lunch at some place where nobody knows me and I don't have to talk. I have arranged that people-demands be not pressing that day; I can give myself over to the day, contemplating it, "considering it thoughtfully." Noon comes. At sea the navigators are watching through their sextants; they check their chronometers at "high" noon (when the sun is seen to be highest) and know where they are.

Sunset, twilight, dark, the stars—I begin to know a little where I am too. The coming of night, if things are right, is as dramatic, in its special way, as the coming of dawn. Both have their hidden meanings, and both are part of a "day" defined as one complete revolution of the earth.

If the day is the coming of the light, making things that are close reveal themselves sharply and distinctly and separately according to our left-lobe preference, night is the right-lobe time. Subtler things, which were there all the time but could not be seen because of the sun's glare, now come clearly into view, while things close by are no longer so distinct and aggressive but merge softly into a universal oneness with all. Recently I took a motor trip across the country, sometimes driving all night. One night, which happened to be a moonless one, I stopped to stretch, and walked a little way into the desert. There I saw the night part of the day in a blazing sharpness I had never seen before. The air was perfectly clear, without cloudiness, murkiness, or haze. Though I have been an amateur astronomer of sorts and have done a lot of telescope observing, I had never seen anything remotely resembling this glory—millions and billions of swarming stars I had never seen before, great sky roads of luminous star carpets threading their way along the now resplendent Milky Way, which this night was adorned with ten thousand extra bright and glittering star spangles. When I reflected that this same wonder had been there in its full glory all day long, but I had not been able to see it, I was driven into speculation. How much else is there of life that is wonderful, that is there all along, but which we have never seen because we have not made the special arrangements required for it to reveal itself?

On the day I have chosen for contemplation of a day I usually don't think much about people, but try to turn my mind toward the vast dimensions of the creation, the infinity of space, toward earth and sun, daylight and dark. When thoughts of people intrude, I try to choose the people of whom I will think. I select solitary, day-conscious people—the psalmist lifting his eyes to the hills, "from whence cometh his help," the young prophet walking in the desert and encountering his destiny, the poet considering the lake country of England, the mathematical physicist perceiving that space is curved. All are people who have made the big discovery of day-contemplation, that man's rational poise is related to his awareness of time and place, earth and sun.

Sometimes on such a day there is a great moment of full knowledge of being an earth-creature and *belonging*, of being wholly content. This delicious instant may be partly described as a sudden intuition that the whole of creation is offered for you, that you are part of a cosmic creative current, perfectly and universally secure. You are not always so lucky as to experience this moment on the day you have chosen. But a moment comes when you know it's time to return, deeply refreshed, to people and the hurly-burly, to see what contribution you can make.

Whether or not you decide to give over a whole special day

to it, there is a special power in such transpersonal day-contemplation. It can lift you into a dimension that lies above and beyond the usual, far out of reach of the ups-and-downs of your moods. Our ordinary days, of course, vary according to our involvements. Now and then, however, it's rewarding to take a day, strip it of all its nonessentials, and get acquainted with it in its pristine essence.

A practical philosophy has developed around this essence that some people call the "Twenty-Four-Hour Plan." The substance of the plan is to become aware that practical living is closely geared to the turning of the earth. We do our living in twenty-four-hour cycles, one day at a time. If you take on the job of living in daily parcels, say devotees of the plan, you'll make it. People break down for the same reason animals in psychology laboratories do, from too great expectations, followed by too great disappointments too often repeated. Living in one-day parcels cushions such shocks by encouraging us to gear our hopes to what may reasonably be expected of the day in which we find ourselves.

The plan is wonderfully flexible. It encompasses, and places in new perspective, whatever we may conclude is the major need of any given day. It provides for the rhythmic, pulsating quality of life that Arnold Toynbee, acknowledging that one day's need may not be the same as another's, has called the principle of withdrawal-and-return. Sometimes the difficulty is not too great involvement with people but its opposite, too much of being by ourselves. The plan divides life into livable, manageable, daily units. Even if you find nothing else in this book yet grasp this simple life-plan and apply it, your later years, taken one day at a time, will be a richer and more tranquil time.

I do not mean that everything will be rosy. Some days we get nowhere, and when they're over, about all you can say for them is that we have ground through another day. But even that is an accomplishment when you're passing through one of the valleys. Some wounds are healed only by time, administered in twenty-four-hour doses. The things most likely to ruin any twenty-four-hour stretch are fear and resentment. To enjoy life we have to control these plagues. Other people will sometimes blunder or make mistakes and hurt us. But we cannot afford to let this make us afraid, or angry, or resentful. When we try to tidy up our own behavior, daily sweeping out the ashes of burned-out grudges and opening the door to fresh living opportunity, we find that perhaps the words or actions of others have not really hurt us after all. Resentments, replaced by new interests, die from lack of nourishment.

Part of the plan is to keep alert for those small and totally unexpected delights that life is always tossing our way. "The art," wrote John Cowper Powys, "lies in the embrace of those elemental accompaniments of existence which as a rule are taken so stupidly for granted."

These "elemental accompaniments of existence" are the very touch of infinity itself. They are its signals that it is there, and is on your side.

# 20 The Strange and Lovely Transformation of Whole-Seeing

ONCE HAVING EXPERIENCED the oneness of the universe through a solitary excursion into a day and a night, our eyes are opened to how hard we have been trying to make ourselves *separate*.

The illusion of separateness requires a tremendous output of energy to maintain itself. This energy is always frustrated, since it is trying to maintain an illusion that is not true. The illusory structure is always falling apart and calling for a desperate rush of repairs, a pointless and utterly exhausting procedure.

Once we have recognized the oneness of the universe, we can end this charade. The released energy can give us a new vim and buoyancy. Accepting ourselves as part of the universal oneness, one with it and with all, there is no longer any need for the frantic effort of trying to prove we are separate. We can relax. We can let peace come to us, peace of mind. We can take the deepest philosophical pleasure in the peaceful acceptance of our oneness with all.

The fact that most people are not aware of this oneness does not affect its reality. Nor does it affect our peace, in *its* reality.

Other people must eventually come to see what is obviously true. We can afford to be patient with them. Remembering how long it took *us* to see what is obviously true, we can wait patiently while they come to see for themselves what we have at last come to see for *our*selves. This patience is the beginning of wisdom.

Although we have been speaking of an external, visible universe, the cosmic peace we have found is essentially an inward thing, steady and firm, even if hell itself may be raging outside. After experiencing a day and a night of the peace of outer space, we are ready to explore *inner* space. This exploration is the quiet restoration of the view from the suppressed right hemisphere of the brain, and a clear road of access to infinity.

Exactly *how* does one have a "quiet time"? This was a question I asked myself in real exasperation about twenty-five years ago. I got a book on the subject and did the best I could. The results were so encouraging that I wanted more. I took courses and lessons, but they left too many questions unanswered. I traveled the world looking for answers, and finding them. It is out of this background that I answer: "How does one enter contemplation? Follow the directions in Chapter 17 on relaxation techniques. Then, instead of concentrating your attention on bodily relaxation, focus your mind on melting trustingly into the Universal, peacefully turning yourself over to the care of all that is."

For effective results you must first decide why you are meditating. What do you expect to gain? Peace of mind? More frequent intervals of relaxation and tranquillity? Lower blood pressure? Improved general health? More money? Psychic abilities? Spiritual elevation? Self-purification? All these are legitimate motives. Ask, and, if you ask consistently, you will be given what you ask for. But you must know *what* you are asking for. The technique is the same, whatever the motive. The *results* vary according to the individual's motivation. It may change over the years with growth and development. At all stages of life, early or late, it is important to have clearly in mind what you are after. Then you simply apply the technique of relaxation and contemplation already discussed, which is, and has always been, the temporary elimination of the habitual thoughtstream.

Men and women were engaged in quiet contemplation before recorded history began, and they knew exactly why. The practice has entered all the higher philosophies of all countries and cultures, and always for the same reason: "Enlightenment."

Enlightenment means simply learning something. Some confusion has arisen because the word has also been given another and special meaning as the *ultimate* learning.

Some of the swamis like to tell a famous story emphasizing this point. It seems that there was a very elevated yogi who was surrounded by worshipful disciples. One day, while the yogi was in a long and profound meditation, his disciples saw evidence in his facial expression that he was experiencing a great ecstacy. When he had finished his meditation they gathered round to question him. One asked, "Was it beautiful, swami?" The swami nodded. Another asked, "Did you see a great light?" Again the swami nodded. Another asked, "Did you see God?" And when the swami again nodded, the disciple asked, "What did he say?" The swami sat in silent thought for a time. Then he turned to the disciple and replied: "He said to me, 'You talk too much.' "

The point, of course, is that enlightenment in its highest mystical sense cannot be talked about. Those who have experienced it know; those who have not cannot realize it through words. Even if an individual is granted an elevated illumination, this must be taken not as an end but as a beginning. Unless the illumination is lived out in a more generous life, it is lost.

Such elevated experiences are not restricted to Orientals, or to people who elect Oriental costumes, vocabulary, rituals, and chants. The distinguished psychologist, the late Dr. Abraham Maslow of Brandeis University, demonstrated this fact. Convinced that most current psychologies are those of sickness, he concluded that the only truly sane psychology would be a psychology of health. He began to ask people, not about the times when they felt especially bad, but about the times when they felt especially good. He found that practically everybody had experienced intervals of an all-pervasive joy and peace. He gave it the name "peak experience." It is unquestionably the same thing as the "samahdi" of the yogis, the "ecstacy" of the saints, and the "bliss" of the meditators. The experience comes in an even greater intensity at death, which, research has shown, is a most profound ecstacy. The "peak experiences" Maslow recorded all came spontaneously, without inner or outer conditioning and without drugs. Think back. You have had peak experiences yourself. These are conscious contacts with the certain, sure serenity of all that is: Jung's "infinite." The advantage of the daily quiet time is that it makes such contacts more frequent.

Enlightenment in the ordinary sense of "learning something" *can* be talked about, and very much needs to be. It is important to achieving the maturely rounded life. Such learning is of cosmic dimension, and can be gained in only two ways: spontaneous inspiration and planned reflection. Some of us feel it's a good idea to utilize both.

Here the water has been muddied a little with loose talk of being "blissed out." It is true that a pleasant euphoria sometimes comes. You can have euphoria (be "blissed out") without having any enlightenment at all. And you can have important enlightenment without experiencing ecstacy. The point needs to be emphasized: *The objective is not feeling; it is enlightenment.*

Meanings arising in deep quiet come from another source and embrace deeper and broader dimensions than any available to our busy day senses. There are elements of solemnity, devotion, and gratitude. These elements lead to a clear understanding of a necessity to break longstanding lower-order habits of thought and action. What one is actually doing is retiring the temporary tribal ego to allow the emergence of the cosmic Self. This "Self" is that element within us that knows itself to be in harmony with the universal source and cause.

The difficulty arises from the fact that the temporary and local tribal ego does not take kindly to being retired. It can set up such a clamor as to bring about the "dark night of the soul," so often experienced by major and minor saints in the course of their development. Tribe-centered ego is a false friend; things are better when it goes. It is essential to remember that the universe is accepting you exactly as you are at this moment. It does not wait until you attain some other and superior condition, or have fewer flaws. It is accepting you now, as you are. And as you open yourself to the acceptance of the universe, as you are, you are accepting yourself as you are. You are in harmony with yourself and with the universe, just as you are, and just as the universe is. You are at peace.

# 21 Together If Possible, Alone If Necessary

AGING IS SO MUCH A matter of changed self-image and altered concept of role that, as we approach the finish line, we almost don't recognize ourselves. Flash Sweeney, the All-American Tense Driver, the Hard Guy, becomes the Genial Old Man, Father of Fathers, Senior Father to All. The All-American Great Lady, the Glamor Girl, becomes the Mother of Mothers, the loving, nurturing Mother to All. And both sexes may find themselves being a little of both.

The perceptive reader will notice that this chapter is mostly in the first person plural. I am in the same boat you are, and it's a big boat with a lot of other passengers. We sink or stay afloat together.

Our problems are the same: various forms of immersion in negativism and sloth. These are in large part brought on by our taking much too seriously the scorn, condescension, and shunning with which our society tries to put us down and keep us there. Our defensive strategy is first to mobilize our own resources to deal with our personal problems. Then, together, we will ultimately bring about a massive change in our national attitude toward aging. But we don't have to wait for a laggard nation to catch up on its thinking. We can begin right now to realize that joyous fulfillment of a rounded life that is the birthright of every sane citizen.

The cosmically intended natural condition of every human being of any age is happiness. The achievement of happiness is mainly a matter of clearing away joy-killing personal habits of thought and action. Happiness, being natural to us, then bubbles up of itself and overflows.

A great physician, the late Dr. Franz Winkler, made some helpful comments about breaking habits. "In every critical period of life," he wrote, "an inner birth process takes place. Whether its fruits will ripen and mature, or die, or grow into monstrosities, depends on wisdom and free will. Evolution requires a supreme effort of will. Even the best of habits are no more than casts into which one's will may flow. Essentially rigid, and truly useful only as props for routine affairs, they must be overcome—regardless of whether they are good or bad—whenever an act of free will is required."

A victory is something you *win*, not something that is just handed to you. This is especially true of the victory over age. To win it we must pull ourselves together for the "supreme effort of will." The reward is supremely worth the effort. Many have gone this path before us, the path of determined evolutionary striving. None have trod it but assure us that the peace and joy that begin to unfold are far beyond any pleasures known to the earlier years.

"A thousand-mile journey," runs a Chinese proverb, "begins with a single step." You have already taken that step, you have read this book. Now take the next one. When troubles seem so numerous we don't know which one to take on first, we arbitrarily single out *one* for a starter. When that one is in hand we have more confidence to take on the next, and the next.

The method is, in general principle, the same for all cases: first *acceptance*, then *affirmation*. Once we have made our diagnosis we have completed the acceptance part: "This, and exactly this, is what's wrong with me." The obvious next step is to add: "and *this* is what I'm going to do about it." This is the affirmation. We're no longer going to take whatever it is lying down. We're going to act.

What do you think your problem is? Smoking? Overeating? Drinking? Anger? Depression? Loneliness? Longing for the old conditions of job, family, residence (homesickness)? Worry over money, health, past, present, or future circumstances? Boredom? All of these?

We simply follow the standard procedure of breaking an undesirable thought-action pattern and putting a new and creative one in its place. The effort of will may begin with hardly more than a wish. We *wish* we'd get busy and do something about the thought-habits that are wrecking our happiness. Continued wishing eventually enlists the help of the emotions, so we find ourselves *wanting* to do something about it. At this point our higher intelligence can enter the scene and convert wishing and wanting into *willing*. When we begin to *will* something good, some nice things begin to happen. The old joy-killing monsters begin to fall away and the lovely face of true happiness begins to show itself.

Habits are broken the same way they are formed—by repetition. And what we are repeating is a new and elevating thought to take the place of the old and degrading one. The thought we are repeating is not one conceived by somebody else as good for us; it is a thought of our own choosing. We, and nobody else, are our own doctors, and we write our own prescriptions. And we get better—very, very *much* better.

It is an enormous help to us when we can find friends who are just as interested as we are in discovering new, more meaningful, more exciting dimensions in life. We *support* one another instead of pulling each other down. We constitute a *non*competitive society of intelligent simple friendliness to replace the old, competitive gossip-and-trivia society in which we had spent so much of life. In this new atmosphere astonishing things can happen, *good* things.

There are a great number of small, friendly elevation-of-consciousness neighborhood groups across the United States, to be discussed in the next chapter. Many of these are devoted exclusively to mutual support, self-development, and elevation of consciousness. The point here is not to espouse a particular group approach but to stress the importance of some kind of mutuality, some kind of regular and consistent effort in the company of other people. Our lives are important in simultaneous relation to one another and the universe; as totally isolated individuals we dwarf our meaning. Our need for warm relationships with one another is, therefore, not whimsical but real.

Some are better able to absorb and dispense this essential vitamin than others. Some of us are naturally gregarious—joiners, group-organizers, party-givers. At the other extreme are those who vigorously assert that they are "loners," a word used to imply they are persons of independent mind and spirit. These are the saddest of all the culturally deprived. They are also the hardest to help, since they are so firmly convinced that their own way is the only truly correct one. Sometimes there is nothing to do about the pathological loner except let him go his own solitary way. But sometimes, after he has hurt himself enough, he can wake up and learn to help himself.

As a literary man I have been far down the road of the loner. Partly by disposition, partly by occupational necessity, I have long haunted the ivory tower, the meditative silence, the solitude of the figurative mountain cave. I understand the psychology of the loner and, since I have been into it and come out of it, have great sympathy with it.

The indiscriminate, compulsive joiner and the loner are both motivated by fear. The former is afraid of being without the support of mass-mindedness ("I want to be part of the in-thing"); the latter is afraid of being overwhelmed in it ("Nobody's going to brainwash *me*!"). Both deprive themselves of the essential human need for creative interchange with other mature minds.

The compulsive loner can help himself by a simple change of attitude about his loner-ness. Once he begins to see it as a handicap to be got rid of instead of a badge of distinction, the trick is turned. He can then begin to allow himself the social vitamin he needs, if he is going to win the game of life and the victory over age.

Sometimes it helps us loner-types to gain a little understanding of how we got that way. Many of us are the products of parents so assertive of their own opinions over ours during our formative years that the only way we could have a life of our own was to steer clear of them and get by ourselves. We grew up sensitized to mass-mindedness and on guard against it. This, of course, is a healthy enough attitude until it reaches an extreme where it begins to deprive us of essential social relationship. At that point steps should be taken. I know a man who tells me with great pride that he is a confirmed loner, and in the same breath complains about his loneliness!

Like other joy-killers, lonerism is a habit, and is cured by the same means that work for other unwanted habits. They were formed by repeated exposure to inhibiting attitudes and are corrected by repeated exposure to liberating ones. Dr. Fromm's "The time is now" treatment and the repetition technique for habit-breaking are specifics.

Loners do not have to make themselves feel they must endure the company of people whose views are odious to them, or spend tiresome hours over obvious trivia, in order to get the human companionship they have to have. They *do* need the courage to try out a few groups until they find one that makes them feel more or less at home.

The "The time is now" treatment, if diligently practiced, will place past events back where they belong, in the nonexistent past, and end their interference. Whatever past association, whether recent or remote, we felt we had to get away from no longer bothers us once we realize that "the time is now" and we *are* away from it. This may be combined with repeated assertion of an idea that expresses our own truly felt personal need for mature communication with other congenial minds. Repeating something like "I have decided to keep going to group meetings until I find one I want to attend regularly" can relieve even a stubborn case of lonerism, along with the gnawing loneliness that always goes with it.

One thing we can absolutely depend on: There is no way we can escape our essential human endowment as social beings. Dr. Hans Selye, Director of the Institute of Experimental Medicine and Surgery at the University of Montreal, who has spent a lifetime studying how to survive in a pressure-ridden society, says in his penetrating recent book, *Stress Without Distress:* "Most people sincerely wish to be useful to society . . . egotism [is] gradually transformed into a kind of altruism designed for survival." Even Govindananda, who spent a quarter century in complete solitude, followed it with more than three-quarters of a century of active and purposeful sociability. Enjoying people's company need not mean supine agreement with their more scatterbrained ideas, nor do we have to go into a fury to defend our own. Mature socializing, once we get the knack of it, allows us calmly to assert our own beliefs and calmly tolerate those of others. And it keeps us in the social context that is so essential to our ultimate well-being.

There are special cases, of course. There are people who are perfectly amenable socially, but who are alone out of occupational necessity. My own literary work, for example, requires that I be in complete solitude two-thirds of my waking life. And how about the forest rangers, ranchers, personnel of remote satellite-tracking or meteorological stations? Even these need not be deprived of that human essential, exchange of thought with congenial minds. Alcoholics Anonymous conducts a "loner" program by correspondence. The Phenix Society (sponsors of the Phenix Clubs, Box 25, Guilford, CT 06437) provides for "newsletter members" and correspondence reading in philosophy for the middle years.

No one, even the geographically isolated, need be entirely alone in spirit. To remain so in normal society, where social possibilities abound, is to starve oneself in the midst of a banquet. So let us conclude with Robert Browning:

*Grow old along with me!*
*The best is yet to be,*
*The last of life, for which the first was made.*
*Our times are in his hand, who said,*
*"A whole I planned; youth sees but half;*
*See all; trust God; nor be afraid."*

# 22 How Phenix Clubs Are Born and Raised

FIELD EXPERIENCE HAS demonstrated that two of the great killjoys of the later years are meaninglessness and loneliness. By providing regular and congenial companionship on a deeply meaningful level the Phenix Clubs meet both these primary needs. The question now arises, how do you get one going?

The nucleus of a new Phenix Club exists wherever one single person feels "grabbed" by the one single Great Thought: THE PURPOSE OF HUMAN LIFE IS THE CONTINUOUS GENERATION OF MORE AND MORE AWARENESS OF SELF, SPECIES, AND COSMOS. It is this central awakening that transforms mediocrity, electrifies existence, brings meaning, interest, and elevated excitement that continues right to the very end of the individual earthlife and, the evidence strongly suggests, beyond.

There is no question about the profound and enduring truth of this statement. Whole libraries of inspirational literature attest to it, and a long reading list of recent findings in the behavioral sciences confirms it. Beholding consciousness-levels from the sleepy earthworm to human comprehension brings it to direct observation. Without this truth life remains half-hearted and often pointless. With it human life comes to life.

The nurture of this truth is the purpose of the Phenix Club. Once even a single person feels it deeply, a new club may, in a sense, be said to have already come into existence.

In starting a new Phenix Club one first casts about for an acquaintance who might—just *might*—be open to this kind of vision of life's meaning and purpose. In a society that largely taboos serious talk and insists that social conversation be kept trivial and superficial, this may require a little choosing. One may invite the chosen friend to a coffee-break or a dutch-treat lunch. Soft-sell is the watchword: "This thing has recently come to my attention, and I'd like to know what you think of it."

If your friend lights up, your new club now has *two* members. If not, try some others. And don't be discouraged. That marvelous redemptive society, Alcoholics Anonymous, which now has three-quarters of a million members and a group at every crossroads, took four years to get its first hundred members. Phenix Clubs spread coast-to-coast in a single year, but are still numerically small. An idea that reaches deep and pulls strong takes time to penetrate. But have no doubt about the method—the steady, quiet reaching out and testing people. You'll be surprised at some you'll find who, just under the surface, have been desperately lonely for exactly what *you* are looking for. A typical reaction is, "Why, I thought I was all alone in my wish for thinking at a little deeper level. Now I'm finding there are so *many* others like me!"

It is not wise to go after large membership until a companionable feeling has had time to grow up among the founding members. A Phenix Club can, in fact, function on a permanent basis with as few as four members. They run as high as thirty, and have even had special-event attendance up to a hundred, but this is unusual, and the emphasis is never on numbers.

The official definition of a Phenix Club, from our original

1973 "blue brochure," is: THE PHENIX CLUB IS A FRIENDSHIP SOCIETY OF MEN AND WOMEN WHO SEEK THROUGH READING, DISCUSSION, AND REFLECTION TO IMPROVE THE QUALITY OF THEIR LIVES.

Let's look at this definition a piece at a time.

THE PHENIX CLUB IS A FRIENDSHIP SOCIETY. This statement is central. All other aspects of the Phenix Club are secondary to this. We don't take opinions—even our own!—too seriously. We know they will pass and be replaced by others as we learn and grow and turn up new information. Intellectual fads, fashions, and affectations come and go, but a firm foundation of friendliness weathers them all. So long as friendliness is maintained, even the sh rpest clashes of opinion will be dissolved in warmth and humor as the steady, week-to-week broadening of viewpoint proceeds. The important thing is to maintain a place where people can come and discuss their living and reading experience in a friendly atmosphere. This promotes continued personality growth. That is why our prime definition of ourselves is "We are a friendship society."

The next important phrase is: SEEK THROUGH READING, DISCUSSION, AND MEDITATION TO IMPROVE THE QUALITY OF OUR LIVES. The first thing to be read is, of course, the Phenix Club handbook, which contains its basic philosophy. It is called *The Last Third of Life Club*, and is based on the life-experience of the founding members. Critic Frank Tribbe says it's misnamed (the next edition will be called *The Life Club*). "This is a book for *all* ages," Tribbe says. "Some fifteen-year-olds will be adult enough; most twenty-five-year-olds ought to be." (As a matter of record, the youngest Phenix Clubber is twenty-two, the oldest ninety-one. When the mind is expanding, time vanishes.)

The book was originally intended for older persons. But the founders soon discovered something about neglected minds. Unless you continue to learn and grow during the stresses of the thirty-through-forty-seven range, your mind tends to close up shop. You become less receptive to new ideas. In extreme cases you fall into the later-years deathtraps, turning crabby, opinionated, and sour; or empty-headed and giddy in a silly "second childhood"; or sinking into a dull depression; or narcotizing your mind in daily alcoholic stupor.

The book had a will of its own, however, and, though it found substantial readership among those of advanced years, its main support came from those in the mid-life, pre-retirement or empty-nest range. The older members are those who have won their mid-life and pre-retirement struggles and have emerged with open, lively, and curious minds. The younger ones are those who are anticipating these storms by gaining the necessary understanding to weather them successfully.

Those planning to start a new Phenix Club would be well advised to have a number of meetings of the founding nucleus before reaching out for a larger membership. The aim of these early meetings is to study the handbook (each member should have his or her own well-marked copy) and discuss it until some agreement is reached on what the main thrust of the new club is to be.

Phenix Club does not offer a canned or prefabricated philosophy. It invites each individual to develop a personal life-philosophy of his own, and suggests some useful materials. It never says, "You must believe this or that," but only, "Have you considered this? Have you ever looked into that?" Because the finite can never encompass the infinite, all philosophies must be left open-ended, eternally ready to receive new revelation and insight. Thus each Phenix-Clubber, being different from all others, develops a philosophy that enlightens his and her own life. The result is a mosaic of different but harmonious and ever growing life-philosophies.

"The serious problems of life," Carl Jung wrote, "are never really solved. . . . The meaning of a problem seems not to lie in its solution, but in our working at it incessantly. This alone preserves us from stultification." During the past few years we of the Phenix Clubs have lived out the truth of this insight. We have had problems indeed, and still have some. But working at them has produced, almost miraculously, seemingly as a by-product, a life that is singularly free from the terrors that once assailed us, full of moment-to-moment meaning and adventure, and even flavored from time to time with joy.

Each Phenix Club is autonomous, making its own arrangements as to meeting place, time and frequency, and programming. Some clubs meet in members' homes, some in church, library, institutional, or community center meeting rooms. No expenses are involved in becoming a member except acquiring a handbook. The small operating costs are covered by passing the hat at the local meetings. The Phenix Society, Inc. grinds no political axes and conducts no lobbying activities. Political action, we feel, is a matter for individual discretion. As in all other departments of life, we share information in this area, but do not proselytize.

Quite a variety of Phenix Club formats has resulted. One club, in Kansas City, makes a festive occasion of its monthly meeting, having dinner parties at members' homes in rotation and discussing the preselected handbook chapter or special book—which all had read in advance—during the evening. Though all but a few clubs are nonsectarian and noninstitutional, there are clubs in Methodist, Jewish, and Episcopal retirement condominiums in Phoenix, St. Louis, and San Francisco, in a Baptist convalescent home in West Virginia, and in a Presbyterian retirement village in Wisconsin. A New York City club meets every *other* Sunday afternoon. The majority of Phenix Clubs, however, are neighborhood groups of general-population, mixed-age, mixed-faith or unchurched people meeting weekly on weekday evenings and sharing a common interest in consciousness-raising and personal development. Members cover the whole occupational range, alphabetically from architect, businessman, and clerk through housewife, physician, salesman, and stenographer, on to teacher, telephone lineman, and therapist, to waitress, writer, and youth counselor.

We do not insist that groups using our materials call themselves Phenix Clubs. Preexisting organizations, such as church and library reading-and-discussion groups, and Alcoholics Anonymous "Over Forty" groups frequently use our materials while continuing their original identities.

Phenix Club meetings have flexible content but a fairly firm external structure based on long experience with group work. Meetings should begin promptly at the announced time and "nominally" close exactly ninety minutes later. To keep a steady purpose in mind, this preamble is often read in opening each meeting: "The Phenix Club is a friendship society of men and women who seek, through reading, discussion, and reflection, to improve the quality of their lives." Sometimes five minutes of meditative silence or "quiet time" follow. Routine matters, announcements, and the like are quickly disposed of and the meeting itself begins.

In a typical club the twelve chapters ("Conditions") of the handbook and its five appendix chapters are discussed in sequence, one each week, and begun over again every seventeen weeks, with a different member leading each week. The topic leader for each week is chosen and announced the week before. Using the assigned reading passage as takeoff, he delivers a ten-to-fifteen-minute conversation starter, telling of his own experience, parallel reading, and observation concerning the selected passage, perhaps citing one or two of the paragraphs he likes most (or least; conforming with the handbook is not part of the plan; its sole use is as a conversation starter).

After the topic leader's remarks, the meeting is open for general discussion—response to those remarks, agreements or disagreements with them, further and independent reactions to the same reading passage, and so on. Those who have something to say follow the usual custom of addressing the chairman and awaiting the go-ahead signal before taking the floor. Speeches from the floor are normally limited to five minutes, but exceptions can be made. Reticent members are encouraged to comment, even if only briefly. To encourage this the chairman may have a round-robin toward the end of the meeting, asking the views of those not previously heard from.

After an hour and a half the meeting is declared "nominally" closed. This allows those who want to get home early to depart in good grace. If a "hot" discussion is in progress, the participants are free to carry on, with all who choose to stay, in what is known as "after-meeting discussion" (it frequently happens). Cookies and decaffeinated coffee are traditionally served as a goodnight snack.

Two factors work together to make a great Phenix Club—inspired leadership and willingness to do a little work: not much, but a little. Inspiration is something we have to wait upon, it comes or it doesn't. But little hour-long jobs are things we *can* do something about, and they make a tremendous difference. Every group needs someone to take general charge, chair the meetings, guide discussions, help find a meeting place, arbitrate disputes. This someone is called the chairman, and the job can be rotated.

Whether designated as "committees" or just jobs, these things also need doing: treasurer, refreshments, new member outreach (contacts, publicity), old member maintenance ("haven't seen you lately and wondered," etc.), national office contact ("Got anything at Box 25 that might help?"), volunteer rides for those who need them, organizing special events. Volunteering for these minor jobs, and willingly taking one's turn in being topic leader consume little time but add enormously to the effectiveness of the club.

"What do you do after you've been through the handbook two or three times?" is a question frequently asked. The answer is that we do all kinds of things, and it is here that the individuality of each autonomous club shows up most clearly. Two clubs starting the same way at the same time may find themselves following quite different formats a year or two later, depending upon the tastes, interests, and preferences of the memberships. The only general principle followed is this: It's a good idea to touch on a reference to the in-sequence Condition at every meeting, even if it's only a five-minute aside, relating the topic of the evening to the current condition. The reason for this is that the Twelve Conditions have been found through experience to result, if adhered to over a period of time, in a marked upgrading of mental and spiritual state and general outlook. These are the "Twelve Conditions":

1. We admit that death is closer for us who are past mid-life than it is for the average person; that in this respect we are different from the majority of people.
2. We have come to see that, for those who are prepared, the eventual passage from this life can be a glory rather than a dread.
3. We have decided to use our remaining years primarily for the broadening and deepening of our understanding.
4. We assert that the last part of life is given by nature for this high purpose; that it can illuminate all earlier experience in the joyous fulfillment of a rounded life.
5. We have resolved to give over our lives to Cosmic Creative Intelligence as we individually name and experience this divine force.
6. Through regular morning and evening quiet times, we are finding ourselves more and more in harmony with this transcendent Power.
7. Reviewing our past in the company of others has shown us that our earlier life-goals no longer suffice.
8. Through reading, discussion, and reflection, we have humbly attempted to discover and cultivate those higher values that are essential to our new life.
9. Having thus gained a clearer perspective on life's major phases, we have steadfastly sought the wisdom it is the business of life's later years to acquire and preserve.

10. These steps have brought an awareness of cosmic dimensions we had not hitherto explored, and have led us into the realm of deep spiritual experiences.
11. Though aware that the workaday world undervalues spiritual wisdom, we offer what we have of it when asked.
12. As our special responsibility, and as opportunity offers, we carry to others in the later part of their lives the heartening word that seniority can be joyous.

Early in the game, we learned that there are as many ways of reacting to these suggestions as there are human beings. This is why one Phenix Club specializes almost exclusively in inviting (without fee) outside speakers to tell of their specialties and answer questions about them. This group has had a sequence of psychics, world travelers, public servants, astrologers ecolgists, yoga instructors, therapists, and authors to minister to the interests of its members, while the original Phenix Club (founded 1973) remains a small weekly discussion group, breaking formula from time to time only to discuss a book of special appeal.

Other neighborhood groups supplement the Phenix Club program with outside intellectual fare, such as continuing education courses and lectures. These add spice and zest to the conversation and help the general evolutionary effort and purpose, which is simply to keep the mind alive, nourished, and growing. In West Palm Beach and Fort Lauderdale, Florida, the Phenix Society has cooperated with the local Retired Senior Volunteer Program (RSVP), United Way, and community colleges in presenting free day instruction in topics of interest to thoughtfully mature people not effectively served by undergraduate curricula. RSVP provides the faculty by drawing on the free volunteer services of retired college professors, United Way takes care of the printing, mailing, and publicity, and the community college furnishes the lecture rooms. It's the nearest thing yet seen to Jung's dream of "colleges for the second half of life," and will unquestionably spread to other states—it makes such obvious sense!

Condominiums, retirement villages, and high-rise senior apartments, which have been said to "isolate" the aging, can be turned into a special blessing when friendliness and social life are found in groups of people who seek to improve the quality of their lives by reading, discussion, and thought. "I've lived here nine years," said one retirement-settlement dweller, "but didn't know what friendliness *was* until our Phenix Club was started."

Acknowledging that people have special needs, preferences, and interests, we founded Phenix Society, Inc., a non-profit continuing-education foundation, to provide a variety of materials. By writing the Society at Box 25, Guilford, Connecticut 06437, the clubs can obtain these supports. Besides the handbook, tapes, and help in furnishing publicity materials and obtaining speakers, the Society sponsors The Wisdom College, a kind of book-a-month club, with study guides, of consciousness-raising substance designed to help meet the need Dr. Jung cited for "colleges for forty-year-olds" (and over). It also publishes a regular newsletter of book reviews, club news, correspondence, meeting suggestions, and inspirational essays, and sponsors an annual conference that draws hundreds of people from coast to coast.

To sum up what Phenix Club is about, we cannot do better than quote the original descriptive pamphlet, the "blue brochure":

> It grew out of a deep discontent with some of the mistaken beliefs a youth-oriented society tries to impose on those past the middle years. These include the odd notions that those past mid-life are no longer fit to contribute significantly to the needs of the world; that youth will solve the world's problems; that the best thing for the aging to do is amuse themselves with entertainments and nonproductive hobbies until it's time for them to die; that death requires no preparation, since it simply ends the individual personality in oblivion.
>
> The Phenix Club rejects these widespread but uninformed conclusions, and exposes them as ignorant, unhygienic, and

unscientific. It asserts that the fiction that youth can save the world is transparent foolishness, since youth has shown itself so often unable to save even itself. The Phenix Club asserts that the later years are not a useless appendage but a joyous fulfillment of earth-life's ultimate meaning.

# 23 Four Clues to Radiant Living

CARL JUNG HAD some important reflections about people who discuss inspirational subjects. He wrote:

> There are all sorts of attempts being made to appeal to people's idealism, enthusiasm, and ethical conscience. Characteristically, one forgets to apply the necessary question, "who is making the idealistic demand?" Is it someone who jumps over his own shadow to talk on an idealistic program that promises him a welcome alibi . . . cloaking a very different inner world of darkness? One would first like to be assured that the man who talks of ideals is himself ideal.
>
> But to be ideal is impossible. Most of the ideals that are preached and paraded before us sound rather hollow and become acceptable only when their opposite is openly admitted to. Without this counterweight the ideal becomes incredible because of its humorlessness, and degenerates into bluff.
>
> Bluff leads to no good. Recognition of the shadow, on the other hand, leads to the modesty we need in order to acknowledge imperfection. Human relationships are not based on perfection, for this only emphasizes differences. It is based rather

> on imperfection—on what is weak, helpless, and in need of support. The perfect has no need of the other, but weakness has. It seeks support, and does not confront its partner with anything that might force him into an inferior position.

As a summary and conclusion to our discourse on the victory over age, we are about to present four clues to radiant living that we have found helpful. These did not come in unasked inspiration from on high; they rose out of ignorance and desperation. They were born in mistakes and raised in the suffering that results from mistakes. We were indeed "weak, helpless, and in need of support" when we came across them. The fact that we are no longer in that pathetic condition is due entirely to the circumstance that we *did*, willy-nilly, come across them.

Life-crises are recognized more widely today than twenty years ago, when these clues first germinated. But even that long ago there was something we urgently needed to know about this mysterious impasse. Why does the spiral sometimes go down-down-down to despair, sometimes up-up-up to beauty and ecstacy?

This called for a little looking into. I personally had a wretched mid-life crisis. I didn't do *all* the wrong things about it, but I did enough of them to make a pretty good catalog of what *not* to do. Like so many men and women, I considered my situation at mid-life and thought (you aren't supposed to say this out loud; it would wreck the system if many people did *that*; but you are permitted to *think* it): "Here I have all the things they told me would make me happy, the things they call 'success'—professional status, wife, home, kids, couple of cars, boat, airplane, enough to indicate that I was supposed to be a 'winner.' But I'm not happy. I am, in fact, miserable. What went wrong? Why does life seem so empty and meaningless?"

There's a gentleman's agreement in our American cocktail hour and other social occasions that you never talk about anything that deeply matters to you. To introduce anything serious into the conversation is to break the social code. You go to a party and say something like, "I wonder what the *Creation* is all about, anyway?" or "I wonder what human life is meant for, especially my own life?" and you've committed a misdemeanor. "Death?" Hush! We don't talk about death. That's depressing, and we don't talk about anything depressing, only about cheerful things. The understanding is that if you don't like our party you can get out and go home, but don't come here and wreck it. "This is for *fun*; even if we're all miserable, it's *fun*, dammit!" We've all run into this tribal taboo against discussing ultimates.

We reached a point of desperation where we didn't care any more. When I say "we" I mean a few men who were going through the same spin-dry cycle I was. For over a year there were just two of us; we agreed to have lunch once a week to talk about ultimate meanings. Our number once got up to seven. Then one died, one moved away, I moved away, and we started over with two, and three, and so on.

After twenty years of this we had made some discoveries. One was Jung's axiom, previously referred to, which applies to all who cross the threshold of thirty: You can't live the *second* half of your life by the same standards as the first half. Don't be surprised by this, he said. This is the way it is. You just can't do it. If you try, you're only going to come to grief.

By now we understood that thoroughly. We had tried the impossible and *had* come to grief. The old get-up-and-go, beat 'em, lick 'em, be-top-dog stuff didn't work any more. Oh, sure, we could still beat down a wounded opponent as we did in the old days if we wanted to, but there was no longer any satisfaction in it. It was not giving our lives the *meaning* they needed.

So it was a matter of okay-doctor-you're-right-so-far-now-what-else? The what-else was simply that the later years have a special and transcendently important meaning of their own, quite separate from, and philosophically far above, the meanings of life's first half. Mankind used to know that higher meaning and utilize it. But modern man has lost it. In the early human tribes it was the older person's job to become wise, and he took his responsibility seriously.

If he didn't, he just didn't belong in the council of elders. If somebody said he was a stupid old man, it was heresy. No old man was *allowed* to be stupid. He jolly well *had* to learn the tribal wisdom and be able to give it out. The same held for women, of course. The matriarchs were expected to be deeply knowing, and most of them were, and were respected for it.

I hear many older people today complaining that young people "don't have respect" for us. Well, why should they? Too many of us have allowed ourselves to become stupid, placidly accepting the "done-for" role our work-obsessed, youth-nutty tribe has agreed to thrust upon us.

We refused to play this stupid game. We did not at once become wise, of course, but at least we *now* knew the name of the game. We had mapped out a course of our own and were no longer vulnerable to the tribal hokum. In our reading and our suffering and our looking and our knocking around we had acquired, not wisdom, but something else. The wisest, we had learned, never claimed wisdom. We followed Socrates, who said, "They call me the wisest man in Athens, but all I ever do is ask questions."

While asking questions, and finding some beautiful, exciting answers, we did become, if not wise, at least a little bit less foolish. And when we note the colossal follies humanity is committing daily, that's progress. Now, after more than twenty years, completely assured that we'd found something that *worked*, and having had the joy of seeing so many others share these same blessings, we were ready for Dr. Jung's next proviso: "We have colleges, universities, professional schools to prepare us for the first half of life. But are there schools for forty-year-olds to prepare them for the much more difficult second half?"

The answer is, of course, no. There are none. We couldn't

start a college right off (though we will eventually), but we could do the next best thing, which was to put together this book on how we did it, and how anyone else can do it who wants to strongly enough. A few people are smart enough to see this thing coming as early as their late twenties and early thirties, and start doing something about it. They have a very good chance of avoiding the mid-life hell that we had to go through. Wisdom, once perceived as a value worth working toward, has an attraction that overarches the generations.

With, I hope, the humility Dr. Jung says is essential, I think I have made clear that this movement was born in weakness. We seek, and just to the extent that we seek, we find. We never arrive at a position where we are able to tell somebody else how to be radiant.

But we have *clues.*

A "clue" is something a smart detective gets hold of and then follows through in an effort to solve his case. We have very definite clues, which, if followed persistently, will solve the case. The whole message of this book is encapsulated in these four clues.

And just *which* case is it that needs to be solved? For me, mine, and nobody else's. For you, yours, and nobody else's.

**First clue:**

It's an old one, going all the way back to Socrates, but it still has a modern ring: KNOW YOURSELF.

To know yourself is to understand your pain, and the reason for it. It's terrible to suffer and not know why. To carry around a dead weight of inner psychic misery and not know what it is, or what to do about it, is an agony, and an unnecessary one.

Finally we found out. *Life-crises are nature pushing you to a new level of awareness.* "Only at a point of genuine bankruptcy of hope," writes Joseph Chilton Pearce, "can we glimpse beyond our cultural [tribal] conditioning and sense a possible turning." This "genuine bankruptcy of hope" is life-crisis. The "possible turning" is the beginning of our "glimpse beyond" the tribal cul-de-sac, into new dimensions of knowing.

The nub of it is that if you *go along* with your natural maturation, work *in harmony* with it, and with the Cosmic Creative Intelligence, the thing may not instantly go away, but it will greatly diminish and become bearable because it now has meaning. The meaning is that your pain is a growing pain. It is the evolutionary thrust in you demanding to be heard.

Another, more mundane and everyday aspect of knowing yourself is covered neatly by one of Jung's observations on our social scene: "Most people confuse self-knowledge with knowledge of their conscious ego-personalities. They measure their self-knowledge by what the average person in their social environment knows of *him*self. Thus individual judgment becomes increasingly unsure of itself. Responsibility is shuffled off on a corporate body—the government—that has no reality of itself, and is only a camouflage for those who know how to manipulate it. Yet it usurps the decision-making function of the real life-carrier, the individual self."

Part of knowing yourself, then, is to return the decision-making to your true self, as distinguished from the tribal-survival-role self that has brought you to such an impasse. Much of this is psychological excess baggage from early efforts to please and placate Mommy-and-Daddy's generation and may now be jettisoned, since "the time is now." It is not so much a matter of whether these ideas are morally right or wrong as it is the bald fact that they simply DO NOT SUFFICE for the new life-situation. So, *know yourself.* Or, since that may turn out to be a lengthy journey, let us say rather, *Make a resolute beginning toward knowing yourself.*

And just what is the ultimate goal of this clue? What is there concealed that is so exceptionally valuable? Happiness. "The Kingdom of Heaven," which, according to scripture, is "within you." Our problems are triumphantly solved by the full development of our own inner resources. So don't run to far places or to the bank or the government for the solutions. When you take up the clue of knowing yourself, and follow it through, the solutions begin to appear.

**Second clue:**

BE yourself. Here we touch upon that vast area of truth that is able to distinguish between the *immediate* world of ACTION and the *universal* world of BEING.

We Americans know very little about this universal world of being, hence find it very hard to BE ourselves. We are an action-oriented people. Unless you're *doing* something you're considered to be doing nothing, and that is not allowed. The watchword is, "Don't just sit there, *do* something." What we're saying now is: "If you want to become acquainted with the world of being—that is, the part of it that is being yourself—take a little time each day to *not* do something. Just sit there."

A good way to begin to grasp the concept of being is to think of a flower. What does a flower *do*? It doesn't do anything, it just *is*. And yet it is a beautifully ornate, distinctly valuable, entirely legitimate participant in the total scene. Its job is just to BE.

You are infinitely more complex, more glorious, more ultimately valuable than any flower. And yet the sole job of one of the most important parts of you, the *universal* part, is just to BE.

**Third clue:**

TRUST the Infinite.

After his two meditative years in his cabin at Walden Pond. Thoreau remarked, "I think we can safely *trust* a great deal more than we do." Trust what? The universe that sustains us, of course. There are a thousand names for it in languages around the world, and a hundred variations in our own tongue. The universe likes to be acknowledged, but is not fussy about the label attached to it so long as it is respectful. Some call it the cosmos, some call it God. Whatever name you give it, get to know it a little, and trust it. It will, Dr. Jung believed, be rewarding if you do. "Just as a social being cannot live without a tie to his community," he said, "the individual self will never find his own spiritual and moral autonomy anywhere except in an extra-mundane principle. For this anchor, the individual needs the evidence of inner transcendent experience, which alone can protect him from his otherwise inevitable submersion in the mass."

And how do you "trust" the universe? The same way we trust anything else—by watching it, experiencing, knowing it, seeing what it does with us, and, finally, trusting it. Is it worthy of our trust? To me the mere fact of your reading this book is evidence enough of its trustworthiness. It is a truly incredible thing, when you think of the circumstances of this precarious planet. Think back on the ups and downs, the vicissitudes, the narrow escapes you've been through. Yet miraculously, you're *there*, sitting and—enjoyably, I hope—reading a book! In this experience of trusting, one finds oneself related, not just to family, business, and tribe, but to the world and all that's in it, to the cosmos, to the universe, visible and invisible, to infinity and to eternity. So . . . trust the Infinite.

**Fourth clue:**

EVOLVE. If none of the other clues reach you instantly, this one alone will pull you through if you stay with it, for it leads to that single essential without which a life cannot be lived, meaning.

We have among us a certain category of chest-beater and hand-wringer who loves to go through these antics and carry on somewhat in this fashion: "Oh God—if there *is* a God—why don't you reveal to me your mysterious purpose, so that I may know and cooperate?" Yet all the while the purpose and intent of the creation is being written in plain sight, in mile-high letters, for all to see. There is much in the universe that is forever mysterious, but its general intent is glaringly clear. It is written in the fossil records, in the archaeological records, in the historical records, in the personal and individual experience of life, and in plain, broad daylight observation.

The purpose of the total creation, and of the human life that flourishes within it, is the evolution of consciousness. The whole evolutionary progression, from gas, to stone, to crystal, to one-celled organism, to plant, to animal, to man and beyond, reveals its single-pointed striving toward ever higher, broader, and deeper capacities for being *aware*, that is, for consciousness.

Mankind today has given itself over to crisis-living, making a great to-do in its press and media about this "crisis" or that—the petroleum crisis, the atomic bomb crisis, the population crisis, the food crisis, and so on. Actually, humanity has only one crisis, a crisis of consciousness. To the most highly evolved minds of our species, ready solutions are already known to all the "crises" just named. The only thing that prevents their adoption is the fact that the mass human mind is still operating at too low a level of awareness to accept them.

The human evolutionary thrust, Julian Huxley has reminded us, has long since passed beyond new experiment in bodily form. Physically, we have been about as we are for many millennia, and will probably remain so. The evolutionary spearhead has transferred its thrust to *psychosocial* evolution, the evolution of consciousness. Therefore, if you want to realize the ultimate meaning of your life, awaken your sleeping faculties to a new effort of exploration of new ideas, new thoughts, new dimensions, new possibilities. Expand your consciousness, that is, *evolve.*

As soon as you begin to take even the first few tentative steps in this direction, a new self-confidence flows into your being. Self-confidence is not having everything hunky-dory, or "having it made." It is not an easy chair where you sit or a house you live in. Self-confidence is a road you travel. It is knowing you are on the right track, regardless of what way stations labeled "success" or "failure" you may pass along the way. When you are evolving your consciousness you are in tune with the all-encompassing purpose of the cosmos; you are in harmony with the universe. You are on the right track. Knowing this, you live with a new self-confidence, *élan*, and poise.

We have already noted that a clue is something a detective uses to solve a case. But clues need to be followed up or the case remains unsolved. These, then, are the four clues to radiant living:

1. **Know Yourself.**
2. **Be Yourself.**
3. **Trust the Infinite.**
4. **Evolve.**

This book has suggested some of the things the good detective might do to follow up these clues.

# Appendix A.
# The New Gerontology: A Note to the Healing Professions

What heals?

We are slowly emerging from an era when it was widely and earnestly believed that healing was accomplished mainly by chemical and surgical intervention. The state of the individual's morale was regarded as secondary.

We are seeing this order gradually being reversed. More and more, the state of the individual's morale is being seen as the primary factor.

Nowhere is the need for hastening this change more pronounced than in gerontology. Morale is low and ebbing. Hypochondria and drug (alcohol) abuse are rampant. A new professional outlook is overdue. The new gerontology is, in fact, already making its appearance. These are some of the ways in which it differs from the old:

The old gerontology tended to look upon aging as a degenerative disease, and upon the aging person as a patient who is ill. The new view is that age is a normal and healthy part of life having its own special and potentially joyful functions.

The old gerontology tended to see aging as something that began to need attention during the sixties. The new insight requires that the psychic realities of the whole life be clearly understood in the thirties if the later experience is to be optimum.

The old gerontology often isolates the aging person in a mental climate emphasizing decaying tissues as the factor of primary significance, and sees no redeeming element in death. The new gerontology declares that a healthy mental and philosophical attitude is primary and that acceptance of an eventual death-transition is part of a healthy attitude. It notes that such an attitude brings an improvement in physical health.

The old gerontology questioned the competence of older people to deal with their own problems. Observing that the main problems of aging are inward attitudes, the new gerontology suggests that they can be ultimately solved only by the aging individual, who is fully competent to do so once the true meaning of aging becomes clear.

The old gerontology tends to look upon the post-retirement years as a decline into senile childishness, where childish pastimes are appropriate. The new perspective reveals that these years are the arena for life's most important work.

The old gerontology saw death as an end; the new gerontology sees it as a graduation.

# Appendix B.
# Some Mind-Expanding Reading

A variety of books, some old, some new, on a variety of subjects, having only one thing in common: They expand the awareness.

Bucke, Richard M., *Cosmic Consciousness.* New York; Dutton

Castaneda, Carlos, *The Teachings of Don Juan.* New York: Pocket Books

Commoner, Barry, *Science and Survival.* New York: Viking

*A Course in Miracles*, 3 vols. New York: Foundation for Inner Peace (1 West 81st Street, Suite 5D, N.Y. 10024)

Dean, Stanley, *Psychiatry and Mysticism.* Chicago: Nelson-Hall

De Beauvoir, Simone, *The Coming of Age.* New York: Warner Books

Eiseley, Loren, *The Firmament of Time.* New York: Atheneum

Ellison, Jerome, *The Last Third of Life Club.* Philadelphia: Pilgrim

Frankl, Viktor, *Man's Search for Meaning.* New York: Pocket Books

James, William, *The Varieties of Religious Experience.* New York: Mentor

Johnson, Raynor, *The Imprisoned Splendour.* Wheaton: Theosophical

Jung, Carl G., *Modern Man in Search of a Soul.* New York: Harbrace

——— *Memories, Dreams, Reflections.* New York: Random

——— *Man and His Symbols.* New York: Dell

Kubler-Ross, Elisabeth, *Death, The Final Stage of Growth.* Englewood Cliffs: Prentice-Hall

May, Rollo, *Love and Will.* New York: Dell

——— *Man's Search for Himself.* New York: Dell

Meadows, D. L. & D. H., *The Limits to Growth.* New York: New American Library.

Norton, Suza, *Yoga for People Over Fifty.* Old Greenwich: Devin-Adair

Pearce, Joseph C., *The Crack in the Cosmic Egg.* New York: Pocket Books

Sandweiss, Samuel, *Sai Baba: The Holy Man and the Psychiatrist.* San Diego: Birth Day

Schumacher, E. F., *Small Is Beautiful.* New York: Harper & Row

Selye, Hans, *The Stress of Life.* New York: McGraw-Hill

——— *Stress Without Distress.* Philadelphia: J. B. Lippincott

Thoreau, Henry David, *Walden.* Garden City: Doubleday

White, John, ed., *The Highest State of Consciousness.* Garden City: Doubleday

——— *What Is Meditation?* Garden City: Doubleday

White, Stewart Edward, *The Unobstructed Universe.* New York: Dutton

# About the Author

The two-million-circulation magazine *Modern Maturity*, commenting on various efforts of aging persons to handle their situation creatively, remarked in its November, 1975, issue that "Jerome Ellison is way out in front." Seventy this year (1977), he has long been known to the reading public as novelist, essayist, critic, reporter, editor, and educator. Born in Maywood, Ill., in 1907, he graduated from the University of Michigan in 1930 and went to New York to participate in the literary and artistic life of the city. He has published ten books—fiction and nonfiction—and "innumerable" essays, articles, and short stories. He was managing editor of *Collier's Weekly* and associate editor of *Reader's Digest.* In World War II he was with the Overseas Branch of the Office of War Information in Washington, New York, and London. He is a world traveler and is nationally known on the lecture platform. "At mid-life," he writes, "I left the metropolitan rat race for the relative calm of the groves of Academe." He has been professor at Indiana University, at New York University, and at the University of New Haven. He is founder of the Phenix Clubs, and president of a nonprofit educational corporation for the study of aging called the Phenix Society. Its address is Box 25, Guilford, Connecticut 06437.

Behavioral Genetics:
Simple Systems

# Behavioral Genetics: Simple Systems

A SYMPOSIUM
HELD MARCH 4-5, 1971, AT THE
UNIVERSITY OF COLORADO

James R. Wilson, Editor

Colorado Associated University Press
Boulder, Colorado 1973

This symposium was supported by
NIMH Training Grant MH-11167 to the
Institute for Behavioral Genetics


ISBN: 87081-037-5
Library of Congress Catalog Number: 72-96157
Designed by Dave Comstock
Printed in the United States of America

# Contents

# Preface

THESE papers were presented at a symposium sponsored by the Institute for Behavioral Genetics, University of Colorado, Boulder, during 4 and 5 March 1971.

The background for the symposium should perhaps be explained. The Institute for Behavioral Genetics (IBG) is an interdisciplinary research and training unit of the graduate school, but is not degree-granting. Institute faculty have joint appointments in an academic department, such as Aerospace Engineering, Biology, Pharmacology, or Psychology. Institute students are regularly enrolled in one of the academic departments and complete both departmental and Institute requirements. Most are recipients of a fellowship from the training program in behavioral genetics supported by a grant to IBG from the National Institute of Mental Health.*

This training program has provisions for inviting consultants to present lectures and seminars about their work. Many eminent scholars have been kind enough to travel to Boulder for this purpose, and have considerably broadened and enriched the training program by their participation. It occurred to us that a symposium format for the consultants might be even more valuable. The participants would then be able to interact with each other, as well as with Institute personnel. It was decided to try this format, and a student committee, chaired by trainee (now Dr.) Lynda Uphouse, was appointed to begin the necessary planning. In a short time a topic was chosen, speakers were

*NIMH Grant MH-11167.

invited, and preparations were well underway. The later success of the symposium was largely due to the efforts and personal involvement of the trainees. The symposium was open to the public, and was attended by people from several departments and from other universities.

The substance of the symposium was, of course, the eight fine individual presentations made by the invited speakers. These papers and the discussions which followed were recorded during the presentations, and transcripts were furnished to the authors for their use in preparing the version which appears here. Some authors deliberately retained the flavor of their oral presentations, and made only those changes necessary to correct errors in the transcript or to insure understanding. Others chose to revise moderately and thereby present a somewhat more formal paper for this volume. Each technique has its advantages, and the versions included here are faithful to the authors' desires; they include only minor editing — chiefly to remove accidental repetitions of words or "stage directions" or other amenities of the occasion. The result is felt to be in the spirit of the symposium, since one of the delightful aspects which emerged was the great variety of the presentations — both in style and in content. It seems only fair to pass on to the reader at least some of this variety in the present version.

Though a number of people made important contributions to the symposium and to publication of the proceedings, several should be singled out for personal thanks. These include Professor Gerald McClearn, Director of IBG and Director of the NIMH training program, whose enthusiastic endorsement made the symposium possible; Agnes Conley and Lois Johnson, Administrative Assistants, and Rebecca Miles, Secretary, who coped with many administrative details and especially with transcribing the recorded presentations; and Dr. Uphouse, who gave so much time to insure the success of the symposium. Thanks are also due to the faculty and students of IBG for their interest and support. The greatest debt, of course, is owed to the invited participants, who took the time from their busy research and teaching activities to come to Boulder to share their findings and thoughts with us. It is hoped that making their work available in this form will partially compensate them for their efforts.

James R. Wilson
Boulder, Colorado 1973

# The Application of Genetics to the Study of General Laws of Behavior

ROBERT M. MURPHEY
*University of California, Davis*

WHATEVER historical reference point that might be chosen for beginning this discussion would be an arbitrary one. As good a place as any might be Calvin Hall's chapter in the 1951 edition of Stevens' *Handbook of Experimental Psychology*, which appeared nearly a decade before the field of behavioral genetics was "officially" established by Fuller and Thompson's (1960) publication of their textbook by the same name.

In his essay, Hall (1951) lamented the traditional estrangement of psychology and genetics. Even though the two disciplines had achieved zygosity at about the same time in the mid-19th century, genetics had concerned itself almost exclusively with morphology, while psychology's professed interest in genetics had been naive and superficial. He predicted, though, that things were soon to be different.

> A real genetics of behavior promises to emerge . . . as psychologists continue to adopt the procedures of modern genetics, and as geneticists turn their attention more systematically to behavior. This encouraging trend will ultimately give status and stature to an interdisciplinary science of *psychogenetics*.

In the twenty years that have intervened since Hall's writing was published, it is clear that psychogenetics (as it is called in

England); or behavior genetics, behavioral genetics or behavior-genetic analysis (as it is variously termed in the United States – take your pick) has indeed merged as an active area of inquiry in its own right. In attempting to follow the contributions of genetics to general laws of behavior, I shall briefly mention a few accomplishments that behavior genetics has achieved, and then take up some considerations regarding its future.

## Psychology Discovers Heredity

By now most of us are familiar with the origins, antiquity, and outcome of the nature-nurture pseudoquestion; it is a narrative of how sound experimentation and modern biological thinking sought to overcome the forces of darkness, radical (to use the term Skinner used in 1963) empiricism as represented by John B. Watson, and rigid instinct theory as championed by William McDougall. Subsequent to the ascendency of behaviorism over instinct theory in American psychology, it was probably Tryon's (e.g., 1942) successful selective breeding experiments on "maze learning ability" in rats, and his quasiclinical findings relating mental deficiency to genetic endowment (e.g., Jervis, 1939) that were most influential in stimulating a trend toward acknowledging inheritance as a justifiable variable in behavioral events. More detailed accounts of the period are provided elsewhere (e.g., Broadhurst, 1966; Fuller & Thompson, 1960; Hirsch, 1963, 1967), and no more need be said about it here. Instead, let us return to Hall's (1951) chapter, which in retrospect seems to have been highly influential in setting the course that behavior genetics would follow over the succeeding twenty years. In it, Hall listed four aims for facilitating the study of behavior-genetic relationships:

> (1) To discover whether a given pattern is transmitted from generation to generation, (2) to determine the number and nature of the genetic factors involved in the trait, (3) to locate the gene or genes on the chromosomes, and (4) to determine the manner in which the genes act to produce the trait.

The decades of the 1950's and 1960's saw an exciting proliferation of behavior genetic research activity which concerned itself principally with the first three of Hall's objectives; his fourth, in a strict sense, has received considerably less attention. In fact, two recent attempts to specify the purposes of modern behavior genetics (Bovet, Bovet-Nitti & Oliverio, 1969; Maier & Maier, 1970) quoted or paraphrased Hall's first three goals, but

omitted the fourth altogether. Of course much of the current terminology used to describe behavior-genetic research differs from the language expressed by Hall. Many of us are averse to any mention of "the inheritance of behavior" or "how a gene produces a trait" on the grounds that behavior is a phenotype, not a genotype, and so forth. Even so, to a large extent (though certainly not entirely) the area's literature can be characterized as demonstrating time after time that inbred and selected strains, and systematic crosses between them, differ on almost any behavioral trait that one might choose to measure, and that, with respect to human behavior, certain phenotypes tend to be related to consanguinity — all of which can be considered as *prima facie* evidence for genetic involvement in behavior (e.g., DeFries, 1967).

This state of affairs prompted Hirsch (1963) to remark:

> When different strains within a species are compared, it actually becomes a challenge *not* to find differences in one or more behaviors. When strain comparisons are followed by appropriate genetic crosses, genetic correlates of behavioral differences are demonstrated . . . with the wisdom of hindsight, we can ask why so many demonstrations were necessary.

Later on, Hirsch (1967) reckoned that it must have been the result of environmentalism's reluctance to admit heredity as a valid issue in the study of behavior, and for that reason,

> far too much effort has been spent proving the trivial points that this or that behavior shows a genetic component or a genotype-environment interaction, or in chasing single genes in order to have a more clear-cut case. Behavior geneticists have been so preoccupied with the defeat of environmentalist opposition that they have had little time for the more important task of a critical analysis of their own work . . .

Other writers are in essential agreement with Hirsch's assessment, except that some — McClearn (1970) for example — view in a somewhat more benign light the considerable amount of information that has accumulated, our increasing scientific acuity, and the greater emphasis upon interdisciplinary efforts. Furthermore, few of us as individuals would readily agree that we have been particularly remiss in the critical examination of our own research, although most of us are willing to entertain the possibility of that fault in the work of others.

In any case, American psychology has finally owned up to the familial legitimacy of the half-breed retainer in its household, as evidenced by the title of Richard J. Rose's paper, presented

at the most recent (1970) American Psychological Association Convention: *Revised Perspectives on Clinical Research and Training now that we know Heredity exists.* No modern textbook on behavior is complete without its chapter on behavior genetics and/or its catch phrases regarding its importance. On the basis of what has been said so far, however, one would be hard pressed to point out a specific instance of genetics' contributing to general laws of behavior.

What behavior genetics does seem to have accomplished is to present convincing evidence that valid, important laws of behavior are not likely to be discovered unless both environmental contingencies and individual differences in biological endowment are taken into account. That is, rather than furnishing general explanatory principles with respect to how behavior comes to vary as it does, behavior genetics has been a reliable source of concrete testimony in behalf of the importance of genetic endowment in the intricate webs (not chains) of events that intervene between the fertilization of eggs and behavioral variation (e.g., Murphey, 1973). Although the point that genes have something to do with behavior may strike us as a trivial one now, the influence of that point upon behavioral science is not; it represents a perhaps primordial but necessary step in the transition from findings to interpretations to understanding. As was indicated earlier, most would agree that these successes are quite modest and that more needs to be done. The question is, of course, what?

It would be outrageously pretentious of me to attempt to specify the nature and kinds of contributions that genetics can or will offer to the general laws of behavior in the future. The direction in which behavior genetics might go now is highly conjectural, and in talking about it, one is placed in a multiple choice situation similar to that faced by Dorothy in L. Frank Baum's (1909) *Road to Oz.* You may recall that, upon being confronted with so many roads that she found herself counting some of them twice, Dorothy desperately informed the Shaggy Man that they were lost. " 'That's nothing to be afraid of,' he replied, . . . 'Each of these roads must lead somewhere, or it wouldn't be here. So what does it matter?' "

Hopefully, behavior genetics will take a great many roads in the future; diversity of interest is after all the safest way to enhance the chances of survival of the discipline. But before leaving, whatever the road, there are several preparatory matters that are worth taking under consideration.

## General Laws of Behavior

*What do we want to know?*

A first question that might be asked has to do with the kinds of knowledge that we are seeking; just what is it that we are trying to find out about behavior? Or to state the question in a hardly less ambiguous way, what constitutes a sufficiently general and lawful behavioral relationship so as to be an acceptable scientific goal? Most experimental psychologists who were brought to academic eclosion and who studied for their Ph.D. qualifying examinations during the last decade or two learned from the writings of the late Kenneth W. Spence (1951) that behavioral theory should emulate physics as much as possible, and that there are two kinds of variables between which we are to find functional relationships. *R-variables* (dependent variables) are measurements of what organisms do. They can include "simple" responses (actones), complex ones (actions), and more general response characteristics (traits, abilities, and the like). *S-variables* (independent variables) are measurements of past and present physical and/or environmental conditions under which the responses of organisms occur. The task of behavioral science, as Spence interpreted it, is to state the quantitative interrelations between independent and dependent variables in the form $R = f(S)$. He saw the problem as being twofold: "1) to discover what the relevant $S$ variables are, and 2) to ascertain the nature of the functional relations holding between the two groups of variables."

In the quest for $R = f(S)$, the preferred procedure was to introduce operationally defined, hypothetical constructs (e.g., habit, strength, and drive) as intervening variables which have reference to the components of $S$, and then to manipulate the appropriate independent variables experimentally in an effort to discover how $R$ varies as a function of them. Spence accused schools of thought such as Kurt Lewin's Field Theory, Gestalt Psychology, and "Ultra-Positivism" (as represented by Woodrow and, by implication, Skinner) of leading to *R-R* instead of *S-R* connections; *R-R* laws are those which merely specify correlations between dependent variables rather than independent-dependent variable contingencies. After a while, I shall contend that one of the most adaptive directions that behavior genetics can take over the next few years is to carry out a systematic search for some form of *R-R* relationships. But at the moment, let us return to the *S-R* approach.

Within the traditional framework, $R$ is seen as *the response* of animals to some set of circumstances. Translated into the real world of data, *the response* of animals turns out to be a typologically conceived entity which reflects some sort of "typical" behavior within "the limits of error" – i.e., an arithmetic mean ($\bar{R}$). An important task of the practitioners of $S$-$R$ strategy, which is still characteristic of behavioral science even though different terms may be employed nowadays, is to minimize the deviations of "errors" around $\bar{R}$ by adding more variables and/or achieving better control over those that were being manipulated. Implicit in the process is the notion that animals within (if not between) species are fundamentally very much alike. This has been called the "uniformity assumption" (e.g., Hirsch, 1963). Counterfactual though it may be, the postulate has been with us at least since the time of Descartes, who stated in his *Discourse on Method* that ". . . there are differences in degree only in the *accidental* qualities, and not the *essential* qualities or natures of individuals of the same species." At this juncture, genetics' contribution of the importance and ubiquity of individual differences to behavioral theory becomes more clear. Whereas most of experimental psychology is looking for laws something in the form of $\bar{R} = f(S)$ — and $S$ may include morphological variables — behavior genetics is more interested in individual differences in responses as a function of *variation* in its "determinants" or $\sigma^2_R = f(\sigma^2_S)$, where $\sigma^2_S$ is partitioned into ever smaller components, beginning with genetic variance ($\sigma^2_G$), environmental variance ($\sigma^2_E$), and variance associated with interactions between genetic and environmental factors ($\sigma^2_{GE}$). Research interest, as expressed in behavior genetics, is not necessarily distributed equally over the three components of variance just mentioned. Vale & Vale (1969), for example, have made a good case in favor of restricting the hunt for general laws to the study of organism-environment interactions, while others (e.g., Jinks & Fulker, 1970) invest much in estimating values for the various components of particular phenotypes, computing heritabilities, and what not.

At any rate, there does seem to be general consensus among behavior geneticists about the centrality of diversity being the proper subject matter for their discipline. This attitude — moving from type concepts to population concepts — is of course consistent with current trends in other areas of biology as reflected in the writings of Ernst Mayr (e.g., 1963, 1968), who has said that the replacement of typological with population thinking is

possibly the greatest conceptual revolution in the history of biology, and of J. J. C. Smart (1959), who spoke of the profound misconception of the biological sciences (he included psychology) "that biology is a science of the same general sort as physics, just as chemistry is." The idea of viewing the phenomena with which we work as distributions rather than qualitative types is so fetching these days that a currently fashionable way to insult a colleague is to charge him with being a typological thinker; and because our linguistic restrictions are such as they are, of course the accuser is always right — never mind that the notion of typological thinking is itself a typology.

Given that behavior genetics is in the diversity business, it further seems that one major aspect of its interest is to link genetic variation with behavioral variation. In this regard, it is once again appropriate to bring up the question of just *what* it is that we are trying to find out. According to Hirsch (e.g., 1967), behavior genetics as it ought to be ("behavior-genetic analysis") is an attempt "to relate the phenomena of behavior to the knowledge that exists at other levels of biosocial organization and in other branches of science." Indeed, when we examine the literature that reasonably falls under the general rubric of "behavior genetics" there is a wide variety of interests with respect to levels of analysis, ranging from the study of the simple movements (taxes) of lower organisms to complex social systems of humans – from simple neurological lesions to intelligence and schizophrenia. Implicit in most of these is the hope that, eventually, the research program will be able to present an account of how specific genetic mechanisms (and perhaps environmental ones as well) contribute to individual differences in the behavioral phenotypes under study — somewhat in the genre of Hall's (1951) fourth objective, "to determine the manner in which the genes act to produce the trait" (one wishes that he had used the word "influence" in place of "produce"). Two problems inherent in arriving at this kind of goal spring to mind immediately. The first complication has to do with reductionism and the second with an empirical leap from genetical information encoded in DNA molecules to behavioral phenotypes. Let us consider reductionism first.

*Do we need to reduce?*

In light of the current level of popular concern about holistic and reductionistic approaches in biology, it is worth mentioning

that, superficially at least, reductionism can evidently present a barrier to understanding behavior, though at the same time it is the major avenue by which most of us get our work done. The problem is especially relevant in the context of this symposium: "Simple Systems."

As has been mentioned, the bulk of behavior genetics assumes that individual differences in phenotypes are attributable to the joint effects of a number of variables and their interactions. In our newer textbooks, whether on animal behavior or human personality, we commonly see platitudes to the effect that the nature-nurture problem was resolved when it was realized that both nature and nurture play a role in behavioral development, or (in the more intricate treatments) behavior is a phenotype, and genotypes — not phenotypes — are inherited. In their commendable essay, Vale & Vale (1969) persuasively contended that "Able men . . . have had no difficulty in demonstrating that neither environmental nor organismic variation alone can account for behavior. Rather, several lines of thought point to the conclusion that a successful psychology must come to grips meaningfully with the problem of the *interaction* of organisms and environments."

Nor is it unusual to find an adroit analogy: "just as architecture is more than the study of building materials, so is biology more than the study of macromolecules" (Mayr, 1968, citing Michael Ghiselin). In one of the more thoroughly anti-reductionistic statements to come out recently, Polanyi (1968), a master of analogy, has created likenesses between living mechanisms and several non-living ones in an effort to show that life is hierarchical, where the various strata of organization, while dependent upon lower levels, are nonetheless irreducible to them. As one illustration of his point, Polanyi used spoken literary composition: "The voice you produce is shaped into words by a vocabulary; a given vocabulary is shaped into sentences in accordance with a grammar; and the sentences are fitted into a style, which in turn is made to convey the ideas of the composition. Thus each level is subject to dual control: (i) control in accordance with the laws that apply to its elements in themselves, and (ii) control in accordance with the laws of the powers that control the comprehensive entity formed by these elements." And a few paragraphs later, ". . . living beings form a hierarchy in which each higher level represents a distinctive principle that harnesses the

level below it (while being itself irreducible to its lower principles."

The usual counterargument, of course, is that we fail to distinguish higher principles from the lower ones only because we have not understood enough of the lower ones to a sufficient degree and/or we have not been clever enough to see how they are put together so as to form a higher level. Moreover, it is not easy to imagine understanding a building or a piece of prose in much more than an aesthetic, descriptive sense without some knowledge of the component parts. That is, it may be that general, ostensive descriptions can get by without recourse to lower level principles, but general understanding cannot. If this point is not self-evident, try to construct your own building without the aid or knowledge of appropriate materials, or try to comprehend a foreign language without knowledge of its components.

Apart from that, we simply have not developed techniques suitable for holistic studies. At present, most of us tend to choose some limited domain of behavior and its correlates, and then exploit it insofar as we are able. The preponderance of our efforts to perform feats holistic has wound up being relatively impotent statistical models. From the arguments of Polanyi and others (e.g., Grobstein, 1962; Rosen, 1968) it is clear that the issue is not *whether* lower level knowledge is useful, but *how to use* lower level findings intelligently and effectively. Polanyi went on to provide a conciliatory path for achieving understanding at more than one level: "We can generally descend to the components of a lower level by analyzing a higher level, but the opposite process involves an integration of the principles of the lower level, and this integration may be beyond our powers." This brings us back to the problem of relating genes to behavior.

*What does the gene say?*

Let us accept Polanyi's reasoning as being correct. If it is true that lower levels need to be understood, and that they can be comprehended through the workings of higher levels, where does behavior genetics stand? At the moment we appear to be working almost exclusively at the higher levels. In other words, we are not in fact relating gene action to behavior. To my knowledge none of my colleagues in behavior genetics has ever seen a gene. What we are doing is manipulating behavioral

phenotypes and their correlates, or vice versa, and, by way of analyzing phenotypic differences and certain antecedent conditions (e.g., who mated with whom), we are *inferring* that in so doing we have demonstrated genetical involvement in the phenomena under study. That is a long way from finding out *how* genes contribute to phenotypic variation.

At present, about as close as anyone has come to that ideal is seen in two examples from the literature of *Drosophila* behavior genetics. In the first example, Kaplan & Trout (1969) were able to isolate strains of *Drosophila* that are disposed to seizure-like states (including leg shaking) when subjected to particular kinds of environmental stimulation, such as etherization or passing a shadow over the dorsal surface of the fly. Various cross-breeding regimens combined with chromosome mapping techniques suggested that the phenotypes are associated with single-gene mutations. Two "hyperkinetic" strains of unusual interest, $Hk^{1P}$ and $Hk^{2T}$, differed from one another with respect to the magnitude or vigor of the response, and the genes associated with the difference appeared to be allelomorphic, i.e., they occupied the same locus (30.9 $\pm$0.6 for $Hk^{1P}$ and 30.4 $\pm$0.7 for $Hk^{2T}$) on the *X*-chromosome. It should not be forgotten that the gene loci were inferred from crossover frequencies, which were in turn inferred from phenotypic combinations involving the trait of interest and "marker" traits, the locations of the genetic correlates of which had been inferred through a similar process (it is realized that certain *Drosophila* markers can be related to physical material on the chromosomes). Supportive evidence as to their being alleles was found when $Hk^{1P}/Hk^{2T}$ crosses produced hybrid offspring which were phenotypically intermediate between the two parental strains. Kaplan & Trout found also that the response magnitude increases positively with the age of the fly. Further work by Ikeda & Kaplan (1970) made use of sex mosaics (gynandromorphs) which were produced by eliminating one of the two *X*-chromosomes from daughter nuclei early in mitosis. Thus, a fly may be partially "male" (*XO*) and partially female (*XX*); the procedure is described in more detail by Hotta & Benzer (1970) and by Morgan & Bridges (1919). Certain breeding manipulations made it possible to compare mutant with nonmutant sections in the same fly, enabling the researchers to relate phenotypes (and by implication, genes) to more or less delineated areas. Ikeda & Kaplan found that hyperkinesis is associated with individual legs and, probably, specific lesions of the thoracic

ganglion. Simultaneously and somewhat similarly, Hotta & Benzer (1970) were able to relate several *Drosophila* eye mutants to morphological traits and their correlates( including electroretinograms) through the use of mosaic subjects. These very interesting findings and their interpretations can be regarded as empirical illustrations supporting Polanyi's argument that it is possible to characterize a lower level by analyzing a higher level, insofar as they concern gene action. Moreover, the studies elucidate a remarkable series of relationships from behavior to lower level, physiological functions. The missing step, of course, is how the information encoded in the genes is translated within a variety of environmental contexts so as to influence morphology and its behavioral outcomes. Perhaps the point is being belabored (and the intent of what is being said is not pejorative in any way), but it is still the case that even the very eloquent and advanced research of the kind just mentioned continues to lay away more and more evidence that genes have *something* to do with behavior without saying *just what* they have to do with behavior over and above what is discovered with respect to morphological and/or environmental correlates.

## Ancillary Contributions

*New hope for the R-R relationship?*

If we are stuck in the snows of an impasse we can either try to dig our way out or else try to keep warm and busy while waiting for the thaw. The latter is probably the optimal strategy.

Let us grant for the moment that it is important to find out how genes influence behavior. At this time, the task of discovering how genetic information is translated into differential protein formation can be carried out much more profitably and efficiently within the context of developmental genetics, which has problems of its own even though astonishing progress is being made (e.g., Brown & Dawid, 1969; Fristrom, 1970; Ursprung, 1967). Few workers in the behavioral sciences are equipped to deal with the issue; and besides, there are other useful things that we can do which are potentially more consequential for our own purposes.

But it may not be very important to find out how genes influence behavior. It is not at all clear that knowledge of specific gene-phenotype relations can provide sufficient generality so as to elucidate the general nature of behavioral variation. For example, suppose that we knew the biochemical steps that intervened between a particular codon and, say, a morphological trait

which was associated with a particular actone in a particular sample of organisms. It is not obvious that such knowledge in and of itself would have predictive and explanatory efficacy *vis-à-vis* other genes, other morphological features, or other behaviors in other organisms. Nor is it evident that the mechanisms of individual gene action, by themselves, have much to say about norms of reaction in contexts of other genes and a wide range of developmental and proximal environments.

The question is, can there even be a meaningful discipline of behavior genetics if gene action is not central to our research? What appears to be happening today is that the kinds of behavior genetic experiments to which we are accustomed are providing information about morphology-environment-behavior relations, where the independent variable (say breeding regimen or strain) serves as an ancillary technique for imposing homogeneity upon the subject variance. It has been asserted repeatedly in this discussion that experimental methods, as they are ordinarily put to use in behavior genetics, are not closing the gap between gene action *per se* and behavior. Furthermore, it may not be worthwhile to do so just now. Because we do not have a general, unifying theory of behavior, and because we do not as yet seem to know what it is that we are trying to find out about behavior, individual "behavior genetic" experiments tend to be specific and not immediately amenable to integration into general laws. At present and in the foreseeable future, the role of genetic manipulations as independent variables can continue to serve as valuable accessory tools for behavior-genetic analysis, but they are not apt to promote general understanding if they are made central to our efforts. Under the circumstances, possibily the most immediately valuable contribution that we can make is to involve ourselves in a systematic search for *R-R* (or, more precisely, $\sigma^2_R$-$\sigma^2_R$) relationships.

*Variation: Love it or leave it?*

If behavior genetics is to be so broad and bold as to try to relate behavioral phenomena to the knowledge that exists at all levels of biosocial organization, as Hirsch (1967) has suggested, then we must take what Köhler (1959) has called a "wide view" of behavioral phenomena. In looking about us, the view is so wide that we can hardly decide which of the phenomena ought to be singled out for study. That is to say, behavior genetics for the most part has not progressed beyond a stage of guessing as

to what the relevant variables are. The question as to what we want to find out about behavior has been brought up several times in this paper, and you may have noticed that in each instance I neglected to offer any more of an answer than a vague digression to another topic. Moreover, the terms we use in reference to the observations that we do make are lacking in linguistic precision. As Hirsch & Erlenmeyer-Kimling (1967) have pointed out, "To be a part of science, the results of observation must be formulated in terms that will place them in the public domain." "Conflicting results" and "failures to replicate" are not unusual in our literature, but the variables under scrutiny are often so imprecisely stated that it is not at all clear whether or not the incompatible findings are even related to one another, except that similar labels are used.

I shall return to these problems momentarily, but owing to the clinical training in my background, I always feel some need to construct a case history of a morbid condition before trying to cure it. The study of behavior cannot get away with the "young science" excuse for very much longer. A major source of our trouble was in the formative years. Others (e.g., Hebb, 1960; Hirsch, 1963, 1967) have already faulted us for fixating on dualism, cathecting physics, and regressing to Lockean empiricism; there is no real need to repeat the charges here. Something mentioned less often is early deprivation. Instead of undergoing the usual prolonged period of naturalistic observation, description, and classification in accordance with how the observations seemed to be related to one another, as have so many other disciplines, our precocious science rushed into the laboratory and began experimenting almost *de novo.* While some amount of therapy is being applied — the influence of ethology (an outgrowth of biology, not physics) has been healthy in this regard — its effects have been mostly propaedeutic. Methodical probing for $\sigma^2_R$-$\sigma^2_R$ relationships could serve not only as remedial exercise, but if successful, might furnish considerable information as to what variables ought to be studied and, incidentally, help us to decide upon what we want to find out about behavior. Unfortunately, the $\sigma^2_R$-$\sigma^2_R$ approach is itself encumbered with several puzzling features, the solutions to which require that some artful decisions be made.

Verplanck (1970) has articulated the need for better descriptions (a primary way of communicating observations) in the behavioral sciences very nicely. The immediate purpose of de-

scriptions, according to Mayr (1964), is diagnosis, that is, segregation and identification of the phenomena of interest as a prelude to classification. Hinde (1959) has provided some theoretical and practical notes on how to go about describing behavior, and the present paper is not the best occasion for pursuing the matter further. Instead, let us examine what behavior genetics can do with descriptions once they are obtained.

First of all, objects and events are not identified and described without relating the descriptions to those of other objects or events. Descriptions acquire scientific meaning by a strategy of systematic contrast. This notion has currency at both a theoretical and an experimental level of analysis, as suggested by Marler & Hamilton (1966): ". . . at no point is the inference drawn that a particular trait in a given individual is inherited; rather a certain difference between the traits of two individuals is shown to be inherited." Likewise, in Barnett's (1967) opinion, "One does not say that the possession of two eyes is inherited. An animal belonging to a two-eyed species may have one eye. . . . In this case we say that the difference from the normal type is genetically determined. Similarly, . . . we do not say that learning ability is inherited, though it is possible to find genetical differences in learning ability." Perhaps the point is so obvious as to be trite, but it does us no harm to be reminded of it occasionally.

Moving now from the relatively specific situations depicted by Marler & Hamilton and by Barnett, there is no clearly manifest way to find suitable means for integrating large amounts of descriptive material — an obligation which is of course imperative if we are to contribute to general, as opposed to limited domain, laws of behavior. A good place to go for our lessons in this regard is to the accounts of similar undertakings in other branches of biology — especially botany and zoology — as they continue their efforts to impose order upon inherently variable phenomena. The topic at hand is taxonomy.

The accepted meaning of "taxonomy" — I shall use Simpson's (1962) words — ". . . is the theoretical study of classification, including its bases, principles, procedures, and rules," which is to say that classification is the subject matter of taxonomy, and descriptions make up the subject matter of classification (see also, Hirsch & Erlenmeyer-Kimling, 1967). The case for making classification central to behavior genetics is a straightforward one: (1) by grouping similar phenomena, it is easier to detect the magnitude and quality of dissimilarities between

and within groups; (2) isolation of groups facilitates the search for correlates, between and within groups, of the traits being classified; (3) knowledge of trait correlates between and within groups aids in generating hypotheses about the factors giving rise to differences between and within groups; (4) tests of the hypotheses can a) help us to construct more precise hypotheses than we had before, and/or b) tell us that we are looking at trivial characteristics which have little or no predictive power.

In addition to the advantages just listed, classification can work wonders toward improving the denotative meanings of the terms that we try to use in labeling the phenomena with which we work, thus enhancing the likelihood that different investigators can replicate one another, or at the very least, talk about the same topics. It should be made clear that classification is an exercise in some form of correlational (though not necessarily mathematical) analysis, and by its very nature, classification in and of itself does not provide causal understanding. It can bring cognitive structure to behavioral variation, which at times seems to be such a chaotic and unfortunate nuisance that it interferes with our study of the real issues — but behavioral variation *is* the real issue. Behavioral variation is the theme of our discipline.

Regrettably, the process of classification is attended by a number of perplexities that cannot be resolved here, but which ought to be mentioned before leaving the topic. To begin with, as Cowan (1955) has so aptly observed, "All classifications are subjective, and, like religions and political opinions, have a large element of aesthetic unreason about them. A classification consists of two elements: the objects to be classified, and the subject who feels the urge to classify them. There cannot be a classification without a classifier or taxonomist, and no two taxonomists will approach objects in the same way." Given the amount of differential slip that is allowed, naturally there is no correct or incorrect solution to the three interrelated problems that are about to be brought up (your attention is invited to Colles (1967), Constance (1964), Ghiselin (1966), Hull (1967), Rollins (1965), and Sokal & Camin (1965) for more thorough treatments of the issues referred to here). Each of them will be presented in a rather biased way, reflecting to a large extent the influence of Sokal & Camin (1965) and Sokal & Sneath (1963).

1. *Should behavioral classification be guided by theory?* Most of the biological classification systems with which we are

familiar make use of a phyletic (phylogenetic) approach to taxonomy. That is, since the time of Linnaeus (whose work led from "artificial" to "natural" classification; see Benson, 1957, for some subtleties of this) and Darwin (who postulated inter-specific continuity), taxonomists have in general tried to classify their phenomena *vis-à-vis* presumed phylogenetic or evolutionary relationships. The circularity of the approach is obvious:

Why are groups *A* and *B* phenotypically similar?
Because they are phylogenetically related.
How do we know that they are phylogenetically related?
Because they are phenotypically similar.
Etc.

There are some striking parallels between this way of thinking and the concept of reinforcement in learning theory. To the extent that theory is employed in guiding classification, it can be argued, the output will be biased by how the classifier thinks the phenomena ought to be related. For the sake of this discussion, let us ignore the likely impossibility of ever evading any form of theory whatsoever, including private, intuitive ones. An alternative of phyletic taxonomy is phenetic or empirical taxonomy, which seeks to generate groupings in a more or less mechanical way and without relating the structure of the system to evolutionary or any other theory (e.g., Heywood, 1967). Furthermore, phyletic taxonomy has by tradition attributed more importance to some traits than to others, in part so that the classification will be compatible with the theory. Phenetic taxonomy consciously tries to allow the traits to determine their own weightings by virtue of their predictive value. Finally, it is news to no one that behavioral science has no unifying theory, analogous to evolution, to which its classifications can be related *a priori*, although such a theory might emerge once the relevant relationships have been identified empirically.

2. *What should constitute groupings?* This question is divisible into at least two parts. The first concerns establishing criteria for recognizing similarities and differences in descriptions; i.e., how much similarity is needed before two or more descriptions can be included in the same grouping, how much dissimilarity will be allowed, and how much of a difference is justification for segregated groupings? A reasonable answer to the points just raised will have to be found somewhere other than in this paper, except to cite Vale & Vale (1969) statement that ". . . the psychologist who would find general laws must search somewhere *between* idiography and average organism."

The second part of the larger question has to do with how we conceptualize groupings once they are made. Shall we follow the tradition of binary taxonomy and think of the groupings as types (e.g., genera and species) or shall we conceive of them as distributions? An answer to this is already available. Right now, the difference between *R-R* and $\sigma^2_R$-$\sigma^2_R$ relationships has more meaning. *R-R* constructs are inseparably typological whereas $\sigma^2_R$-$\sigma^2_R$ relationships are population oriented. If variation is the central theme of behavior genetics, then we must deal with variation as it exists and not as we would prefer it to be. Because most biological and especially behavioral variation appears to be quantitative in nature (e.g., Parsons, 1967), we must prepare ourselves for quantitative analyses (as is being done already; e.g., Cattell, 1959; McClearn & Meredith, 1964), including nonlinear multivariate techniques (but see Vale & Vale, 1969, who would probably disagree with the spirit of this sentence). Let us not create types out of convenience, but when those rare qualitative differences do present themselves, let us rejoice in our good fortune.

3. *What should be classified?* It has been seen — *1.*, above — that theory, and for that matter individual differences in classifiers, can bias the kinds of descriptions that are deemed important enough to be classified. At another level, the question encompasses the range of variables to be classified. We know that behavior is related to genes, morphology, experience, and by definition, stimuli — all of which are used here in their very broadest possible sense. Are we to attempt to classify every variable that occurs to us as a latently relevant subject? Given that we do not yet know in a general way what it is that we are supposed to be finding out, the answer has to be *yes*, with the understanding that we are at present incapable of processing all of the potentially available material simultaneously. That is, it will be necessary to contrive hierarchies of integrative taxonomies after some fashion. At this point, the magnitude of the task appears overwhelming. And it may be so, but if we are serious about understanding behavior in any holistic way, it is something that ought to be attempted.

## Literature Cited

Barnett, S.A. 1967. Instinct and intelligence: Behavior of animals and man. Prentice-Hall, Englewood Cliffs, N.J.

Baum, L.F. 1909. The road to Oz. Reilly and Lee, Chicago.

Benson, L. 1957. Plant classification. Heath, Boston.

Bovet, D., F. Bovet-Nitti, and A. Oliverio. 1969. Genetic aspects of learning and memory in mice. Science 163:139-149.

Broadhurst, P.L. 1966. Behavioral inheritance: Past and present. Cond. Reflex 1:3-15.

Brown, D.D., and I.B. Dawid. 1969. Developmental genetics. Annu. Rev. Genet. 3:127-154.

Cattell, R.B. 1959. Personality theory growing from multivariate quantitative research. *In* S. Koch [ed.] Psychology: A study of a science, Vol. 3. McGraw-Hill, New York.

Colles, D.H. 1967. Phylogenetic fallacy. Syst. Zool. 16:289-295.

Constance, L. 1964. Systematic botany — an unending synthesis. Taxon 13:257-273.

Cowan, S.T. 1955. Introduction: The philosophy of classification. J. Gen. Microbiol. 12:314-319.

DeFries, J.C. 1967. Quantitative genetics and behavior: Overview and perspective. *In* J. Hirsch [ed.] Behavior-genetic analysis. McGraw-Hill, New York.

Fristrom, J.W. 1970. The developmental biology of *Drosophila*. Annu. Rev. Genet. 4:325-346.

Fuller J.L., and W.R. Thompson. 1960. Behavior genetics. Wiley, New York.

Ghiselin, M.T. 1966. On psychologism in the logic of taxonomic controversies. Syst. Zool. 15:207-215.

Grobstein, C. 1962. Levels of ontogeny. Amer. Sci. 50:46-58.

Hall, C.S. 1951. The genetics of behavior. *In* S. S. Stevens [ed.] Handbook of experimental psychology. Wiley, New York.

Hebb, D.O. 1960. The American revolution. Amer. Psychol. 15:735-745.

Heywood, V.H. 1967. Plant taxonomy. St. Martin's, New York.

Hinde, R.A. 1959. Some recent trends in ethology. *In* S. Koch [ed.] Psychology: A study of a science, Vol. 2. McGraw-Hill, New York.

Hirsch, J. 1963. Behavior genetics and individuality understood: Behaviorism's counterfactual dogma blinded the behavioral sciences to the significance of meiosis. Science 142:1436-1442.

Hirsch, J. 1967. Behavior-genetic, or "experimental," analysis: The challenge of science versus the lure of technology. Amer. Psychol. 22:118-130.

Hirsch, J., and L. Erlenmeyer-Kimling. 1967. Behavior genetics. *In* H. Helson and W. Bevan [eds.] Contemporary approaches to psychology. Van Nostrand, Princeton, N.J.

Hotta, Y., and S. Benzer. 1970. Genetic dissection of the *Drosophila* nervous system by means of mosaics. Nat. Acad. Sci., Proc. 67:1156-1163.

Hull, D.L. 1967. Certainty and circularity in evolutionary taxonomy. Evolution 21:174-189.

Ikeda, K., and W.D. Kaplan. 1970. Unilaterally patterned neural activity of gynandromorphs, mosaic for a neurological mutant of *Drosophila melanogaster*. Nat. Acad. Sci., Proc. 67:1480-1487.

Jervis, G.A. 1939. The genetics of phenylpyruvic oligophrenia. J. Ment. Sci. 85: 719-762.

Jinks, J.L., and D.W. Fulker. 1970. Comparison of the biometrical genetical, MAVA, and classical approaches to the analysis of human behavior. Psychol. Bull. 73:311-349.

Kaplan, W.D., and W.E. Trout. 1969. The behavior of four neurological mutants of *Drosophila*. Genetics 61:399-409.

Köhler, W. 1959. Gestalt psychology. Mentor, New York.

Maier, R.A., and B.M. Maier. 1970. Comparative animal behavior. Brooks/Cole, Belmont, Cal.

Marler, P.R., and W.J. Hamilton. 1966. Mechanisms of animal behavior. Wiley, New York.

Mayr, E. 1963. Animal species and evolution. Harvard, Cambridge, Mass.

Mayr E. 1964. Systematics and the origin of species. Dover, New York.

Mayr, E. 1968. The role of systematics in biology: The study of all aspects of the diversity of life is one of the most important concerns in biology. Science 159:595-599.

McClearn, G.E. 1970. Behavioral genetics. Annu. Rev. Genet. 4:437-468.

McClearn, G.E., and W. Meredith. 1964. Dimensional analysis of activity and elimination in a genetically heterogeneous group of mice (*Mus musculus*). Anim. Behav. 12:1-10.

Morgan, T.H., and C.B. Bridges. 1919. The origin of gynandromorphs. Carnegie Inst., Washington.

Murphey, R.M. 1973. Genetic correlates of behavior. *In* G. Bermant [ed.] Perspectives in animal behavior. Scott-Foresman, Chicago.

Parsons, P.A. 1967. The genetic analysis of behavior. Methuen, London.

Polanyi, M. 1968. Life's irreducible structure: Live mechanisms and information in DNA are boundary conditions with a sequence of boundaries above them. Science 160:1308-1312.

Rollins, R.C. 1965. On the bases of biological classification. Taxon 14:1-6.

Rosen, R. 1968. Some comments on the physico-chemical description of biological activity. J. Theoret. Biol. 18:380-386.

Simpson, G.G. 1962. Principles of animal taxonomy. (2nd printing) Columbia, New York.

Skinner. B.F. 1963. Behaviorism at fifty: The rapid growth of a scientific analysis of behavior calls for a restatement of the philosophy of psychology. Science 140:951-958.

Smart, J.J.C. 1959. Can biology be an exact science? Synthesis 2:359-368.

Sokal, R.R., and J.H. Camin. 1965. The two taxonomies: Areas of agreement and conflict. Syst. Zool. 14:176-195.

Sokal, R.R., and P.H.A. Sneath. 1963. Principles of numerical taxonomy. Freeman, San Francisco.

Spence, K.W. 1951. Types of constructs in psychology. *In* M.H. Marx [ed.] Psychological theory: Contemporary readings. Macmillan, New York.

Tryon, R.C. 1942. Individual differences. *In* F.A. Moss [ed.] Comparative psychology. Prentice-Hall, Englewood Cliffs, N.J.

Ursprung, H. 1967. Developmental genetics. Annu. Rev. Genet. 1:139-162.

Vale, J.R., and C.A. Vale. 1969. Individual differences and general laws in psychology: A reconciliation. Amer. Psychol. 24:1093-1108.

Verplanck, W.S. 1970. An "overstatement" on psychological research: What is a dissertation? Psycho. Rec. 20:119-122.

## Discussion

*Dr. Witt:* I was very struck with what you said and thought that maybe if you had spoken an hour longer I would have started to understand what I do, which I never did up to now. It seems to me, if you talk about classification and so on, that there are people who think they know what we should do and then there are other people that do but haven't the slightest idea what they should do. I have been asked frequently what I do and I've always answered with a question and that, I realize now, was a very cowardly way out because I really don't. People have said why do you study spider webs, what good do they really contribute to our knowledge of human beings? And the answer has been, we have to study spider webs directly because the study of human beings has not contributed to the knowledge of spider webs up to now. It merely means to say that we charge — with our "simple" systems, and "simple" is in quotes — and do something, and then somebody comes along and says I *THINK* I know what you tried to do, and I say I haven't the slightest idea what I did but I did just enjoy myself after doing it. So, I had a feeling that we, with our simple systems and Dr. Best and so on, are keeping warm and busy and you are trying to dig us out, Dr. Murphey.

*Dr. Murphey:* I am as evasive as anyone else.

*Dr. Witt:* I'm really not sure about that. But the question naturally is: What is the bridge between the gene and behavior?

But you then come naturally to the problem, what is the bridge between the gene and developing morphology, and then morphology and behavior. I mean there are so many steps in between, and we talk glibly about central nervous systems regulating things, but we haven't the slightest idea how the central nervous system does anything. We know it has some electric sparks and we know some ions rush around across the cell membrane and we know some transmitter substances go from one nerve cell to the next, but that doesn't explain why an animal moves its arm in this way and not in another way at a particular moment. There are so many steps between the gene and the behavior. But naturally what you said, if I understood you right, is that you take your behavioral traits and you try to describe them as well as possible and do some kind of statistics on them.

*Dr. Murphey:* They are not necessarily statistics — we try to relate them. We try to get them all together and see what is related. A great deal of the process of relating them will have to be mathematical.

*Dr. Witt:* But let me say we take — I can't think of anything but spider webs, I don't know why — spider webs just as a simple system. We do measure these webs and try to group the webs by the individuals that built them — were these siblings or non-siblings — and compare the variation. Is that a useful procedure?

*Dr. Murphey:* I should think that would be one way of generating data which could be used in the kinds of relational undertaking that I'm suggesting.

*Dr. Witt:* That's what I thought.

*Dr. Best:* One of the things that I'd like to hear you comment on is the traditional view of something like the analysis of variance, which has been that you have two parts to happenings — that one part is the determinant part and the other part left over is error. You know, a sort of stochastic element. Now to the extent that most experimental psychologists that I know can reduce it to a determinant state of affairs, they feel they've been successful to the extent that there is a stochastic element left over; they feel that this is the difference between men and gods. There is, I think, a reasonable amount of both rationale and empirical evidence presently to suggest that the generation of random behaviors itself is a thing that may actually be programmed in. The best answer to solving many of life's problems may not always be a deterministic solution. If you push an animal

hard — such as a rabbit that's being chased by a coyote or by an owl, the rabbit is hopelessly outgunned on any kind of deterministic strategy that it can come up with. An owl can fly faster than a rabbit can run, and a coyote is smarter than the smartest rabbit is ever going to be, so that there is probably a point in its survival struggle at which it pushes the panic button and says "go random." Now if you take a look at the van Neumann-Morganstern theory of games, you discover that in a two-person game in which you have two opponents and one opponent has all the marbles — he's got more information, he's got more guns, he's got more speed, he's got more of everything than his other opponent — if the player without any marbles tries to follow any deterministic strategy, he's had it, and the only way he can get himself off the hook is to go random and that is what the rabbit does. Now if you proceed to analyze this from the conventional point of view then you say: This is error. And this is what I don't know about the behavior. That's probably what the god of evolution intends for the rabbit's behavior to do — to have that stochastic element to render it indeterminant, and that in itself is a sort of behavior.

*Dr. Murphey:* If your analysis is true, I would consider that to be a very useful description — a very useful observation and description — to plug into this thing. In fact, in the case of jackrabbits — watching how they run — they do use a more or less random observation pattern. . . . There is an old film by Walt Disney called "Grita the Greyhound" which was on about 6 or 7 years ago. There was a magnificent helicopter shot of the greyhound. The plot of the thing was that the greyhound wasn't good at racing so the owner threw it out on the desert. So the film involved how the greyhound survived, and it had some shots of the greyhound chasing a jackrabbit over some sand dunes. You could see the trails there. The greyhound is a magnificently fast animal. Unfortunately he runs in a straight line. He's incredibly fast — much faster than a jack rabbit — and you could see the rabbit alternating, more or less at random. The greyhound would miss. He got on a particular trajectory and sometimes he would literally fall over trying to make the turn, and roll down a sand dune. That was an interesting and important observation. It could be described and plugged into the system.

*Dr. Witt:* Could you go random without choosing a random table? I just wonder how you can do it. You know, there's no plan in the thing. But, even random is a plan.

sensitive mutants. It was a genuine experience. As I walked in, that first day of mine at Caltech, I spied Jean Weigle, his back pressed against the near wall pitching pennies against the far wall, and two graduate students throwing darts at a nearby dart board. The scene was completed by Bob Edgar standing on the laboratory bench trying to pull down a rather sizeable boa constrictor that had wrapped itself around a steampipe in anticipation of striking at the laboratory's pet rat Ajax. I asked, a little stunned, "Why all these animals?" Bob replied, "It's a biology lab, isn't it?" That statement really struck home. I had almost forgotten that "normal biology labs" have animals and plants.

Now my laboratory looks quite biological indeed with a great variety of different snakes and plants growing on such exotic nutrients as potatoes. But, here I am talking about behavior. Surely, my laboratory is interesting, but a psychology laboratory it is not. You might think, because of some of my recent snake work, that I am going to talk about the behavior of snakes, but I am not, because a snake does not have a really very interesting behavior. Not that snakes are not interesting, and not that they do not do some fascinating things, but they have a rather limited behavior. They sit around a lot and flick their forked tongues. Now one may ask what constitutes an interesting behavior? My answer would be a system that demonstrates a variety of defined behaviors as a result of a series of defined stimuli. If a variety of non-specific stimuli always resulted in identical behavior or the identical stimuli caused a variety of behaviors, then in either case it does not appear to me to be a very interesting system. In the latter category would be a system like a single stretch receptor in vertebrates. It responds by showing an increase in spike frequency as a result of stretch. One can design a multitude of experiments showing the rate of stretch, temperature dependence, etc., yet the resulting parameter is always the same — the increase or decrease of spike frequency as a function of stretch. I would call this system just a simple input-output situation, and not use the word behavior.

In recent years there has been a large effort in numerous laboratories to study and try to understand behavior by examining the *neural basis* of behavior. This does appear to be a reasonable approach because there seems to be no doubt that behavior is a direct result of neural activity. An exemplary study would be of the gill withdrawal reflex in Aplysia (Kandel, 1970). If the snail's siphon is gently touched it contracts and the gill withdraws

# Phycomyces and Its Behavior

RUSTEM IGOR GAMOW
*University of Colorado, Boulder*

As THE years go on, I suppose I am beginning to feel more and more like a biologist, but I wonder if I will ever begin to feel like a psychologist. My early training was indeed biology — the study of flowers and beavers and such. But as I continued into Graduate School I became interested in molecular biology. So I spent a few carefree and happily innocent years working on a virus which measured no more than 950 Å from stem to stern; one could put $10^{14}$ of these into a test tube. During this time I really did not know what "sensory physiology" was, and "behavior" was just a term given to children depending on how they interacted with adults. This is a little strange in retrospect since I was then publishing my first paper on the nature of the specific response of the tail fibers of bacteriophage T4 as a result of a stereospecific chemical stimulus (Gamow, 1969). In general I did get used to thinking of "life" as being microscopic (T4 bacteriophage can only be seen with the electron mircroscope). Well, after my degree I was off to Caltech to learn about sensory physiology from Max Delbrück. The division of biology at Caltech is really more or less in one place instead of being spread out over a large area with a great variety of complicated sounding names. Thus, adjoining the labs of Delbrück were the labs of Bob Edgar. Now, for anyone having worked in the area of molecular biology, just to be in "The Edgar Laboratory" was quite a moment. This is the laboratory which came out with the famous amber mutants and the equally infamous temperature-

*S*, or to look only at variations. They are independent manisfestations of the same continuum. They both may be right. They probably both have something to contribute. But they tell you different things. I think that most of us here who call ourselves behavioral geneticists are simply more interested in variation than we are in the mean. It doesn't necessarily imply that the people who look at *R* as a function of *S* are wrong in any sense — that they aren't going to find out anything useful.

*Dr. Kates:* I was interested in some grapevine information I got that the institute's sub-branch in Hawaii is called the Institute for Behavioral Biology. I wonder if that isn't the crux of what we're getting at. You may hang a little shingle that says "Behavioral Geneticist" and then find that more encumbering than helpful.

*Dr. Murphey:* You can say the same for any label.

*Dr. Kates*: You addressed yourself to the point that, as behavioral geneticists, what are we to do? I wonder as behavioral biologists, if we don't get a bigger viewpoint on the same story.

*Dr. Murphey:* No. I prefer that title and I use the term "behavioral geneticists" to refer to people who purport to be studying relationships between genes and behavior. That's the only restriction I put on it.

*Dr. Capretta:* Concerning classification systems, what success do you think there has been if you look at some of the systems ethologists — Lorenz or Tinbergen or Scott — developed. Do you think there has been much success or do you think they have given us any guidelines for development?

*Dr. Murphey:* No. As you know, their approach to behavior has not been to study variation. They have provided us with a number of very interesting observations and descriptions, but to have called those things instincts and fixed action patterns and to have otherwise labeled them is to imply some amount of immutability that does not seem to be in the same spirit. They have provided us things that ought to be studied and ought to be related and relatable to other phenomena.

*Dr. Murphey:* I have some phenomenological feeling for randomness. Beyond that I can't really say what it is.

*Dr. Kates:* I think that this has been elegantly described by Ken Roeder, for the moth and the bat. The bat was chasing a moth and is programmed, and the program says when the sound is at one level and one set of neurons has been turned on, you fly the hell out of there. But then the sound gets stronger, and any strategy that's logical has got to go. So, in fact the moth — and this has been demonstrated — randomly takes a course with gravity, and cannot be pretracked by the bat. This brings up in my mind the first equation you wrote on the board, *R* is a function of *S*, and that's what I'd like to talk about for a minute.

$\bar{R}$ as a function of *S* is a very disturbing concept to me. I think more and more people in several fields of behavioral analysis are getting a little perturbed with the whole concept of a mean. For instance, in slow wave analysis-evoked potentials in the brain, you look at the slow wave. You don't look at one slow wave, you look at 14, or 211, and get the mean. You say *THAT* is the evoked potential for this stimulus. You can quantify the stimulus and you can then give the mean of the response. Now I believe that an organism — and I don't think this is a belief which is unfounded — could have *N* degrees of freedom for coming up with a response, so that the mean may be absolutely no representation of what the organism ever does. But, it seems to me, the approach of looking at *R* as a function of *S* can be a valid one, and if you can define *S* because you put it in under the most stringent controls, and you can make some definition of each individual *R*, then the last point of the equation to look at is the *f*. *f* is never addressed. We just say *f*, a function. Now what is that function? Everybody in this room will agree that this function is some other function of the genotype plus the environment. I wonder if a definition of the function term in that equation isn't a very good task to undertake.

*Dr. Murphey:* In response to both your questions, I have a personal dislike for talking about "the determinants" of behavior — I prefer to talk about correlates. With respect to your comments on looking at *R* as a function of *S* and looking at variances, in statistical theory, means and variances are independent of one another except that they're moments of the same distribution. I think we could carry the analogy there into behavioral science — that it's neither wrong nor right to look at *R* as a function of

into a cavity under the mantle shelf. This study has led to the further understanding of two important phenomena — plasticity of neuronal activity and habituation.

A second example I like to give is the work of J. G. Nicholls (1968) who has chosen the leech as his simple system. The leech (Hirudo) contains a tail ganglion and a chain of 23 ganglia. One can identify individual neurons in such a ganglion and then trace out individual sensory and motor pathways. One has identified both electrophysiologically and morphologically three nonoverlapping sensory inputs, (a) touch sensitivity, (b) pressure sensitivity, (c) noxious stimuli such as pinching or squeezing.

As a third example, A. D. D. Willows (1971) has been working with a close relative of Aplysia, *Tritonia diomedia.* This animal has huge (1 mm) easily identifiable brain cells (shape and color). When touched by a sun star (a multiarmed starfish) it undergoes an escape reaction. It churns and contracts to avoid the grasp of the sun star, though the behavior is very stereotyped. Specifically, Willows found that stimulation of a single cell near the central commissure (stimulation of ¼ sec.) resulted in a complete *escape response* which lasted up to 30 seconds.

This work on the leech, Aplysia, and the Tritonia is certainly very exciting and exemplifies systems that are bound to extend our knowledge concerning the neural basis of behavior. On the other hand, are these not just extensions of the stretch receptor mentioned earlier? Of the examples cited above I believe only the Tritonia comes close to true behavior — behavior being a series of contractile motions that resemble very much the "natural" escape response.

In the study of behavior there appear to be two basic approaches, the general neurophysiological approach of studying a system from the inside out, and the psychophysical approach of studying a system from the outside in. Later on I will suggest that a psychophysical approach should and could be applied to simple systems, and even applied to systems that indeed may not possess what is normally called a psychology.

In the work on Aplysia, Tritonia, and the leech, the experimental procedure was to change something on the inside and then observe what happened to behavior on the outside. One may well argue, and I would tend to do so, that tracing a sensory pathway through to its motor output such as the gill withdrawal is not studying behavior. Physiology certainly, but not behavior. I guess I feel that behavior, to be called behavior, must be studied under

reasonably natural circumstances. The problem is that the kind of answers that one obtains by observing behavior are rather more psychological than mechanistic. I remember some beautiful work done on the play behavior of deer (Muller-Schwarze, 1969). These deer went through the motions of play every morning for about ten minutes and then again every evening. The question was asked if they were inhibited in their playing in the morning, would they play longer in the evening? The answer was, in fact, no.

Now I would like to ask the question whether one could observe a series of complex natural behaviors and be able to obtain information on the comprehensible level (mechanistic level) as one obtains by looking at an organism from the inside out. For instance, if one tries to determine the behavior of a mouse that has received a flash of green light, a foot shock on its right front foot, and a short blast of a whistle at a certain pitch, most of us would agree that it is firstly difficult to obtain reproducible data and, secondly, hard if not impossible to interpret what mechanistic steps occur which finally result in behavior.

Now it is my impression that people have studied complex behavior in complex organisms and simple "behaviors" in simple organisms in hopes of understanding complex behavior in complex organisms. Perhaps there have not been enough psychophysical (if we can use the word) experiments studying complex behavior in simple systems in order to gain more information about complex behavior in complex systems.

I gather, after talking to a number of people actively working in the field, that what is lacking in brain research is a concept of "how we think, what is consciousness, how memory is stored and recalled." Lacking this concept, we carefully examine the architecture in hopes of understanding how the brain functions. It is the old biological technique of trying to relate structure to function. When I say the concept is lacking I am really saying a theory is lacking. In general most biologists maintain that our understanding of even the simplest biological system is so primitive that little or no theory is possible. Of course physics had almost always been the other way. Theory always rode far ahead. Paul Dirac first predicted the positron which was then found many years later. General relatively was verified by experiment after the theory. Now I am willing to concede that it may be a long time before theory gallops ahead of experiments in biology, but it seems clear to me that they may now ride side by side. A case in point is the now famous lac operon model of

*Escherichia coli* that culminated in a Nobel prize award to Jacob and Monod (1961). The model they formulated was based on a number of faulty assumptions and thus it was even a greater theoretical feat in that their theory, which was based on some experimental evidence, was verified in every detail after the model was constructed. In one of his many enlightening seminars, Günther Stent marvelled that anyone can arrive at a correct model using the correct assumption, but that it takes real genius to arrive at a correct model using faulty assumptions. What he is really saying, I believe, is that one must be clever in just the same way that Dirac was clever and Einstein was clever. Perhaps it is too early for the people studying the brain to rely on theory, just as it is for people studying the behavior of complex organisms. But what about theory of behavior in simple systems? May we not discover a lac operon type of system of behavior? Is it not possible that an experimental-theoretical analysis of a simple system may bring forth a *concept* that will guide us in trying to understand more complex systems such as the brain?

The system I wish to describe to you is the sporangiophore of the fungus Phycomyces that elicits behavioral responses to a gaseous substance, to a change in gravity, to a variety of light intensities and light distributions, and to mechanical stretch. The behavior we observe involves a variety of distinct growth patterns, all of which have certain recognizable characteristics. Now we ask, can we do the mouse experiment described earlier? For instance, what kind of behavioral response would we see if we gave Phycomyces a ten-second stimulus of unilateral light followed immediately by three g's of gravity and then a defined mechanical stretch? I do not know the answer to this question but my guess is that it would be reproducible! Now, perhaps if we repeat the experiment but at a lower temperature we will find that the "stretch transducer" will fail to contribute to the total behavior. The behavior would be the same as if the stimulus were never given. This could tell us something about how one kind of stimulus receptor is physically separated from another. An experiment similar but less complex in nature will be described later on in this paper.

Recently an excellent review article appeared (Bergman et al., 1969) concerning the sensory and genetic aspects of Phycomyces. In this review there is an amusing but pertinent description concerning the natural history of Phycomyces:

"During the last 150 years, it (Phycomyces) has been found

growing on the walls and timbers of an oil mill in Finland, in oil mills in Saxony, on litter under beech trees in England, on refuse on a warehouse floor in Hull, England, on a decaying wasp nest in Ireland, on dead wood in Urbana, Illinois, on human feces in a cave near Rome, Italy, and on peaches in a refrigerator in Cold Spring Harbor, New York; it has been isolated numerous times from mouse, rabbit, and horse dung."

As a neurophysiological system it is rather unique for the following two reasons:

1. It is a plant.
2. It is one cell in which the receptor is also the effector.

Thus, we will not be dealing with the usual neurons, synapses, transmitter substances, groups of ganglia and so on.

Phycomyces may never become the phage of sensory physiology, which is in fact a very dubious fate for any organism. To unravel the secrets from any beauty one must appreciate the beauty; to attempt to strip out the secrets without a constant marveling of these secrets appears to me almost sacrilegious. I would like to make note of two scientists who have contributed more to the understanding of Phycomyces' secrets than all other workers put together: E. S. Castle of Harvard, who has been contributing to our knowledge of Phycomyces since 1928, and Max Delbrück of Caltech. Delbrück chose Phycomyces probably for a variety of reasons (I am sure beauty was not an uninfluencing factor), but primarily because it showed a behavioral response to an easily measured stimulus, it was large and easily grown and maintained, and it has a long latency period between the stimulus and the response. He felt that this allowed one to study in a more leisurely fashion the physical-chemical events occurring between the stimulus and the response. Figure 1 is a photograph (a time exposure) of the mature Phycomyces sporangiophore (stage IV b). Figure 2 is a time exposure once every twenty minutes demonstrating how rapid the head (the sporangium) grows as it matures. The sporangium contains on the order of a hundred thousand spores from which new generations arise.

**Figure 1. A time exposure of Phycomyces, stage IV, sporangiophore. (photograph courtesy M. Delbrück).**

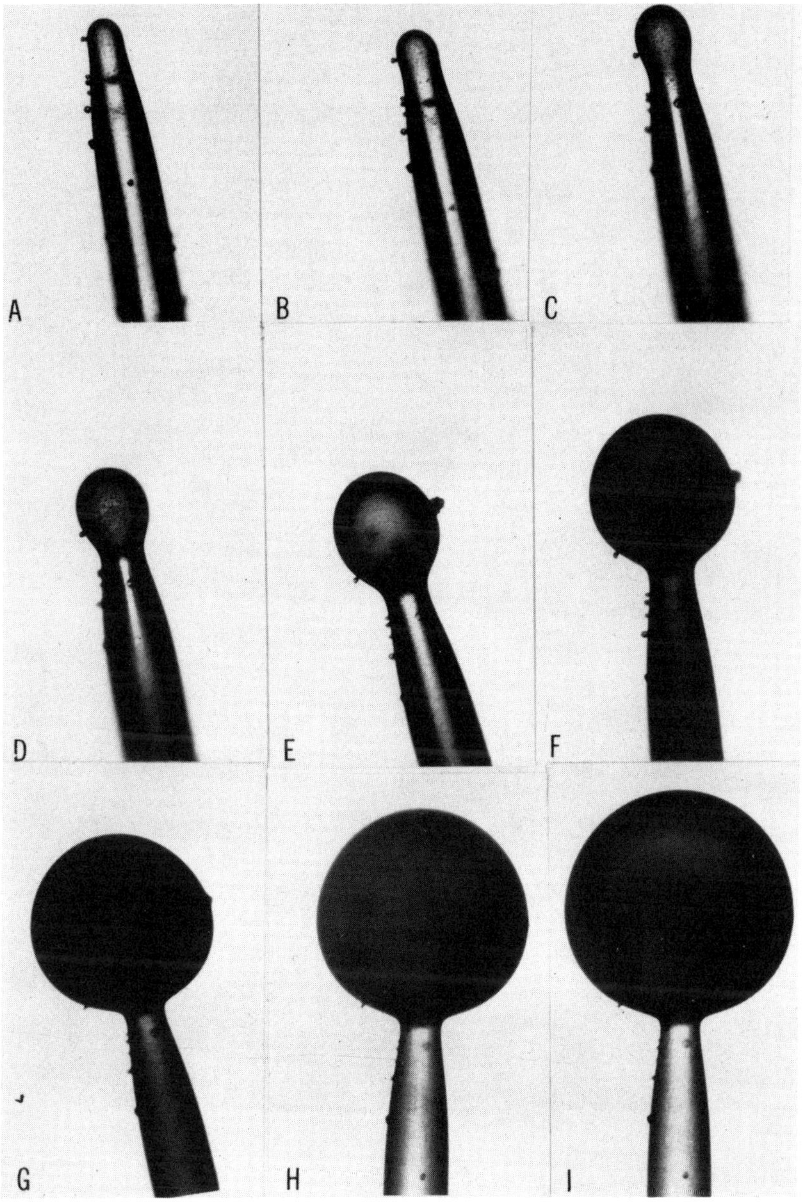


Figure 2. Development of a sporangiophore from stage I to stage III. Each exposure was separated by 20 minutes. Figure 2A, late stage I; Figure 2B-2H, the sporangium is steadily increasing in area and is thus defined to be stage II; Figure 2I, the sporangium has stopped growing and is thus called Stage III. Stage IV (Figure 1) is reached when the region below the sporangium begins to elongate.

## Habituation and the Avoidance Growth Response

Phycomyces grow away from any solid object, presumably because they emit a stimulating gas that tends to be concentrated between the organism and the barrier (Johnson and Gamow, 1971) (see insert 1).

Insert 1

Growth away from barrier with dots representing the stimulating gas.

Insert 2

Increase in growth parallel to a double symmetrical barrier.

It follows from a single barrier experiment that if the stimulating gas were uniformly distributed around the sporangiophore all sides would attempt to "avoid the barrier" and a symmetrical increase in growth would result (insert 2). Insertion of a sporangiophore into a double barrier should give a growth response and, in point of fact, it does exactly that — a response now termed the avoidance growth response. The question we now can ask is, if one sensory pathway (the light pathway, let's say) was so saturated from a previous high light stimulus that no additional light would cause a response, would a second type of stimulus, such as the avoidance growth response stimulus, give a response? In slightly different language, if one sensory pathway is habituated by a saturating stimulus, will the organism (remember we are still dealing with a single cell) respond to another type of stimulus? In fact, we did just that. We saturated the light growth response with a high intensity laser and found that, although the organism would not respond to the laser, it gave a normal avoidance growth response (Ortega and Gamow, 1970). This suggests that light responses saturate at some point before the avoidance growth response enters. If the failure to respond to light is a depleted energy pool, it is clear that there must be different pools for the different receptors.

## The Cold Storage Experiment

Margaret Petzuch and Max Delbrück recently published a very interesting and charming paper (Petzuch and Delbrück, 1970). They asked the following question: If a Phycomyces sporangiophore is cooled down to a point that it will no longer grow, and then given a light stimulus, will it "remember" this stimulus and react accordingly after it is warmed up? The stimulus they gave was an asymmetric light stimulus, thus they needed only to observe the direction of bend. They found that it does "remember" the stimulus and it will remember it up to at least two hours. Thus it appears that the initial photoreaction is rather insensitive to temperature and in addition that the product of the photoreaction must stay around for a reasonably long time. In this regard it would be interesting to test whether other types of stimuli are also stored in the cold, for instance the stretch stimulus. To my knowledge that experiment has not been done.

## The Light Growth Response

Over a very wide range, a symmetrical increase in light intensity results in a transitory increase in growth. This increase lasts for about ten minutes after which the growth returns to its normal rate of about 3mm per hour. This return to basal growth rate is a consequence of adaptation. The sporangiophore responds only to a change in flux of light either above or below the level of light to which it had been previously adapted. Until saturation, which is about thirty times the intensity of light over background light, it gives a graded response. The response, as many sensory responses, is a linear function of the logarithm of the stimulus. The situation is little different from that in our retina when we turn off the overhead light. As a function of time after turning off the light our retina becomes more and more sensitive; in other words, it will respond to lesser and lesser light intensities. Similarly, after a saturating light step-up stimulus the sporangiophore will fail to give a full response until it has been adapted to the new light level. The kinetics of adaptation have been described by Delbrück and Reichardt (1965) and appear to follow equation (1) which, as they note, is similar to the one describing the charging of a capacitor. The majority of the mathematical concepts concerning adaptation have been worked out previously by Delbrück and Reichardt, but in a slightly different form than presented

here. It is convenient for us to present this form in order to be able to use the adaptation equation in terms of predicting behavioral responses based on this equation.

## The Tropic Responses

In addition to responding to either a symmetrical increase or decrease in light intensity, Phycomyces is also positively phototropic; it actively grows towards visible light. The threshold, action spectra, and distribution of growth are the same in the two responses but they differ in the following two ways. First, the latency is a little longer than in the light growth response, and second, it shows no adaptation. In other words it will continually grow and follow an asymmetric light intensity.

The sporangiophore uses a fascinating mechanism to determine the direction of light. The stalk itself is cylindrical and has a higher refractive index than air. The result of this is that light entering the proximal side will be bent and focused on the distal side; because of this there will be a higher intensity of light on the back side, which appears to be the signal to the organism for that side to grow faster. This is known as the lens effect (Buder, 1918; Shropshire, 1962) (see Figure 3). This leads to the prediction that a sporangiophore immersed in a medium of matching refractive index should not display the tropic responses. This is true; a sporangiophore immersed in an oil with matching refractive index will not bend towards light, but grow straight up.

In the remainder of this paper I wish to address myself to the following question. Given the equation describing the kinetics of adaptation (as derived from the light growth response experiments and their response-growth characteristics), could we predict the unilateral roping behavior? For the most part the analysis given here has been the work of Daniel Leon Johnson, and constitutes Part 1 of his dissertation. The methods used for the computer analysis are too detailed to be presented here; therefore only a sketch of the method is given.

Delbrück and Reichardt based equation (1) on direct experimental evidence. Briefly, their analysis consisted of adapting a soprangiophore to a light intensity $A_0$, and then measuring the ability of the light growth response to occur at lower and lower light intensities as a function of time after the adapting light had been turned off. A is defined in units of light intensity, but is clearly measured in terms of the sporangiophore's ability to give a certain predefined growth response.

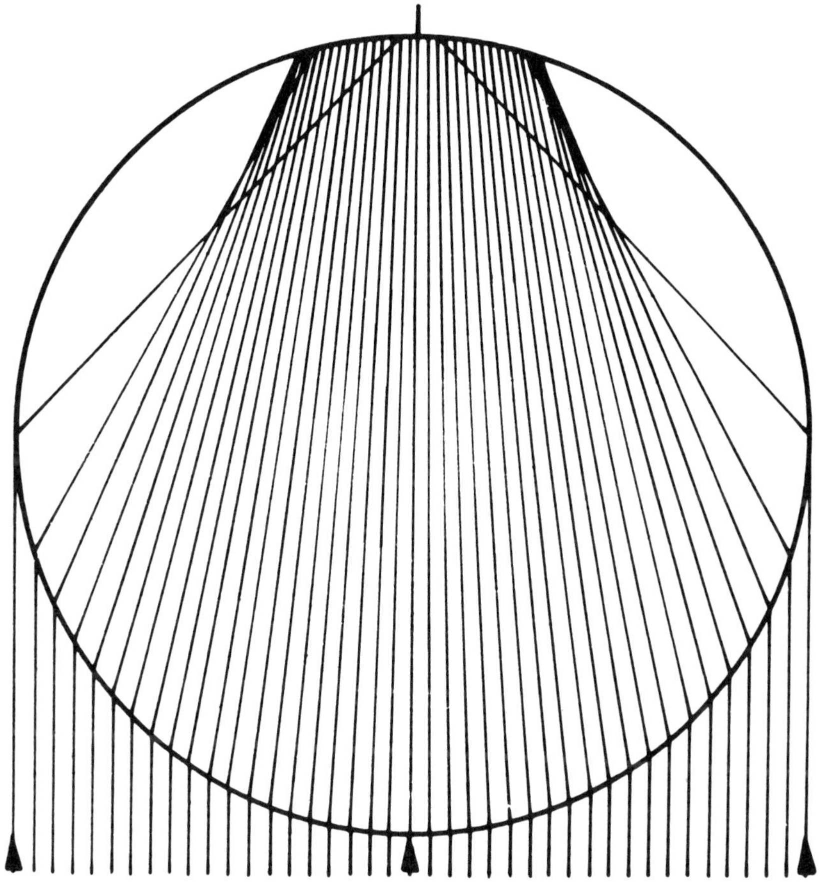

**Figure 3. Graphic representation of the lens effect (drawing modified from E. S. Castle). The angle that the light rays are bent as they pass through the sporangiophore's cross section can be calculated simply from the known refractive index.**

$$(1) \qquad \frac{dA}{dt} = \frac{I - A}{b}$$

where $A(t)$ = level of adaptation
$b$ = time constant
$I = kA_0$ = stimulating laser
$A_0$ = level of adaptation at $t = 0$
$t$ = duration of stimulating intensity, I

In general, the decrease in the magnitude of an electrical output in a sensory receptor after successive stimuli is called adaptation. Although this appears to be a common property of many sensory receptors, the mechanism of adaptation in any one system is still essentially unknown. A second general property of sensory receptors is that the sensory response is a linear function of the logarithm of the stimulus. This relationship was first formalized by Fechner. Its bases, however, were observations made by Weber in which

$$\frac{\text{A change of stimulus}}{\text{original amount of stimulus}} = \text{JND (just noticeable difference)}$$

Generalizing, Fechner proposed

$$\delta S = c\left(\frac{\delta R}{R}\right)$$

where S = response and R = stimulus and c = a constant of proportionality. This can be integrated with initial conditions R = r when S = 0 to obtain

$$S = K \log\left(\frac{R}{r}\right)$$

If we let R be in units of r, we obtain

$$S = K \log R.$$

There have been many criticisms of the underlying assumptions of this development, but for our purposes we will mention just one. This criticism is that time is not an element in the development. Without time, there is no way to relate how the phenomenon of adaptation can mediate a response. Furthermore, this lack of time consideration does not predict the reciprocity law that states that within limits the response of a receptor is a function of the dosage. This reciprocity rule holds for Phycomyces (Delbrück and Reichardt, 1956). Consider a constant dosage D = Rt. Here R is the step-up in stimulus and t is the length of time R is applied.

Now if $t_1R_1 = t_2R_2$

$$r_1 \ln R_1 \neq t_2 \ln R_2$$

$$r_1 \ln R_1 \neq t_2 \ln R_2$$

With this criticism in mind, it appears that the adaptation phenomenon could be intimately related to the logarithmic relationship between the response and the stimulus (details given in the appendix).

Given the boundary condition that

$$A(t) = A_0 \quad \text{when } t = 0$$

Delbrück obtained the following solution:

$$A(t) = A_0k + (A_0 - A_0k)e^{-t/b} \tag{2}$$

Figure 4A is a plot of how $A(t)$ changes when the light intensity is increased from $A_0$ to some new intensity I. It is clear that the sensitivity to the new light intensity I decreases as t becomes large. In order to follow this change in sensitivity, let us define a new function $R_1$, which is the new stimulating intensity I divided by the function $A(t)$.

$$R_1 = \frac{I}{A_{(t)}} = \frac{A_0k}{A_0k + (A_0 - A_0k)_e^{-t/b}} = \frac{1}{1 + \left(\frac{1-k}{k}\right)e^{-t/b}} \tag{3}$$

It is important to note that the initial level of adaptation $A_0$ does not appear in this equation except in terms of k; since $k = I/A_0$, k is then clearly the parameter that defines the stimulus strength. Since the light growth response is a function of stimulus, it follows from equation (3) that $R_1$ is also proportional to the light growth response.

Figure 4B is a plot of the function $R_1$ versus time. At time equal to zero $R_1$ just equals k and approaches the value of one as t becomes large. This allows us to state that a light growth response occurs when $R_1 > 1$, or simply that the light growth response (G) is proportional to the area under the curve $R_1 - 1$.

$$G \simeq \int_0^t (R_1 - 1)dt = \int_0^t \left[\frac{1}{1 + \left(\frac{1-k}{k}\right)e^{-t/b}} - 1\right]dt =$$
$$b \ln (k - k - 1)_e^{-t/b} \tag{4}$$

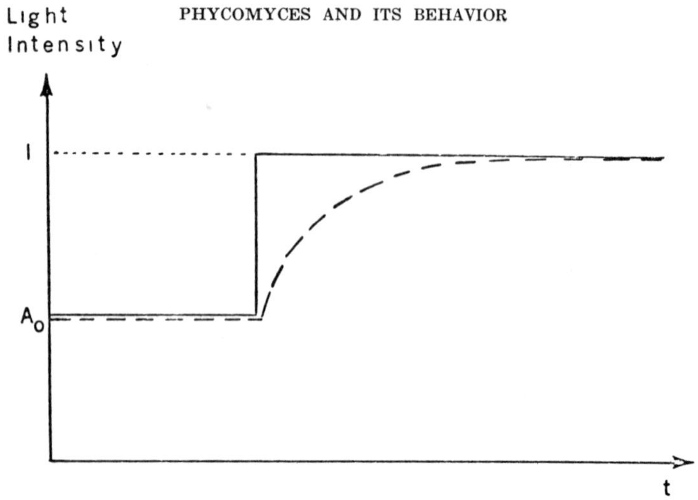


**Figure 4A.** The unbroken line represents the increase in light intensity from $A_0$ to some multiple of $A_0$, I. The broken line represents the increase in adaptation, A(t), as a result of the increased light intensity.

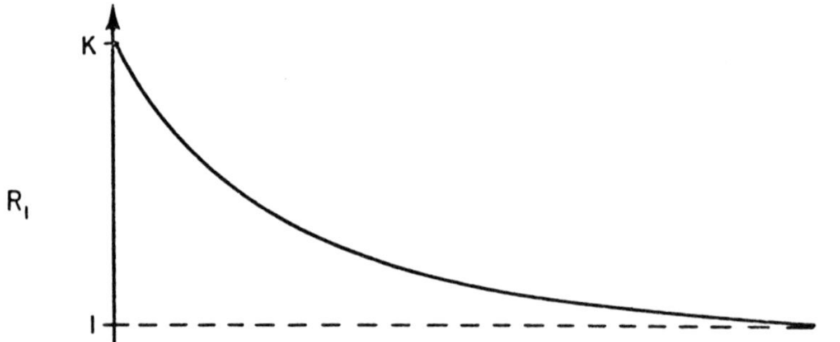


**Figure 4B.** The function $R_1$ versus time is plotted. It is clear that $R_1 = k$ when $t = 0$ and that $R_1$ approaches one as t becomes large.

Equation (4) reduces to equation (5) when t becomes large compared to the time constant b.

$$G \simeq b \ln k \qquad \text{or} \qquad G = B' \ln k \tag{5}$$
$$B' = \text{proportionality constant.}$$

Equation (5) states that the response is proportional to the stimulus k, which is just the expression we were looking for.

An additional property of most sensory receptors is that, within reasonable time limits, their response is proportional to the logarithm of the dosage of the stimulus, where the dosage is defined to be the product of the intensity of the stimulus times the duration of the stimulus. Within the duration range extending from 4 to 60 seconds, Delbrück and Reichardt (1956) verified this dosage relationship in Phycomyces. This dosage-response relationship is known under several names, such as Bloch's law and the Bunsen-Roscoe law, but is generally known as the reciprocity law (Bouman, 1959). Equation (4) predicts that the reciprocity relationship should hold for stimuli durations ($t_0$) much less than b, i.e., $t_0 << b$.

For

$$t_0 << b \qquad \text{then } e^{-t_0/b} \simeq 1 - \frac{t_0}{b}.$$

If this is substituted into equation (4) we obtain

$$G \simeq b \ln \left[1 + \frac{kt_0}{b} + \frac{t_0}{b}\right] \tag{6}$$

for reasonably large values of k; where k still is equal to $I/A_0$ equation (6) becomes equation (7)

$$G \simeq \ln \left[1 + \frac{kt_0}{b}\right] \tag{7}$$

which predicts reciprocity since $kt_0$ is the dosage (intensity x times) term and thus is constant. Therefore (6) is constant regardless of the value of $t_0$.

Although we have looked closely only at this one system we predict that the same relationships will hold for any sensory system for which the kinetics of adaptation follow some simple exponential function. We have done a similar analysis but assumed that the kinetics of adaptation is a simple linear function, and have found that the logarithmic-stimulus response relationship is still approximately true except for small values of k. We conclude, therefore, that it is the general phenomenon of adaptation that predicts both the logarithmic-stimulus relationship and the response dosage relationship.

## Prediction of Tropic Behavior by Means of Computer Modeling

As mentioned earlier Phycomyces sporangiophores undergo adaptation when a change in uniform symmetric light intensity is the given stimulus. An increase in growth occurs when the intensity is greater than $A_0$, i.e., $I/A_0 > 1$, and a decrease in growth occurs when $I/A_0 < 1$. On the other hand no apparent adaptation occurs when the light is unilateral; the sporangiophore continues to grow toward a unilateral light source. In order to explain this lack of adaptation two hypotheses have been proposed and tested.

I. The Twist Model: An interesting and curious fact concerning mature stage IV sporangiophores is that the growing zone twists during growth such that the sporangium itself undergoes one complete revolution every thirty minutes. As can be seen in Fig. 3, one would continuously have new parts of the cytoplasm entering a new intensity of light. This could well explain why adaptation is never achieved.

II. The Uniform Adapted Model: In this hypothesis one does not consider twist but assumes that the level of adaptation throughout the cross section is at some average value $\bar{A}_0$. This means that during continuous unilateral illumination some regions of the growing zone are above the value of $\bar{A}_0$ and some regions are below the value of $\bar{A}_0$.

Both models were tested by means of a computer. A clear result was that twist alone cannot explain the lack of phototropic adaptation. The uniform adaptation model not only predicted the lack of adaptation in the phototropic response, but furthermore predicted the known experimental finding that when a steady-state unilateral light source is either increased or decreased there cccurs a transitory decrease in bending angle towards the light. This response has been termed phototropic inversion (Bergman et al., 1969).

### General Method of Analysis

From Figure 3 one can calculate what the light intensity would be per small unit areas throughout the cross section. In Figure 5, assuming that 10 units of light enter the proximal side, one can then assign an intensity to a series of segments that have been arranged in three concentric rings A, B and C.

In the previous section an equation (equation 4) was given which related a change in growth rate to k, where k was defined to be $I/A_0$ and I is the stimulating intensity. From

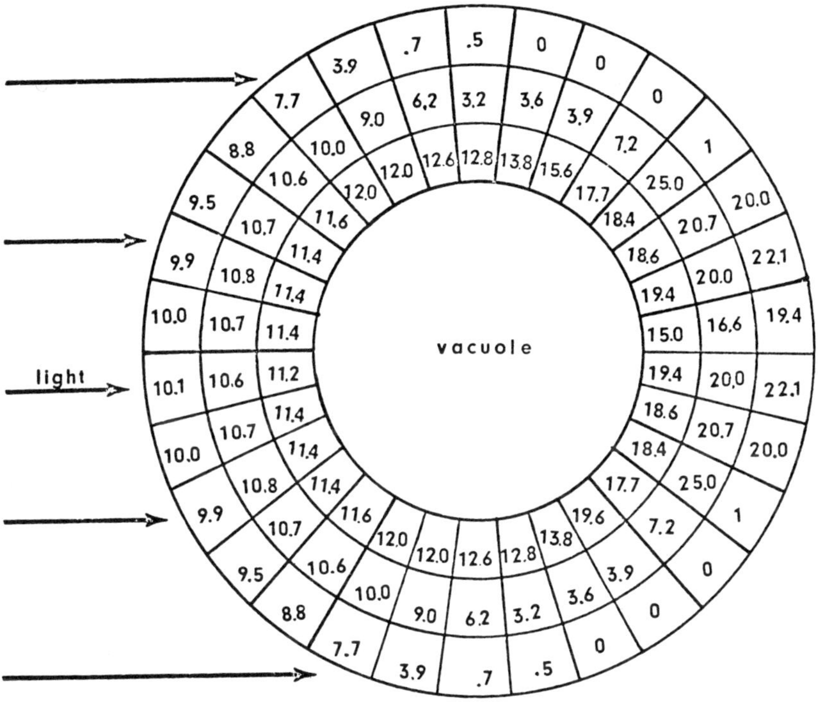


**Figure 5. For the computer analysis it is necessary to calculate the changing light intensity as the light passes through the sporangiophore. Because the rays are converging, segments on the distal side have a higher intensity of light per segment. In the graph the light entering was assumed to be 10 units.**

straight experimental data in which percentage of increase of growth above the basal growth was measured as a function of k, one could now judiciously choose constants such that equation 4 matched the experimental data. This growth function is not a complicated one and some typical values are: $k = 1$ means no change in growth, $k = 2$ means 25.5 precent increase in growth and so until $k \geq 64$, at which time the growth reaches a maximum of 69 percent increase. The equation also gives negative values for rate of growth for k's $< 1$.

When considering the twist model there is a constant change

in the level of adaptation and therefore a constant change in the growth output for any one value of k. In the twist model we have assumed that each new segment is rotating into a new position once every minute. During the minute the adaptation level changes from the initial value $A_i$, to a final value $A_f$.

$$A_f = I + (A_i - I)_e^{-t/b} \qquad \text{for } t = 1 \text{ minute}, \quad b = t \text{ minutes}$$

Given these two parameters it is clearly a difficult but solvable problem. Although a great deal of effort was exerted to fit the experimental data, it was not possible to come closer than an order of magnitude to the experimental data. Furthermore it did not predict the phototropic inversion.

When the uniform adapted model was tested, entirely different results were obtained. Figure 6 is a computer print out of what one would expect if one only considered one of the three rings and then all three rings together. Bending is highly significant, coming very close to the experimental 3° per minute.

In other words, it tells us that the segments on the distal side have an advantage over the proximal segments both because of lever arm advantage and a greater number of them receiving a larger $I/A_0$ stimulus. We can now further ask if this model would predict phototropic inversion and, as can be verified by Figure 7, it does. Both an increase in unilateral light intensity ($k > 1$) or a decrease ($k < 1$) causes a temporary decrease in bending angle. This test in fact demonstrates that other factors are involved, since in the experimental data the bending angle is temporarily reversed instead of just decreased. Twist certainly contributes to the phototropic response but in only a small way, and cannot account for steady troping.

If we now remember that both the adaptation equation and our growth rate equation were obtained strictly from the light growth response experiments, we are indeed predicting an additional behavior as a result of a different set of stimuli (admittedly after the fact, since those behaviors are not new). The fact that they appear to correlate so nicely with the photographic responses certainly is suggestive that we are beginning to understand some aspects of Phycomyces behavior.

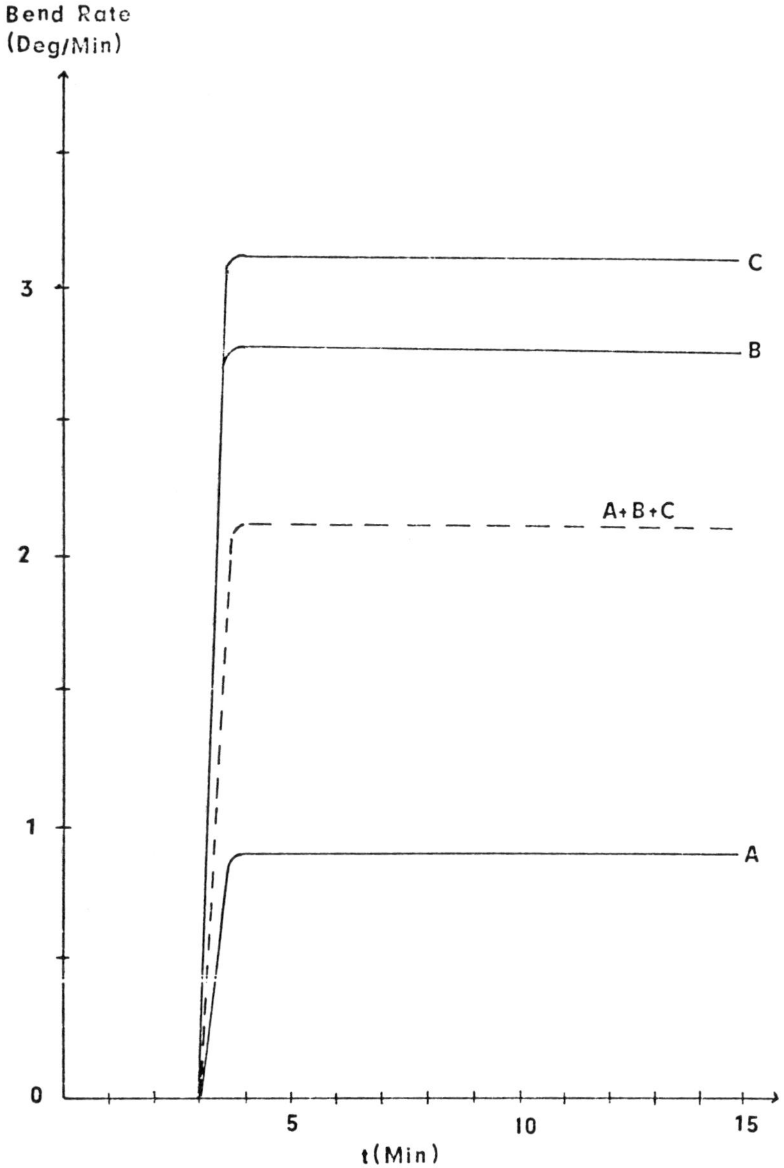


Figure 6. From the growth equation one can calculate the percent increase of growth for every segment. These increases in growth can then be translated into a bending angle. In this graph curve A, B, C represents the bending angle of only the segments in the outer ring, middle ring or inner rings contributing to growth respectively. Curve A + B + C represents the bending angle for all three rings simultaneously. Note that adaptation does not occur.

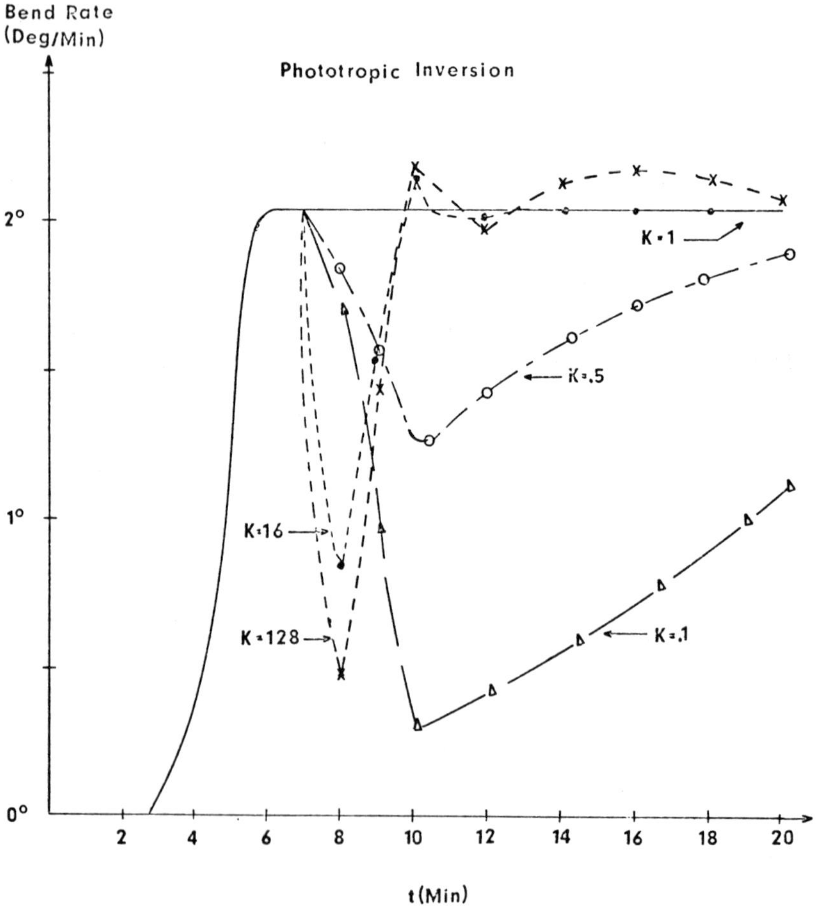


Figure 7. This graph shows the decrease in bending angle when a unilateral light intensity is either increased k > 1 or decreased k < 1. Experimentally the bending angle is actually reversed for a short period of time; this does not occur in our simulation but it certainly tends to go in the correct direction.

## Appendix

The integration of the adaptation equation

$$\frac{dA}{dt} = \frac{I - A}{b}$$

$$\frac{dA}{I - A} = \frac{dt}{b}$$

$$A\text{-}I = C_1 e^{-t/b} \qquad \text{at } t = 0, A_{(t)}, = A_{(0)} \quad \text{then } C_1 = A_0 - I$$

substituting

$$A_{(t)} = I + (A_0 - I)_e^{-t/b} \qquad \text{but } I = A_0 k$$

so $$A_{(t)} = A_0 k + (A_0 - A_0 k)_e^{-t/b} .$$

The area of $R_1 - 1$ in Figure 2 for some defined stimulus duration $t_0$, is equal to:

$$\int_0^{t_0} (R_1 - 1)dt = \int_0^{t_0} \left[ \frac{1}{1 + \left(\frac{1 - k}{k}\right) e^{-t/b}} - 1 \right] dt =$$

$$\int_0^{t_0} \left[ \frac{\frac{-(1 - k)_e^{-t/b}}{k}}{1 + \left(\frac{1 - k}{k}\right) e^{-t/b}} \right] dt = \int_0^{t_0} \frac{1}{\left(\frac{k}{k - 1}\right) e^{-t/b} - 1} dt =$$

$$- b \left[ \ln \left( \frac{1}{\frac{k}{k - 1} - e^{-t/b}} \right) \right]_{\circ}^{t_0} = b \left[ \ln \left( \frac{k}{k - 1} - e^{-t/b} \right) \right]_{\circ}^{t_0} =$$

$$b \left[ \ln \frac{k}{k - 1} - e^{-t_0/b} - \ln \left( \frac{k}{k - 1} - 1 \right) \right] =$$

$$b \left[ \ln \left( \frac{\frac{k}{k - 1} - e^{-t_0/b}}{\frac{k}{k - 1} - 1} \right) \right] = b \ln[k - (k - 1)_e^{-t/b}]$$

The reciprocity relationship follows from

if $\quad t_0 << b \quad$ then $\quad e^{-t/b} \simeq 1 - t_0/b$

so that $\quad G \simeq b \ln[k - (k-1)(1 - t_0/b)]$

$$\simeq b \ln\left[k - k + 1 + \frac{kt_0}{b} - \frac{t_0}{b}\right]$$

$$\simeq b \ln\left[1 + \frac{kt_0}{b} - \frac{t_0}{b}\right]$$

for large reasonably large k

$$\simeq b[1 + kt_0/b] .$$

## Literature Cited

Bergman, K., P.V. Burke, E. Cerdà-Olmedo, C.N. David, M. Delbrück, K.W. Foster, E.W. Goodell, M. Heisenzerg, G. Meissner, M. Zalokar, D.S. Dennison, and W. Shropshire, Jr. 1969. Phycomyces. Bacteriol. Rev. 33:99-157.

Buder, J. 1918. Die inversion des Phototropismus bei phycomyces. Ber. Deut. Botan. Ges. 36:104-105.

Delbrück, M., and W. Reichardt. 1956. System analysis for the light growth reaction in phycomyces, p. 3-44. *In* D. Rudnick [ed.] Cellular mechanisms in differentiation and growth. Princeton Univ. Press, Princeton, N. J.

Gamow, R. I. 1969. Thermodynamic treatment of bacteriophage T4B adsorption kinetics. J. Virol. 4:113-115.

Jacob, F., and J. Monod. 1961. Genetic regulatory mechanisms in the synthesis of proteins. J. Mol. Biol. 3:318-356.

Johnson, D.L., and R.I. Gamow. 1971. The avoidance response in phycomyces. J. Gen. Physiol. 57:41-49.

Kandel, E.R. 1970. Nerve cells and behavior. Sci. Amer. 223:57-70.

Muller-Schwarze, L. 1969. The ontogeny of play behavior [paper presented at the 2nd Annual Winter Conference on Brain Research].

Nicholls, J.G. 1968. Receptive fields, synaptic connections and regenerations, patterns of sensory neurons in the CNS of the leech, p. 3-16. *In* D.A. Baylor and J.G. Nicholls [ed.] Physiological and biochemical aspects of nervous integration. Prentice-Hall, Englewood Cliffs, N. J.

Ortega, J.K., and R.I. Gamow. 1970. Phycomyces: Habituation of the light growth response. Science 168:1374-1375.

Petzuch, M., and M. Delbrück. 1970. Effects of cold periods on the stimulus-response system of phycomyces. J. Gen. Physiol. 56:297-308.

Shropshire, W., Jr. 1962 . The lens effect and phototropism of phycomyces. J. Gen. Physiol. 45:949-958.

Willows, A.D.D. 1971. Giant brain cells in mollusks. Sci. Amer. 224:69-75.

## Discussion

*Dr. Feldman:* I would like to pick out for further discussion a couple of things that impressed me. The most significant thing to me, perhaps, was the repeatability of the system. That is, you have a system in which you can take a stimulus and get a response, and you essentially get the same response all the time; you don't have to worry about seventeen other things happening, sixteen of which are probably unknown. I have a feeling that this is the same thing that Dr. Kates was getting at. It seems to me that if you are trying to study the effect of a stimulus on a given system, then you would like to be able to isolate the response of the organism to that particular stimulus and not have to worry about a number of others. This is clearly what this system lends itself to most elegantly. There are some remarkably elegant experiments that have been done and that Igor is in the process of doing. I have one question about the adaptive significance of these light responses. You did point out that the organism does not photosynthesize, and that there is no basic known light response other than this kind of tropic system. You said the organism grows faster when in the light, or goes toward the light. I would like to know if you have any ideas about why the organism likes to get into the light. Does this light response have any particular significance to the organism itself, or is it some historical remnant from a time when it had adaptive significance?

I have another small question, basically one of clarification. When you mentioned something about a one-to-one relationship between receptor and effector, I wasn't quite sure what you meant by that — that the receptor *was* the effector. I guess I'm not really familiar with those terms. Perhaps I ought to stop there.

*Dr. Gamow:* Well, the adaptive significance, as they made fairly clear, is a problem. Let's say we have here a stone with Phycomyces spores germinating under it. The organism needs to get out from under the stone to disseminate its spores. The end

result is that it grows out this way and that, until it is above the ground with the 100,000 spores in its head. I think that the adaptive significance of the response, going toward light, is just getting out. This is why it is so long — up to 5 and 10 centimeters long. If it didn't have the avoidance response, if it didn't have the ability to grow away from objects while going towards the light, it probably would crash into the object. The light response seems to be a mechanism for getting its head (which is very fragile — you touch it and it breaks) out into the open. What is not clear is why, when you give an increase in light symmetrically, you get a growth response. Everybody thinks that that is an artifact of the tropic response. There has been devised something called the "sunrise" experiment. This is one of Delbrück's experiments. If you continue to increase the light intensity, it will continue to grow fast. That is to say, it grows up to 5 millimeters per hour as long as you increase the light.

Let's consider the receptor-receptor idea further. If you stimulate any one segment with a laser micro-beam, you will have a behavior response — a growth response — right there, but not elsewhere. There seems to be a one-to-one correspondence to where the photon is received, and it produces some product that causes an increase in growth right there.

*Dr. Feldman:* I guess this is my difficulty. I don't see what the idea of the one-to-one thing is. Why don't you just say that whatever it's doing, it is just not making a diffusible substance?

*Dr. Gamow:* O.K.

*Dr. Feldman:* I mean, I thought you were implying something mechanistic about what particular molecule is doing what.

*Dr. Gamow:* No. By one-to-one I just mean that the light intensity here at this one region causes a response in this one region.

*Dr. Feldman:* So whatever the product of that light is, or secondary or even tertiary products, they are just not diffusible?

*Dr. Gamow:* Right — or diffuse very slowly. Another thing did occur to me that I think is worth mentioning. This system demonstrates elegantly to those of us working on rhythms that there are an enormous number of effects of visible light on biological systems that have no organized photoreceptor. This is very intriguing. When we talk about clocks, we need to ask what the original adaptive significance of a clock was. Consider Drosophila. It emerges from its pupa case at the time of highest

humidity, and this is good for Drosophila. A prey runs around at a different time of day than the predator. But the original selective value for a clock clearly could not be these divergent kinds of uses to which the clock has been put. There has been a suggestion by a number of people that, in fact, visible light is affecting general intermediary metabolism to a much greater extent than previously thought. There are enormous numbers of colored molecules in cells. The more striking, of course, are in the electron transport chain. I think it is worth speaking about, in terms of effects of environmental stimuli on biological systems, to think that there may be some fundamental effects of lights on systems other than through organized photoreceptors. You may have seen about a year or so ago that Warren Butler had a very nice paper in *Science* which demonstrated that at least one of these effects would produce an oxidation of cytochrome oxidase and literally block electron transport.

So, when one is thinking of the effects of environmental stimuli on biological systems, I think this question of effects of light on general metabolic systems is one that we should keep in mind, rather than always having to think about an organized photoreceptor.

*Dr. Kater:* I didn't understand what you meant by adaptation. Did you also say "habituation"? When you put up the curve, I really didn't know what the experiment was.

*Dr. Gamow:* If you keep an organism, either your eye or a phyco, in bright light, then you turn the light off — completely — you realize that in your retina and in the phyco you have an increase in sensitivity.

*Dr. Kater:* How did you show it?

*Dr. Gamow:* In the case of your eye, there is a decrease of sensitivity such that you don't see a very dim light. In the case of phyco, it does not give you a response to light.

*Dr. Kater:* If you give it a very bright light, . . . it's growing like hell. Is that correct?

*Dr. Gamow:* It's growing like hell. Sure, it's growing up and it's coming down.

*Dr. Kater:* It's coming down again?

*Dr. Gamow:* Even in the light, it is coming down. If you increase the light, it will increase up to a point. Up to a point it is transitory, because then it comes back to the growth it had

before the increase of light — even if you leave the lights on.

*Dr. Kater:* Another questions is, if you are at some intermediate level of intensity, can't you then raise the light level up again and then the thing will take off?

*Dr. Gamow:* That's right.

*Dr. Kater:* This is something I wanted to ask you then about possibly different pools of energy. Isn't that experiment just on the same response an indication that the adaptation, or whatever you call it, can't simply be a depletion of energy, because if that were the case then why would you then get a further response by the same stimulus at a greater intensity?

*Dr. Gamow:* Well, there was a theoretical paper by Castle, who had been publishing on Phycomyces since 1927 (every two years) — a most beautiful elegant paper. He had an idea to which I can't really do justice. You have a certain level, pool level. You give it a saturating stimulus that depletes the pool to a new level to where you can't get any more, a new steady state — consecutive steady states — in between, if you only deplete it half-way. Then you have enough of that pool left for you to go up a little more again.

*Dr. Kater:* But then it would be incumbent on somebody proposing such a model to say that the half-way pool was rate limiting. I mean, you can't have a half-way pool, and have the thing start to come down if you still have half of the pool still there, or something like that.

*Dr. Gamow:* Sure you can. Because it's concentration-dependent.

*Dr. Kater:* That's right. But the pool has to be the critical rate.

*Dr. Gamow:* Yes, yes, yes. That's really elegant, and it's really a neat system. You probably never heard of the very early work on Fucus egg development, which has never been carried to this extent, but which has almost identical behavior. It was picked up by Jaffe, and he showed the electrical things happening in there too, and it was really neat. To expand — free associating for a minute — if you put methylene blue on any old neuron, it becomes a photoreceptor. Methylene blue goes into the oxidative-phosphorylation cycle, and you have made a neuron into a photoreceptor. By the same token you can do this to muscle. You put methylene blue in there and you get a photoreceptor effector. You are looking at receptor-effector systems in a single cell, and I think in terms of the generalizations that we all would like to

see made, this is really a neat place to try to do this.

*Dr. Best:* I'll try one devil's advocate ploy. If I were working in design engineering, for a company that produced operational amplifiers, and we produced this really cruddy one that nobody bought, I might buy all the operational amplifiers on the market and try to find out what makes them better. I would really be puzzled when I got to the ones that were in those little sealed packages that were single units, or maybe a chip, because I had the receptor, the effector, all the resistors — everything in one unit. But if I could get at a tube model, with those little resistors in there and the capacitors and separate things, I think maybe I could help make my unit a little better. This is rhetorical, I'm sure you see, but what I'm getting at is: In your opinion, do you think it is better to study the receptor process in a specialized receptor cell, effector process, and effector cell — an integrative process in a specialized integrating cell — or is it a false analogy that I've made, suggesting that this is a sealed package like a chip or an op amp that has been incapsulated?

*Dr. Gamow:* You are exactly correct. A lot of people have speculated that maybe Phyco is really like a solid-state system. There are people like Reichardt who worked with Delbrück in some earlier experiments. He felt that it would be better to look at a multicellular organism which had diversity of labor than to study one cell that contained within it all the functions performed by the multicellular system. That's one philosophy. The other philosophy is that it may not be simpler if it is all in one cell. It may be more complex. But until somebody works out one system or the other, it is difficult to say which is the better system. There are two different philosophies.

*Dr. Feldman:* There seems to be an assumption that if it's all in a cell, then you can't take the parts apart and, in fact, I think this may not be a valid assumption. Since we are here in a behavioral genetics symposium, in fact, perhaps the single most important use of genetics is that you can get mutants that are lacking one component of the system, and if this were not true, then nobody would be able to study regulation in E. coli, for example, and take apart the components.

*Dr. Kater:* That's the million dollar chip you just threw in. Because the whole thing for this group, as I see it, is not the mutant approach. The thing that struck me is that you say this is invariant. Isn't there variation in the normal population?

Can you look at the variation in the normal population and then attempt genetic selection? I think the mutant approach will tell you about the genetics of the system. Looking at the normal variation, and the genetic complements modifying "normal function," I would agree with you. There is a good chance that if you take a solid state chip and you put it into acid or do this and that to it, there will be ways of analyzing it. I think it is very appropriate for this particular group to think about using that kind of tool in their work.

*Dr. Gamow:* No, the Delbrück group is very much into it. There is variability in the normal population. The experiments are easy enough to do. But you do pull your hair out, and things do stop growing on you and I'm making the best of all possible worlds. Sure. We had some large variability in one experiment and finally concluded that it was a very humid summer.

*Dr. Sharpless:* Two questions about the experiments and the methodology: Does one segment of Phycomyces show adaptation as a consequence of having exposed other segments to light?

*Dr. Gamow:* That experiment we have tried and tried. A half year ago, we used a microbeam laser for this purpose. It turned out to be very difficult to do. The methodology was a little complex and the results were not conclusive. If you shine ultraviolet light on it, it grows away from it. We might ask why. The answer is incredibly simple and very cute. The inside of the cell has many aromatic compounds and the u.v. doesn't get through. So now we have more light coming in the front side than in the back side, and the front side grows faster. When the Phycomyces is illuminated by 304 millimicrons the advantage of the lens effect is just compensated by the absorption and as a result the organism grows straight up. Some experiments done last year by Dave Dennison using u.v. stimuli have suggested that adaptation is not localized. These experiments are still in progress and a definite answer to the question concerning the spread of adaptation is not yet available.

*Question:* So the adaptation diffuses?

*Dr. Gamow:* Yes, there's good indication that it does, based on some experiments with u.v. light and line integrals that Varjú and Delbrück had done maybe five or six years ago. They're very complicated papers to read. It always bogs me down whether they're measuring from the head to this point or from this point. It's a moving system — a relativistic point of view.

*Dr. Sharpless:* The second question: When you do the adaptation experiment, is the entire organism bathed in light, or just one segment?

*Dr. Gamow:* Which one in particular?

*Dr. Sharpless*: The one that gave you the curves, the one you were talking about.

*Dr. Gamow:* Habituation?

*Dr. Sharpless:* Yes.

*Dr. Gamow:* When we performed the adaptation test, nicking the sporangiophore with a laser microbeam, we bathed only the one segment.

# Learning and Neurophysiology of Planarians: Transphyletic Similarities in the Organization of Brains and Psychological Behavior Systems

JAY BOYD BEST
*Colorado State University, Ft. Collins*

THERE has been a kind of unspoken, philosophical tradition that all right-thinking investigators are expected to believe implicity. To be suspected of infidelity to these notions was to expose oneself to the charge of softheadedness and sentimental anthropomorphizing. As far as I can ascertain, there was never any compelling reason for believing, yet these notions have had a profound influence on thinking in those sciences pertaining to behavior, the nervous system, and the attempts to simulate and model psychological behavior systems. These ideas have had a profound influence on the intellectual thrust and direction of groups such as our present one. In the course of this paper, I want to develop some notions which diverge quite sharply from this conventional wisdom and, by doing so, try to present the beginning of a unifying conception of behavioral organization on a transphyletic scale. In order to circumscribe this target of discourse I would like to express this set of traditional notions in explicit form. These notions go somewhat along the following lines.

> Stereotyped behaviors are simple and associated with simple neural systems. Simple behavior is stereotyped and is the expected product of a simple neural system. The psychological behaviors of sentient creatures came gradually into existence from simple, stereotyped, behaviors as a consequence of the evolution of large complex brains from small simple ones. Only complex neural and cybernetic systems show psychological behaviors. Thus, the simple nervous systems of invertebrates show, accordingly, mostly stereotype behaviors, with psychological behaviors being exceptional and probably suspect at that. Higher vertebrates show, conversely, almost entirely plastic psychological behaviors, with little stereotyped behavior, and that probably suspect. Brains are probably similar to large, complex, digital computers. The elements of both brain and computer are simple and stereotyped and most likely binary. Invertebrate nervous systems are similar to small computers with simple stereotyped programs; for this reason, they do not show sentient behavior. Computers do not show sentient or psychological behaviors because they are still too simple. Someday someone is going to build a vast computer, perhaps filling the interior of a surplus dirigible hangar, or the basement of MIT at any rate, and it will exhibit sentient psychological behavior and, according to the cautionary prophecies of the late Norbert Weiner and Kubrick's "2001 — A Space Odyssey," perhaps paranoia and megalomania.

So much for the conventional wisdom.

One way to approach this problem encompassed by the conventional wisdom is to ask what the first brain to evolve was like. Was it more like a little piece of spinal cord that showed only steretoyped behaviors, or was it organized more like a miniature psychological system. We are about a billion and a half years too late to inspect this first brain, so probably the best we can do is to take a look at the brain of some animal like a planarian which, I think I could argue, is probably similar to the first animal really to evolve a true brain. This brain is only about 200 micrograms, wet weight. I might start out by simply asking — does it have something like psychological behaviors? One of the clues to this, indicating that maybe it has something like psychological behaviors, is one we turned up by using electron microscopy. A couple of years ago we discovered that there were dense core synaptic vesicle systems in it, which as all neurochemists and electron microscopists know, is a sort of structural indicator of the kinds of nerve endings that use aromatic amine neurotransmitters like dopamine, serotonin and noradrenaline. We already

knew that they had light core synaptic vesicles in them, but one of the things that is interesting about the aromatic amine neurotransmitters is that, if you look through mammalian brains, you will find them in abundance in the region of the brain that is usually called the limbic system, which is almost, but not quite, synonymous with what is called the rhinencephalon, or smell brain, that is very intimately involved with things like plastic events of learning, with motivation and emotional affect.

If one examines this limbic system, either by stimulating it or by recording from it, it seems as though everything that comes across to the observer as a psychological behavior sends ripples through this limbic system. The limbic system seems to use a lot of aromatic amine neurotransmitters, much more than most other structures of the CNS do. Other structures of the nervous system that use aromatic amine neurotransmitters tend to be radiations, or projections, or distribution systems from this limbic system. This prompted us to ask whether this old conventional wisdom was correct? In order to answer it the way I said I would answer it, I first have to say what I mean by a "psychological behavior system." By a psychological behavior system I mean a system having the requisite properties to capture the interest of a serious experimental psychologist. If you ask a serious experimental psychologist what he studies, he really will not tell you. He will come out with some phrase like "behavior of organisms." This is very high sounding, but it is not true. He does not study behavior of all organisms. What he studies is a very selected subset of behaviors in a very selected subset of organisms.

If one takes him at his word and studies the behavior of experimental psychologists —not taking the verbal content but what they actually do — one of the things one will find is that those systems which seem to have the minimal characteristics to capture their interest seem to have something that comes across like emotion or emotional affect, something that comes across like motivation or drive, and something that comes across like learning or habituation. They may have other characteristics as well, but these three are the minimal qualifying characteristics. Unless they have all three of these, they do not qualify for his attention.

Now I will simply proceed to answer my question by trying to provide experimental evidence that the flatworms do, in fact, have these three characteristics. Among the things we discovered

in their brains were these dense core synaptic vesicle systems. It seemed just possible that they had the neurological hardware to show something like emotional affect. We already knew that they had something that came across like drive or motivation. At least if one did the same kind of experiment that one did on a rat, one obtained very similar curves, with a somewhat different time span, to those one obtains with a rat. This can be seen in Figure 1.

Now go to the next two characteristics. I want to consider them together because I think that when emotion is coupled to the logic of the learning paradigm, one gets very efficient learning. Conversely, in every case where you find very efficient learning, you nearly always find that these sorts of tasks have highly emotionally affective factors associated. However, whenever emotionally affective factors are allowed to enter into a learning experiment on a rat in such a way that they enter in as extraneous variables and are not a part of the logic of the learning task, then they are very disruptive to learning. There is a type of learning situation generally referred to as approach-avoidance tasks. For example, take a thirsty rat and shock him through his metal drinking tube when he goes to drink. This is one of the good single-trial learning paradigms. This general kind of situation falls in the category of these approach-avoidance paradigms. They are rather interesting because one can, by the same experiment, deal with the last two qualifications jointly. This is good. Since emotion and learning probably are connected anyway, it is just as well to deal with this connection explicity. If these emotionally affective factors are important and one pretends they don't exist, then one cannot rationally control for them. If one is going to control for them, and is not to engage in a kind of intellectual hypocrisy, then one has to admit the possibility of their existence.

So we decided to discard the old conventional wisdom in this case, since there was a very real possibility that it might not only be counter-productive but might not even be true. As long as one refuses to admit the existence of emotional factors in the lower invertebrates, then one cannot rationally deal with them in any way. We said, let us just pretend, in view of the fact that we found the associated neurochemical hardware, that they do exist. Now one of the things any good rat experimenter does is to take a rat, put it into the learning situation beforehand, and habituate it. He habituates it to everything first, and then he tries to couple what he wants it to learn to something that

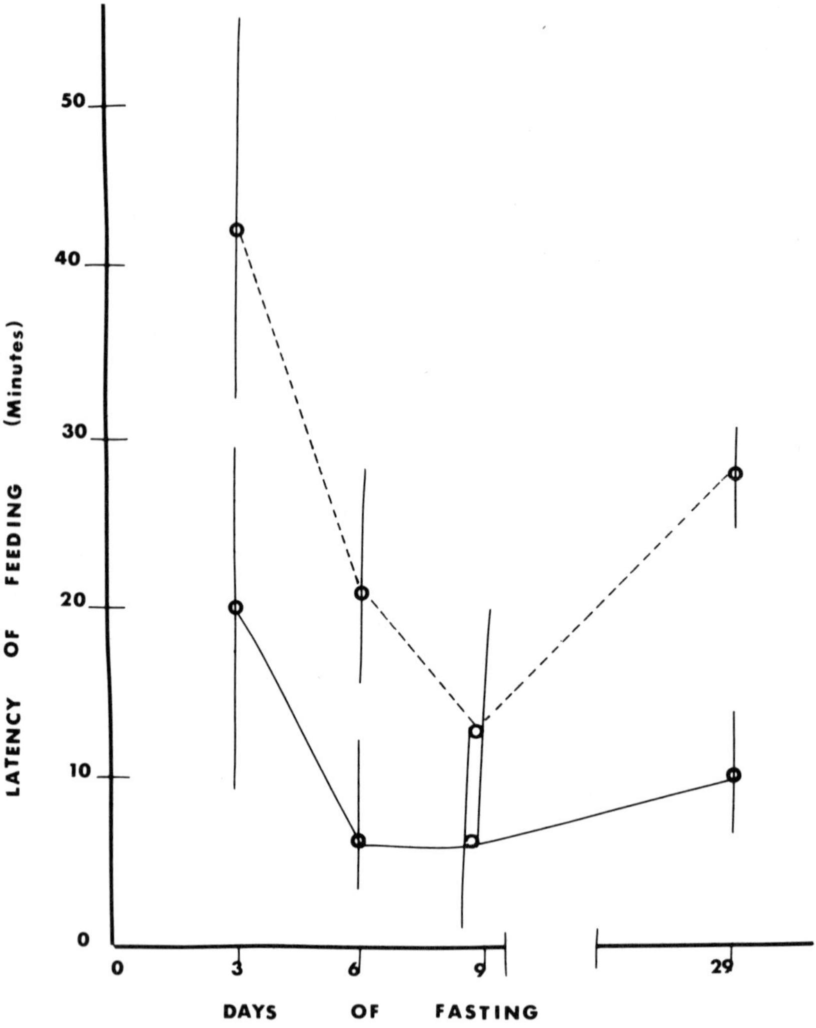


Figure 1. Appetitive drive in planarians. The latency of feeding is shown as a function of the duration of fast for the planarian *Cura foremanii* in a test chamber to which the planarians in one group had been previously habituated (o —— o) and to which those in the other group had not been previously habituated (o- - - --o). For moderate durations of fast the latency decreases with fast duration, but, for long fasts, the feeding latency increases again. Similar results have been obtained with the planarian *Dugesia dorotocephala.* Aside from the time scale, these results resemble those obtained with comparable experiments on rats. From Best and Rubinstein, 1962.

it really hates or really likes. He tries to couple that hateful or delightful thing to the actual paradigm that he is trying to teach the rats.

So I decided to make the same concessions to flatworms that people make to rats, which seemed only fair. Instead of trying to keep the stimulus constant, I said, let us just habituate them to everything we don't want them to take into account. I put 20 worms together in a bowl. Why 20? Briefly, the reason is that they know how many other flatworms are in the bowl. If you isolate them, then they fission. I did not want them fissioning because that introduces changes. Furthermore, I knew this effect was mediated through the brain. There was a possibility that being alone made them "up tight." So I put 20 of them together in a bowl, and had them in the same bowl for three days to a week before I ever did anything to them. I kept the bowls in the very same corner of the laboratory where the learning experiment was to be carried out. I habituated them to the round of light changes that normally occurred in the laboratory. I even habituated them to my sitting there in front of them — everything.

The rule of the game for this particular laboratory room where these learning experiments were carried out was that no one was to do anything unusual. People carried on their business in the laboratory as usual, but nobody was permitted to do anything unusual in that room. The flatworms were never picked up, never transferred to another bowl, never handled. We always kept the same group together.

Now for the actual learning experiment [details were shown at the Symposium by use of a Super-8mm motion picture]. I would fast the planarians for a given length of time. I would then put a little chromium-plated steel washer down in the midst of the bowl, being careful not to hit any of them with this washer, and I would put a piece of liver in the hole of the washer. The planarians would come when they smelled the liver and start climbing over the washer to get to the liver. As they would climb upon the washer to get the liver, I would give them a little localized shock to the tip of the tail. To administer this localized shock, I used a little bipolar probe, which had a microswitch on it, held in my hand like a pencil. Unlike the Thompson and McConnell paradigm, this does not involve the brain at all with primary stimulation. Probably not more than the caudal fifth or quarter of the animal is subjected to primary stimulation from

the shock. So its brain gets information about the stimulus only via the normal routes of nerve pathway conduction. The resulting behavior of the animal is vigorous, but entirely normal. He hoists his tail up out of the way of the shock which lasts only about a third of a second. Only the transgressors are shocked. Since the shock is of such short duration, it is practically over with by the time he begins behavior in which he hoists his tail up out of the way and runs off the washer. In the first several sessions (each training session lasted 30 minutes), these occurred some days apart. Their behavior is modified by this training.

Before the washers are put in each time, they are boiled for about 5-10 minutes and then cooled down and re-rinsed in sterile water. After rinsing in sterile water, they were rinsed in the aged tap water that the flatworms were kept in and then put in the bowl.

One of the things about the situation in which the worms have been shocked is that although they still have not really learned the avoidance very well yet, they gather around the edge of the washer and get as close to the liver as they can without really climbing onto the washer. You see a lot of unusual behavior in the intermediate stages. Sometimes, after they had received a few shock sessions, they altered their behavior and started going on the air-water interface and tried to "fly" over the liver and drop down on it from above, but they got shocked for that too. I think that the proper interpretation of this behavior is that they like liver just as well as they always did, but, in some sense, they know they can get hurt by going onto that washer. In between times they were fed on liver without a washer and, on these occasions, ate readily.

They are often ambivalent in that they reach out and then pull back. Every once in a while somebody will still try to get the liver and then he gets "zapped." I think one of the things that is rather interesting is that if you watch these guys, you get the feeling that instead of being like a stereotyped little Grey Walter mechanical turtle, they act more like a little guy in a worm suit. You see a lot of funny, ambivalent behavior — go across, reach across, turn around and go back. You do not see that in the earlier trials or sessions.

*Question:* Are the worms fed once a day?

*Answer:* No, I usually feed them twice a week.

*Question:* Are they very starved?

*Answer:* They are not very starved. These have been fasted about 5 days, which, in relation to the normal feeding frequency, is not a very bad starvation.

*Question:* What if you drop one of the worms directly on the liver?

*Answer:* I don't know, because you see one of the rules I meticulously follow in these experiments is that I do not handle the worms. The probe, I might add, does not actually touch them and is actually about a millimeter away from the tail. You see, even in rat experiments, and I feel pretty certain that this is also true in the invertebrate experiments, probably the most significant thing that happens to a rat is the mere fact that the experimenter picks him up. I suspect the reason that one gets any sort of sensible learning experiments at all is that white rats are such domesticated animal-vegetables. If you ask people you know why they do not use Norway rats, you hear all sorts of remarks like "intractable" and "armor-plated gloves." I suspect that a lot of wild animals (that have not had all the stuff bred out of them) are so upset about being picked up and placed in a learning situation that they do not know what you are talking about, with your buzzers and lights. Learning is, in consequence, very inefficient. Yet we know from other indications that such wild animals are not stupid.

*Question:* Did any of them actually manage to get any of the liver before they got the shock, or did you always give the shock before they got there?

*Answer:* Well, when the washer is there, I sometimes will let them go ahead to see how far they will go in the later trials. In some of the later sessions, I occasionally let them get on the liver and then I "zap" them.

*Question:* Did I understand that you train one group and you do not train the other group?

*Answer:* Yes. There is a sequel to that which I ought to tell you about in conjunction with the control group. I began thinking about the single-trial learning experiments involving rats getting shocked on their metal drinking tubes. Stop and think about a rat getting shocked on his metal drinking tube and the resulting single-trial learning. A metal drinking tube has been the only thing that the rat has drunk from since he was weaned, and a metal drinking tube is an object to which he is well habituated. It always means water.

I wondered if this extensive pre-habituation to an object that was a source of positive reinforcement was not really one of the essential parts of an efficient learning paradigm. After I had completed these other experiments, I had a control washer group that had never been shocked, and another group that had been shocked every time, right from the very beginning, whenever they tried to climb on the washers. I reasoned that I could use my control group now as a group that was very comparable to that rat-drinking-tube-shock experiment, because by this time the washer was, if anything, not only not an alien object but even an object to which positive reinforcement had always been attached. So I then gave this group a single session in which they were shocked for climbing on the washer.

This session was administered in the same way that I described previously. It lasted thirty minutes. Subsequent test sessions, administered several days afterward, revealed a very dramatic suppression in the number of attempts to cross the washer and feed. These worms, in consequence of this single 30-minute training session in which they had each received an average of only 2 to 4 shocks on the tail, acted like they had received a large number of training sessions. So I think that, if one does the real counterpart to the kinds of conditioned emotional response training that one does with rats, and uses tasks of simple information structure, one gets rapid learning in flatworms too. The rapidity or efficiency of this learning approximates that observed in rats. I should reaffirm that I do not think a flatworm is a rat, but I do not think the essential differences reside in speed of learning *per se*.

A rat has about a gram's worth of brain and that is about 4,000 times as big a brain on a wet-weight basis as these flatworms have. Why is it that flatworms display a behavioral repertoire that comes across like a one-cylinder psychological system? I think the answer is because their brain is like a one-cylinder vertebrate brain. In a Model-T Ford way, it is organized very similarly to the vertebrate brain.

Figure 2 illustrates some of the things I mean. This is the experiment I told you about in which flatworms know how many flatworms they have around, or my basis for contending such a thing. One takes a group of flatworms that have been sorted out into large, medium and small ones, and then one isolates them. They have been previously maintained with something like 40 of them in a bowl and now one isolates them in individual

petri dishes. Then one scores the cumulative number of fissionings in these individuals *versus* the days after this initial isolation. You see the large ones fission a lot more than the medium ones, and the medium ones fission more than the small ones.

*Question:* What is the conclusion?

*Answer:* Well, I will give you the second experiment. See Figure 3. I had 30 worms per bowl. We cut the heads off one

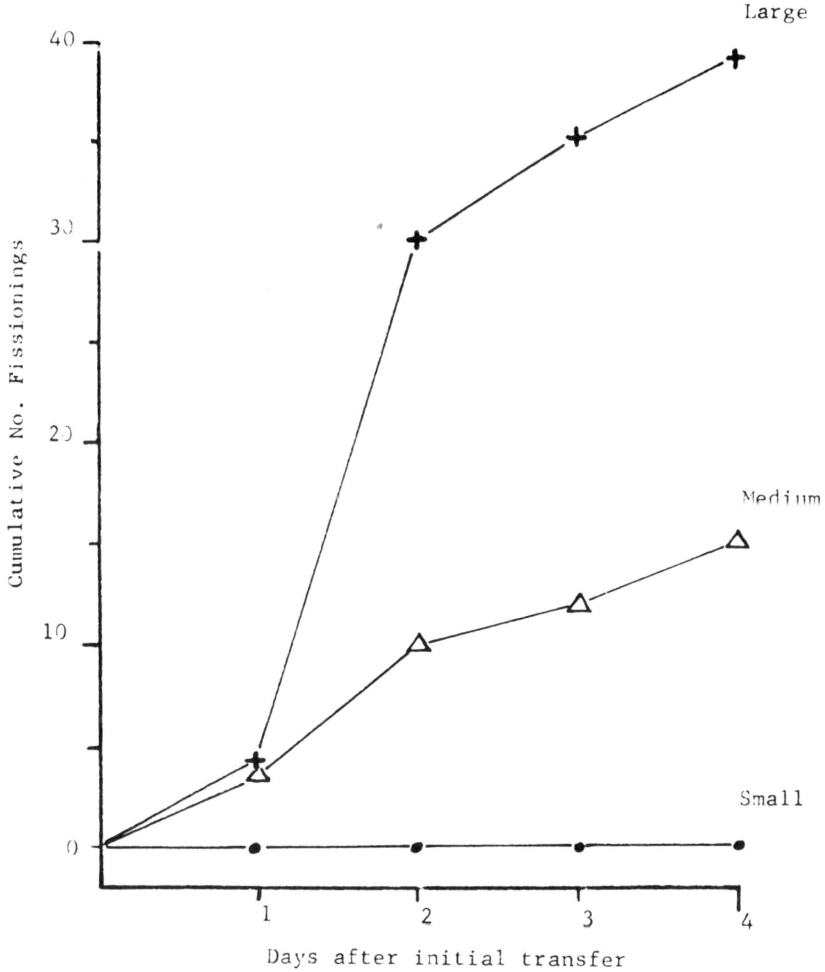


**Figure 2. Effect of social isolation and size upon fissioning in planarians. Data showing the cumulative incidence of fissioning for planarians of the species *Dugesia dorotocephala,* which were placed in isolation at zero time. Sixty planarians were used in each size class. From Best, Goodman and Pigon, 1969.**

group by a transection right at the level of the auricles, across the base of the brain. Then we had a group of uncut normals. We used the same number per-bowl of these as we did in the case of the head amputees. We had a third group in which we had taken the same amount of tissue off the tail. Then we looked at the cumulative number of fissionings as a function of the days after amputation, in each of those groups. You will see

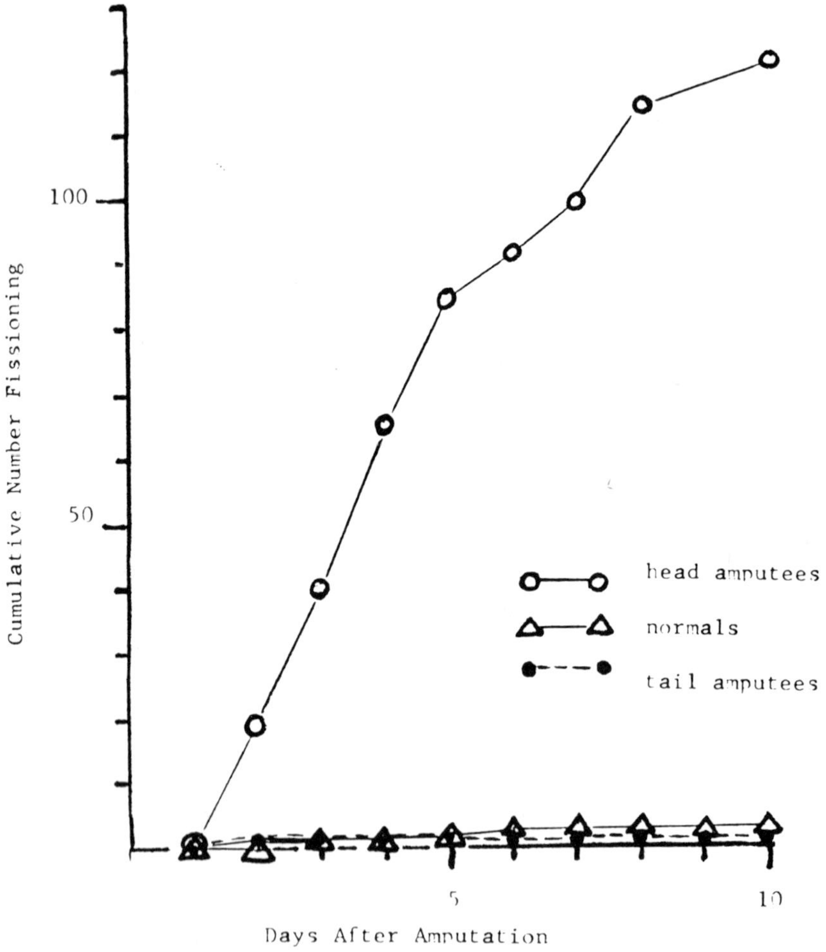


**Figure 3. Inhibition of fissioning by the brain of planarians in the presence of other planarians. Cumulative incidence of fissioning as a function of time for intact, decapitate and tail tip amputee planarians maintained in social groups of 30 per bowl. 180 *Dugesia dorotocephala* were used for each treatment group. From Best, Goodman and Pigon, 1969.**

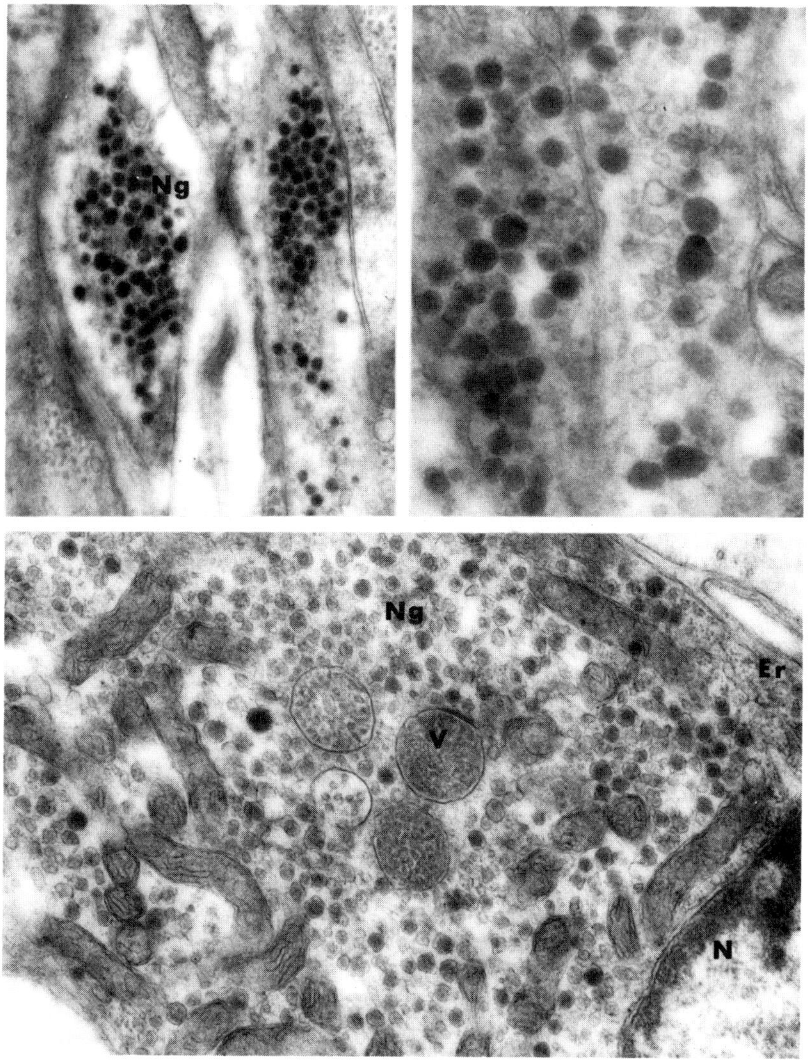


Figure 4. Electron micrograph of neurosecretory granules and a neurosecretory cell in the brain of the planarian *Dugesia dorotocephala.* The two top plates show clusters of two different size classes of neurosecretory granules in the neural fibers of the neuropil of the brain. The bottom plate shows the cytoplasm of a neurosecretory cell in the mature or storage phase. From Morita and Best, 1965.

that the fissioning is completely suppressed in those that had either the tip of the tail amputated or which were intact. But those from which the brain was removed had the fissioning released in spite of the fact that they had a population density that was sufficient to hold the fissioning suppressed in the normals. So the effect is mediated through the brain.

Figure 4 shows what I think is some of the neuroendocrinological hardware through which this effect is mediated; these are some of the neurosecretory cells and granules that we see in this brain. These are some of the larger sizes of neurosecretory granules, and some of the fiber processes. This is an electron micrograph of the cytoplasm of one of these neurosecretory cells in the storage phase where there is not much in the way of endoplasmic recticulum, but there is lots of ER in the neurosecretory cells during the phase in which they manufacture neurosecretory granules. You can see lots of neurosecretory granules, and you can see just a bit of the nucleus. There are other objects too, like multivesicular bodies and mitochondria, but the real thing to notice is the neurosecretory granules.

Figure 5 shows what the learning curve looks like for a single bowl of these guys, if you score the number of attempts and the number of visitations *versus* the session number. Now these large oscillations will probably be a source of distress to some people, but what I contend is that those are marks of authenticity; this actually relates to one of the questions that was raised in the introductory discussion this morning concerning what you miss when you average data out. One of the things you miss when you average the data (whether from rats or flatworms or children) is that extinction behavior does not result in a smooth learning curve. These oscillations in performance during extinction are not random sampling fluctuations. One can perceive this by the frequency with which one obtains sessions with scores lying outside the standard deviation bounds predicted on the basis of a Poisson distribution. Averaging the results from a number of bowls would, of course, give the usual, deceptively simple, monotone decreasing extinction curve that learning theorists present in textbooks.

The anterior portion of the brain of the planarian *Dugesia dorotocephala* contains some oversized neurons and, from the usual interpretations of such oversized neurons, I think almost certainly these neurons are final pathways of convergence to a major motor system. Right up in the most anterior tip of this

brain they feed fibers down into major longitudinal transmission tracts that travel on the ventral surface of the brain.

An interesting feature of this set of large anterior neurons is that there appear to be a number of nerve endings which contain dense-core synaptic vesicles that project onto them, or into their near vicinity. (See Figure 6.) My conjecture is that these are dopaminergic endings. We find lots of diffusely distributed dense-core synaptic vesicle endings in the neuropil around this

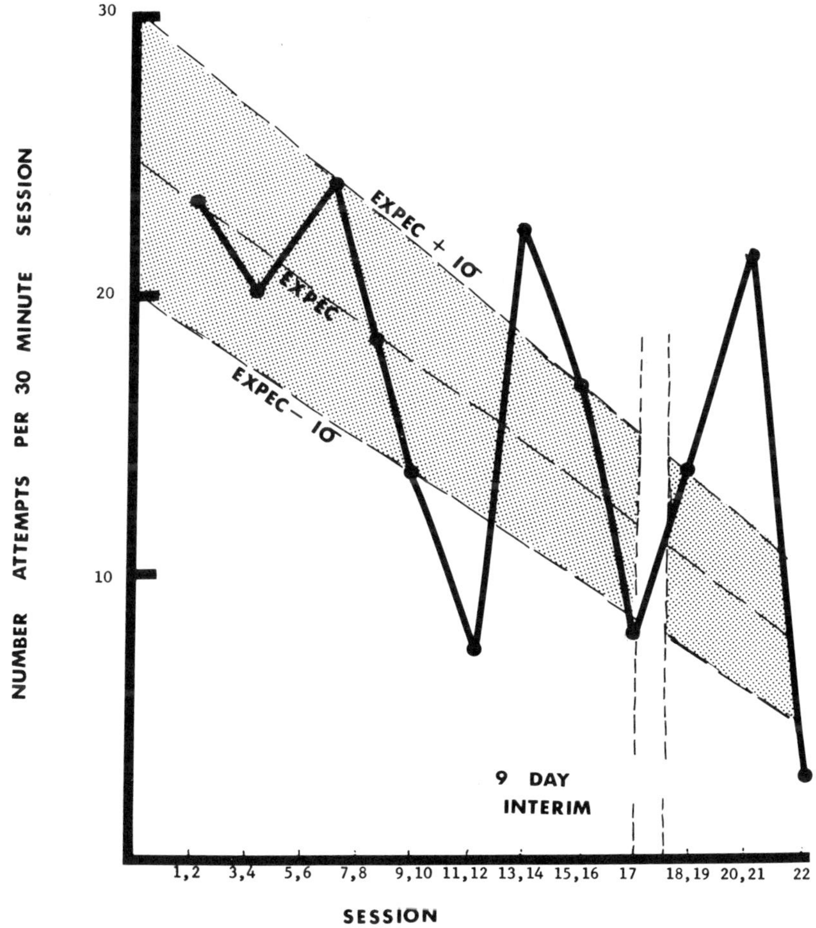


Figure 5. Graph showing the number of attempts to feed (and tail shocks received) as a function of the number of training sessions for a group of planarians of the species *Dugesia dorotocephala.* The boundaries were calculated on the basis of the standard deviation to be expected if the number of attempts per session were Poisson distributed about the linear regression line.

region where it appears that a lot of the chemoreceptor fibers come into this brain. Again I will put my cards on the table. My conjecture is that the anterior motor system is the primordial precursor to what in the vertebrates we call the basal motor ganglion system. I think this region where these chemoreceptor fibers come in and we see all these dense-core synaptic vesicles is a sort of paleorhinencephalon, a limbic system. Both from our chemical analyses and from a fluorescent cytochemical method used by two investigators at Harvard, Welsh and Williams, it appears that there are noradrenalin and serotonin in this brain, and in sizeable quantities.. Since neither they nor we have checked for dopamine, we cannot say whether it is present. But I know there are at least three different size-classes of these dense-core synaptic vesicles, so I suspect that we will find dopamine in there too.

The fissioning control mechanism I described really takes the information from chemoreceptors and holds the worm together. If we isolate the worm, or if we do microsurgery and remove these regions that contain those chemoreceptors, we release the fissioning and then you get a little worm. The mechanism thus controls the size of the worms.

Stop and think about what our own somatotropic control

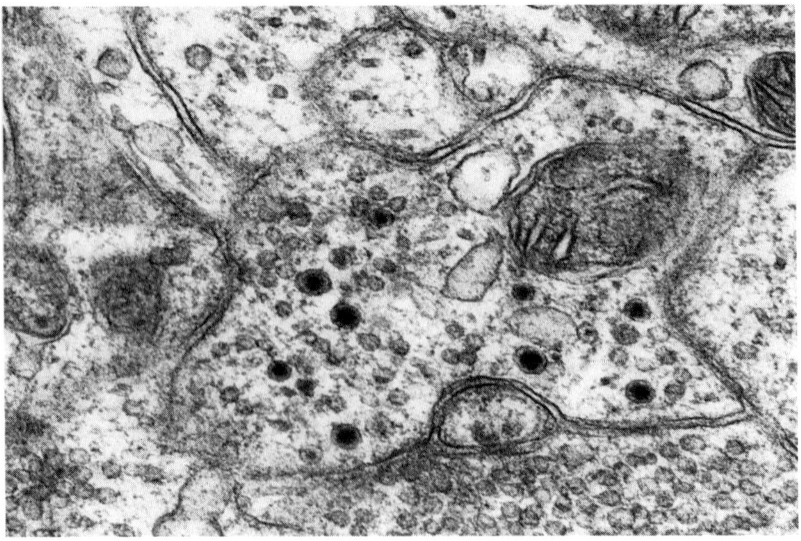

**Figure 6. An election micrograph showing dense-core synaptic vesicles in the brain of the planarian *Dugesia dorotocephala*.**

system regulates. Two major things it regulates are size and sexual state or state of sexual maturation. Already some years ago, Roman Kenk had shown that one can induce gonad formation in flatworms that do not ordinarily show gonad formation by taking the brain from sexually mature planarians and grafting it onto ones that ordinarily did not show gonad formation.

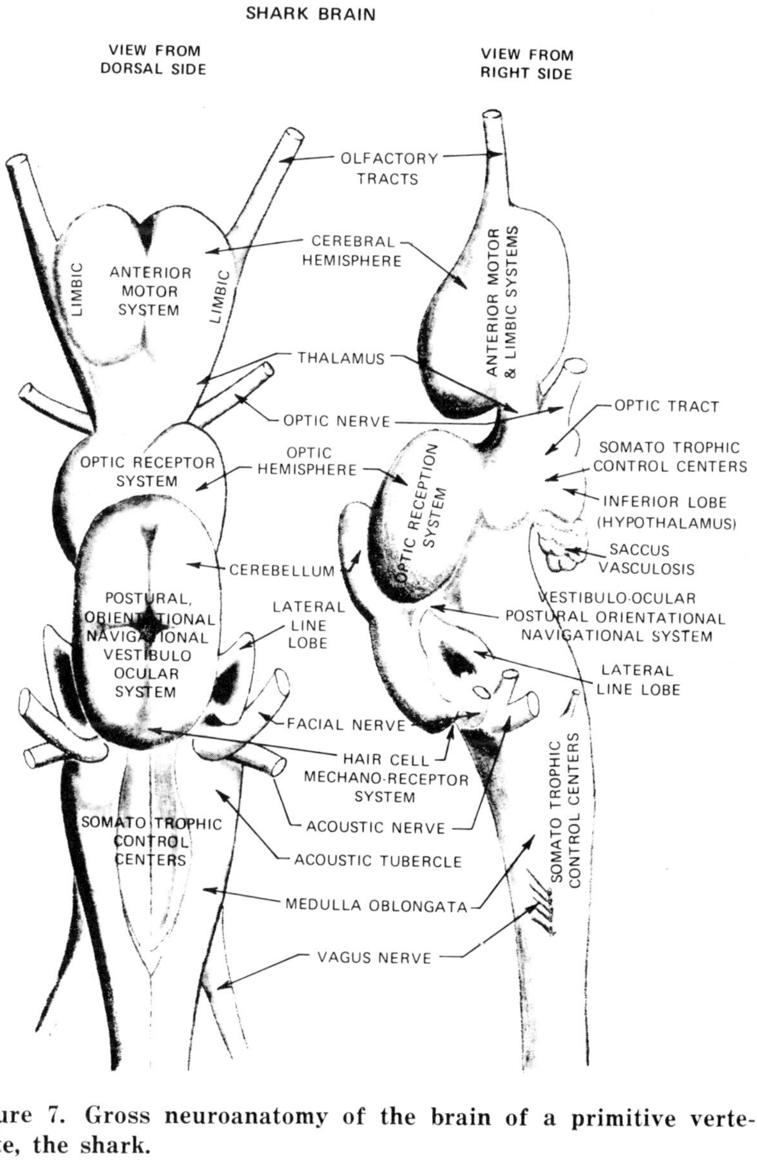


**Figure 7. Gross neuroanatomy of the brain of a primitive vertebrate, the shark.**

Annelids, arthropods and vetebrate creatures like ourselves also have their size and state of sexual maturation controlled from the brain by neuroglandular systems. In planarians there is a large size-class of neurosecretory granule that is primarily concentrated in this region of the brain close to its point of juncture with the segmental nervous system. This location is comparable to that of the medulla oblongata and all of its somatic control centers in the vertebrate brain.

There is probably another system of neuroscretory cells that is located near where the optic tracts come into the brain. There is a similar system in arthropods and there is also a system (the hypothalamic centers) that is located in a very comparable position in the vertebrates.

Some of these relationships can be seen nicely in the shark brain which is all laid out in a linear ensemble. (See Figure 7.) Let us consider what comprises the cerebral hemispheres of a shark's brain.

In this primitive vertebrate, the cerebral hemispheres are mostly comprised of rhinencephalon, or smell brain, and anterior motor system. Sharks have very little cortex, and in fact there is very little cortex really until you get up into the birds. This basal motor ganglion system, or anterior motor system, is the highest center of volitional control for all vertebrates except the mammals. Look at where the optic hemispheres are. You will notice that they are located in a position comparable to where the optic tracts come in from the eyes in the brain of the flatworm. Another thing that is interesting is the location of all the somatotropic control systems. The hypothalmic system lies near where the optic tracts come in and then deflect upwards. Consider this region of the shark's brain. It is concerned with hair cells, mechano-receptors, irrespective of where they are used, whether they are used as lateral line organs, whether they are used as cochlear hair cells, or whether they came from the semi-circular canals. We have not explored it completely yet, but there is, I suspect, coming from the auricles of the flatworm a set of receptors in there that seem to be connected with vibration sense and to be hair cell-like receptors, and would come in through a similar region. What is the cerebellum all about? Well, the cerebellum probably arose originally for integrating this special orientation information from the optic tracts and from the hair cell mechano-receptors. It is therefore an enlargement of the region of neuropil that formed that associational region between the two regions.

Figure 8 is a block diagram of what I call the "universal brain." Now this "universal brain" is presented in the same spirit as one of those metabolic pathway charts. A metabolic chart does not imply that every animal has every detailed pathway, but rather is a framework about which the evolution of these biochemical pathway systems is organized. My conjecture is that this block diagram depicts a similar basic foundation plan for the brains of all bilaterally symmetric encephalized animals on the main evolutionary sequence.

One of the questions you can ask is why would we expect the brains of these diverse animals to have a common organization? Certainly a vast evolutionary gulf exists between the flatworms and animals like us. Both brain mass and body mass have increased by a factor of five million. Yet the ratio of brain to body mass has not changed much. One can still raise the question of why this sort of original plan should be preserved. The probable basis is the mechanism by which the shape of an organ is derived. There are two major ways I can conceive of by which one can get or create a given geometrical shape. One of them is that one can build a mold, which is the usual human way. One builds a mold and pours a fluid into it. The fluid solidifies and one gets concrete blocks and slabs of steel and things like that. If one gets the shape that way, it tells one nothing whatever about the fine structure. All it tells is that what was liquid at one time is now solid. However, there is another way of getting a shape, and that way is probably more relevant here. That is a process like crystallization in which the processes really give rise to the macroscopic shape. In that case, I think that the macroscopic shape tells one an enormous amount about the microscopic structure of the thing. A crystal of sodium chloride is a crystal of sodium cloride all the way down to core. If you think about organogenesis and the sort of work that has been done up to this point in tissue reconstruction experiments there is a lot about such experiments that is reminiscent of a crystallization process. Take a bunch of cells and put them together in a hospitable medium and they slither over one another and "crystallize out." They will "crystallize out" sheets of cardiac muscle or, perhaps, kidney tubules, as Moscona has shown.

I think that probably the main forces that are working here are first the ameboid movement which provides the randomizing process by which one ensemble can find its way into different sorts of configuration. Another process minimizes the interfacial

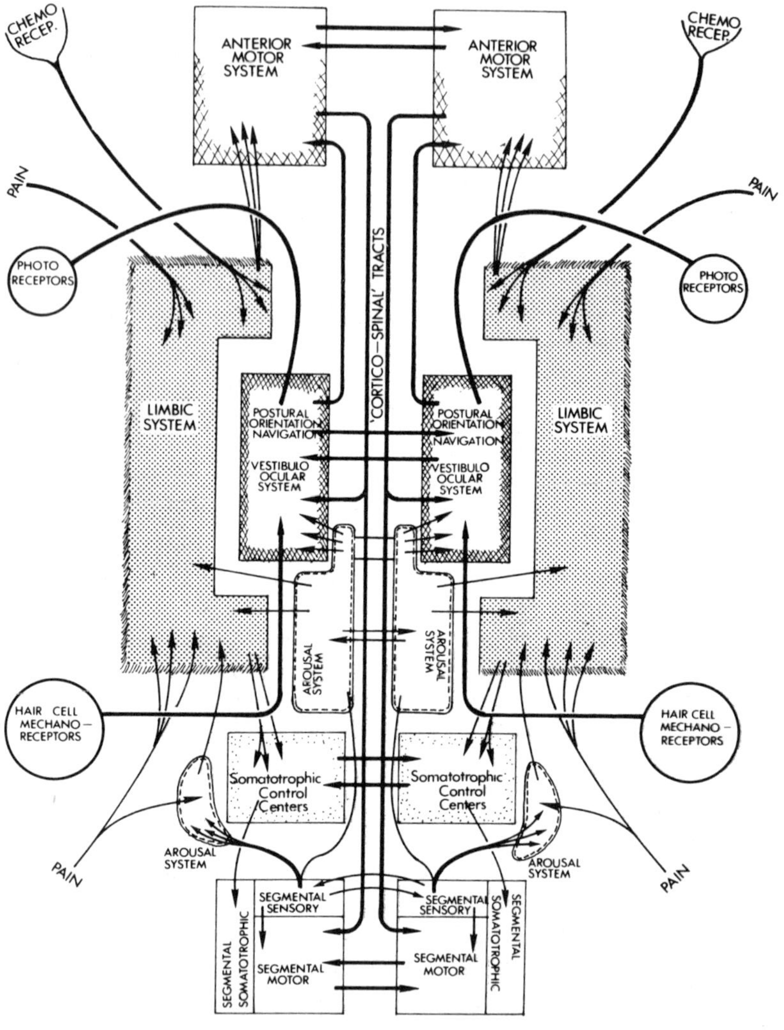


**Figure 8.** The "universal brain." There is an "anterior motor system" for volitional motor control. There is a "limbic system" which mediates emotionally affective processes and reinforcement processes. This "limbic system" is an extension of the chemoreceptive regions. There are two major "somatotrophic control regions," one in the region of juncture of the segmental nervous system and the brain, another in the region of entry of the optic tracts. There is an "orientational control system" which integrates optic and mechanoreceptive information to provide the spatial reference frame of the animal, and which organizes the postural and orientational responses. Stereotyped modular elements of locomotor behavior are organized in the segmental nervous system.

energies of this ensemble of cells — subject to the constraints imposed by the packing of these cells. The genetic code most likely works on the selective stickiness between these surface groups.

Suppose one wanted to modify something constructed this way and did not want to wipe out everything that had been achieved by previous evolution. Probably there is a set of surface adhesion properties that gives the main outlines. One would not want to tamper with those. What one would do is add secondary groupings. I would guess that probably there is some sort of surface grouping that, for example, sorts cells out into an anterior motor system and then a more recently evolved set of groupings that segregates them into groupings like the caudate nucleus, substantia nigra, and globus pallidus.

Neuroblasts are, in many ways, similar to amoebas, and this is not just a superficial analogy. Both neuroblasts and amoebas send out pseudopods and engage in ameboid movement. There are ensembles of microtubules that go out into these pseudopods and into the embryonic dendrites and growth cones as they form. These cellular extensions form trial connections and discriminations that go on inside the brain as these embryonic neurons find their proper place. [At this point a 16 mm film was shown.]

## Discussion

*Dr. Kater:* Dr. Best, you said that you were going to diverge from conventional notions. That was your goal, and I don't think anybody doubts that you diverged from the conventional and traditional notions that people have in mind. I think that one of the most beautiful things about you as an individual is that you are not constrained by a lot of things that were taught in first-year courses, but for which there is very little evidence.

The whole idea that there is a homology that might exist as a continuum from an invertebrate through a vertebrate should be massively insulting to many of us. But it might be true. By the way, have you ever done histochemical fluorescence of your dense-core vesicles?

*Dr. Best:* No, I haven't. Welsh and Williams, however, recently reported a study in which they used the fluorescent histochemical method of Falck and Hillarp to demonstrate serotonin and a primary catecholamine in the brain of a planarian. They think the primary catecholamine may be noradrenaline, dopamine, or both.

*Dr. Kater:* Yes. They didn't use a microfluorometer.

*Dr. Best:* No. Their study was essentially a qualitative one of the distribution of the projections. We haven't used the fluorescence cytochemical method. What I have done — and I have done it because I want to do a particular kind of experiment — is to scale down some of these straight chemical analytical methods about 100-fold so that we can use 50 micrograms of tissue. We have little capillary centrifuge tubes and capillary cuvettes for the spectrophotofluorometer. What I want to do concerns the following: There was a recent experiment by Welch and Welch, in which they took a mouse and just let him see two other mice fighting. He was not involved in the fight at all. They just let him see it. They found this depleted the noradrenaline from his brain stem by as much as 30%. I got to wondering, could one do a counterpart to this with flatworms? Put him in a situation in which he was punished for being there. You've zapped him on the tail or something like that. If one puts him in such a situation, can one show a counterpart to the mouse observations, a change in either serotonin or noradrenaline? So I wanted a method that I could do on one or two flatworms this way, or even a portion of a flatworm brain. That was why I got off onto this microchemical method. In the course of simply working out the method, we determined, as a byproduct, that it had fair amounts of serotonin and noradrenaline in its brain.

*Dr. Kater:* Getting off on a slightly different track for a second, because I think that one of the most startling things of your whole career and your work is the diversity of information that you are bringing to bear on these problems . . . the whole concept which you brought forward with respect to differential adhesivity and Steinberg's work, where Steinberg has been able to demonstrate that in embryogenesis, if you take ectoderm cells and endoderm cells and just put them all together, the ectoderm cells are always going to be on the outside of the ball that forms. You can get mesoderm lines inside, and there seems to be a hierarchical array of very specific surface properties, these being related to calcium and charges on the membranes and this sort of thing. As I've seen your pictures, it becomes clear to me that you didn't get to do something which I think you probably wanted to. You made the comment that the growth of neural processes may in many ways resemble ameboid movement. Do you want to comment?

*Dr. Best:* Yes, here are some pictures [not included here] of embryonic neuroblasts or neurites. This sort of strange-looking object looks like a Druid tree, or something like that. It is a three-dimensional model of the growth cone of a neuroblast. This is from developing rat visual cortex. The student of mine who did this is a beautiful neuroanatomist. It is really an art form with him. He took these rats, raised them from neonates in darkness except for a half hour of dim illumination a day — that was to keep the retina and optic nerve intact — and then at 30 days he put them on a normal light regimen. About 2-5 days after he put them on a normal light regimen, all sorts of things began happening in their visual cortex. In terms of cell number and synaptic count, most of this action is happening in layers 4 and 5. In this 2-5 day period, after he puts the lights on, all these neuroblasts begin moving out. They are preceded by a growth cone, and the growth cone looks like nothing so much as a big pseudopod going out in front. Anyway, he made serial electron micrographs, and then built up a 3-D model of it. I believe what is going on is that the growth cone is like an amoeba pseudopod sniffing its way through grains of quartz, forming temporary adhesions as it goes out. The neurite "tries them on for size" as it is moving out and finding its way to layer 4. That may not be right, but at any rate that is what it looks like.

*Dr. Kater:* I'm sure we will have some questions from the audience at this point.

*Dr. Murphey:* I would simply like to relate a sad story to you. I used to consider myself to be one of the great fruit fly trainers of our times, and I had a situation very similar to yours. My flies were in a T-maze instead of in your situation. They were shocked for going to one side, and they were allowed to do something I thought they would like to do on the other side. I got things that looked like learning curves, and I got things that looked like extinction. Instead of having one group that was untrained and another group that was trained, I ran the flies to both sides of the maze randomly. It seemed to me that in the situation you had it would be possible for your animals to leave some sort of substance on the washer so that another animal would attend to it.

*Dr. Best:* That would be so within an individual session. But what happens between sessions is, at the end of the 30 minutes, I take the washer and liver out. I can interchange washers, which

I do. I can boil the washers, which I do. I showed no within-session data.

*Dr. Murphey:* Let me go on. A few experiments ago I ran the same situation under a different set of circumstances where one group of flies was trained, another group was not trained, but put in the maze for an equivalent amount of time right after the trained group had been in there. Actually, there were two groups of trained flies — one was retrained 24 hours later, and the other 48 hours later. Control group flies were individually yoked to the experimental subjects. They were not trained during the first session. During the second session, either 24 or 48 hours later, controls were trained for the first time, and they did nearly as well as the experimental flies. I tried to make some kind of latent learning interpretation of it, as you did. I'm not saying anything perjorative about what you did. I'm pointing out a potential problem. I tried to postulate latent learning, but could never really get it to work. As I was writing up the article, I went back and correlated the scores between the experimental subjects and the control subjects, and guess what? I accounted for more variance in the experimental subjects' score that way than I had by any of the variables I'd been manipulating all these years. This seemed to suggest that the experimental subjects were probably leaving some sort of alarm pheromone on the incorrect side.

*Dr. Best:* But once again I repeat that I do not see how that can possibly account for what is going on here. Whatever the situation may be with the fruit flies, I don't see how such factors can play a role in this case.

*Dr. Murphey:* When you retrain the control subjects that have not been trained before, they are there in a group. What I would be interested in is what the very first subject does . . . the very first subject that goes to that washer.

*Dr. Best:* I still don't see what that tells you, because . . . you see, they live in there all the time, and the thing they are being trained to do is not to go on a washer which is removed and replaced with another washer that is as alike as Montgomery Ward makes washers.

*Dr. Murphey:* It's possible that I don't understand your paradigm.

*Dr. Best:* I don't quite see why not. They live in the same bowl all the time. They are never taken out, never picked up,

the water is changed in the bowl. I clean it out, but this is not to remove pheromones. It is simply to clean the bowl out. At the onset of a training session, I take a washer with a pair of forceps and place it in the bowl. Then I put a little piece of liver in the middle of the hole in the washer. Then the worms come to this washer. Now, in this other bowl, my controls are run at the same time, to another washer. They are all the same kind of washers. At the end of the 30-minute session, I take the washer and the liver out. During their training, every time they start to go on the washer I shock them on the tip of the tail. Between sessions these washers are not only washed, they are boiled for 5-10 minutes. They're all scrambled up anyway. I can see how you can raise the argument of a within-session effect, but I really cannot see how you can raise that kind of argument from session to session, which is all I showed you the scores on.

*Dr. Kater:* I think I see the dichotomy here. It may already have been resolved by an experiment we don't know about. Is it possible, in a very high affective run, to take the washer from your experimental group and put it in the control group and get normal control efficacy?

*Dr. Best:* I'm not saying that is not a worthwhile experiment to do. All I'm saying is that, irrespective of the outcome, I don't see how pheromones could be playing a role here.

*Dr. Kater:* It would appear that you could have conditioning exactly as you have interpreted it but, perhaps, augmented by an alarm-pheromone type substance.

*Dr. Best:* The ones that are shocked put out a gelatinous mucus. I can look at a washer and tell you whether it has been used in an experiment in which the worms were shocked. But this mucus washes off very readily. In some studies that we did that I never mentioned, we investigated a type of cell in the flatworms that is very similar to a goblet cell. Goblet cells, as the physiologists among you will know, are what we use to make saliva, gut mucus, and things like that. These cells in the flatworm appear to be similar at all stages of maturation to the goblet cells in us. Now if you really look at the early work on conditioning, it was not motor conditioning. Consider Pavlov's early experiments on saliva flow, which is controlled by the autonomic system. The alterations in saliva flow produced in his experiments were almost certainly a kind of conditioned emotional response. The effectors were goblet cells. Pavlov noted that he would get not only dif-

ferent amounts of saliva but also differences in the texture of the saliva, depending on whether it was an aversive stimulus that he was conditioning the dog to or not. To return to the original question on the planarians, I think that there is a change in kind and quality and amount of mucus, but I do not see how it accounts for the particular kind of learning result.

*Dr. Kater:* I may have missed the point. Did you say you did or did not change the water?

*Dr. Best:* I had to change the water anyway after I had fed them, because if there is any residual liver juice in there you get bacterial growth and that kills them.

*Dr. Kater:* Have you ever done an experiment in which you took untreated animals and put them into a situation where there was water in which you had run some trials?

*Dr. Best:* No. Once again one of the things I want to emphasize is that practically all of the arrangements you talk about involve picking the worms up and handling them and this I regard as a far more serious crime than any that might be implicit in anything I have done.

*Dr. Kater:* You said you changed the water, right?

*Dr. Best:* I changed the water but I don't rough the worms up when I change the water. If you take a worm out and put him in another place, then you pick him up.

*Dr. Kater:* But you can take untreated worms and take the water out and put treated water in.

*Dr. Best:* You are obviously concerned about pheromones. Let me give you some information about pheromones. I've told you about this feedback effect of population in which if I put 30 worms in I suppress the fissioning. If I isolate them, the fissioning goes on.

*Dr. Kater:* I want to ask you a question about that.

*Dr. Best:* We did do experiments on that in which we used a cross-perfusion system. We had a high population density here in the upstream chamber, and one worm here in a downstream chamber. Perfusion of the water from the high population density to low population density chamber does not inhibit the fissioning at all. Neither is the fissioning inhibited if you replace a group of worms with a single individual. Now I cannot say whether it is a short half-life pheromone or whether it is not a pheromone

at all. It could be some sort of contact-contact sensing that is responsible for the population feedback inhibition of fissioning. I can tell you that it is not that they see one another, because we did an experiment in which we take a flask and put one worm in it and put this flask down in a beaker. We matched the water level in flask and beaker. We add a little grease so the meniscuses do not curve up and then you cannot tell where one leaves off and the other begins. From the standpoint of the worm, it looks like he is surrounded by bodies, but it doesn't inhibit his fission. We have 30 worms in the beaker and put one worm in the flask and he can see all the worms in the beaker but the worm in the flask fissions anyway. If you take a worm and just cut off his ears, and then put him into a high population density, this suppresses his fissioning in the same way as it does an intact worm. If you leave his ears intact and you then trim a little region off the lateral margin of the head, then his fissioning stays suppressed in the presence of a bunch of other ones.

*Dr. Kater:* The trouble here is you are not only removing a sensing device put perhaps something that he requires to fission. How might you distinguish? He may need those tissues to sense other worms around.

*Dr. Best:* I think he does, as a matter of fact.

*Dr. Kater:* By the same token, it could also be required as an element of the fission process.

*Dr. Best:* Well, on purely logical grounds, and in the absence of other information, I would have to admit you have a point. But I could also tell you another result. I will state the following: the brain in the presence of a high population density of other worms or the information received from the presence of a high density of other worms suppresses fissioning. Removal of the head releases fissioning. Another thing is that if you take a flatworm and subject it to something like social isolation, it usually fissions. Now one will very frequently see that the tail piece will undergo secondary, or tertiary fissioning and give you little micro-worms, whereas the anterior piece never does. This fits in with the idea that the brain is suppressing the fissioning, and this influence becomes attenuated as you go down the worm, and if you get far enough away then the worm gets unstable. There is another aspect of the matter, namely the latency of the onset of the fissioning and the meaning of this latency. There are two ways

one might regard that latency. One is that this is the time it takes the neurohormones to effect the change, like a lag in the machinery. The other is that the brain doesn't realize it is alone until it has had two days. I think it is the latter, because we did an experiment in which we would pulse worms with loneliness. Well, you may say, how do you pulse worms with loneliness? What you do is, you have a high population density and then for some period of time you separate everybody and keep them alone and then put them back together. Now what happens is this. If they don't fission prior to the time that you put them back together, then they don't fission. Whatever recruitment has gone on, all bets are off as soon as you put them back together.

*Dr. Kater:* How long does it take your subjects to extinguish? That is, can you measure learning in trial which is not a carry over from one session to the next?

*Dr. Best:* Everything I've showed you has been a carry over from one session to the next.

*Dr. Feldman:* You are shocking them continually?

*Dr. Best:* No. As a matter of fact if you figure out how many shocks a worm gets in a session, each one probably does not get more than one or two shocks, and each shock only lasts a third of a second.

*Dr. Feldman:* Do you ever have sessions near the end where no shocks are required?

*Dr. Best:* Well, I'll put it this way: In some of the later sessions very few shocks are required. Let's say that we will have a session later on where we may have 25 worms in there and, of course, since they are pretty well matched according to size, I can't tell one from the other, except the one that is trying right at the moment. I know how many worm shocks were delivered within a session. It will drop down to where there are only maybe 6 or 7 shocks for the group of 25, which means that something like two-thirds of the worms are not getting shocks. It's possible that what one is doing is sensitizing them to a fear hormone. Maybe when you shock a guy he throws something out so the others modify their threshold or the way they act to that.

*Dr. Kater:* Do you think you could prove the efficacy of this thing if you could measure individual differences?

*Dr. Best:* Yes.

*Dr. Kater:* Maybe you could get rid of these guys that are slow learners and then you could have ones with no . . .

*Dr. Best:* I think that it is quite possible to breed to it. The reason a lot of the obvious experiments that you might think of haven't been done is because I have run all of these training experiments myself. I don't trust anybody else. This means I have to stick them in at the end of the day after classes and committees and running the rest of the shop.

*Dr. Gamow:* Every experimental organism suffers, maybe. The snake that suffered because Eve gave, or the snake gave, Adam the apple to eat . . . in planaria there is so much bad publicity. A lot of people work on planaria because of transfer of learning. Now, somehow you are convincing us that you can determine whether an animal is learning.

*Dr. Best:* I showed you the kind of thing I do. Perhaps you might do such an experiment to try to see once and for all whether the transfer phenomenon does occur or does not. I think it is one of those long shot things. If somebody wants to grind up worms and feed them to others, well, fine. I was never that strongly entranced by passive transfer, or thought it that probable.

*Dr. Gamow:* But the problem with such an experiment is the criterion of learning.

*Dr. Best:* Yes, that was McConnell's problem, and I can tell you what is going on in his paradigm. I did publish two papers where we analyzed the mechanism of what is happening in the Thompson and McConnell shock-paradigm. You have an ensemble of neurons of different lengths that are being subjected to an electric field, and one of the things you can show — this was a good instance in which the theory predicted the experimental outcome very nicely — is that if a neuron in a volume conductor has certain specific membrane properties, and if the specific membrane properties stay the same as you change the size of the neuron, then there is a direct, predictable relationship between both the rheobase and the chronaxie that is a function of geometry. The upshot of the story is that long neurons have a long rheobase and a long time constant. Little neurons have a high rheobase and a short time constant. So, if you have an ensemble of neurons, and you use different pulse widths of frequency characteristics, you can predict which neurons are going to be

selectively fired off. In the case of an animal like a flatworm, where you have histological studies of the distributions of the different size classes of neurons, you can tell the kind of initial twitch that you are going to get from your shock. What happens is that if you use a long pulse and you turn the gain up on the Grass stimulator just to that point at which you observe the first contraction, then you can pretty well convince yourself that these long duration pulses stimulate the ventral nerve cords. You get a big, long contraction. But the "response" has nothing to do with the normal behavior of the animal. If you go to short duration, high frequency wave-forms, then you selectively stimulate the neurons of the brain. Again what you are doing in this case is you are subjecting this nervous system to what would be the counterpart of the diffuse intercranial stimulation, if you were to talk about it in a larger animal. But it has nothing to do with an ordinary learning experiment. I won't say there are not learning components to it, but what you are primarily doing is stimulating different size classes of neurons depending on the wave-form characteristics. You are selecting different size classes of neurons wherever they happen to be, and I think this is why things were so erratic in the McConnell experiments. It was not a learning experiment fundamentally.

# RNA Metabolism in Response to Neural Stimulation in Aplysia

R. PRICE PETERSON
*University of Pennsylvania, Philadelphia*

It is dogma that the nervous system is functionally plastic, i.e., that it can accommodate itself to handle new information and recognize repeated patterns of stimuli. From this follows a central problem — what is the mechanism of this plasticity? On a cellular level this means, how does one cell induce another to change its pattern of behavior and by what mechanism? Does the stimulus intervene in the neuron's metabolism at transcriptional, translational or purely membrane levels? What would be a reasonable preparation to get at these problems?

The early work of Arvanitaki and Chalazonitas (1949), Kandel and Tauc (1965), Strumwasser (1965), Frazier et al. (1967) all suggested Aplysia as a likely candidate. They have provided a wealth of neurophysiological data on its very large identifiable neurons, where synaptic input can be well regulated and manipulated. Primarily what I wanted to do was to develop a system where nucleic acids and proteins could be studied in a well defined neural system which could also be manipulated neurophysiologically. By well defined I mean one where there could be no confusion as to what macromolecules were neural and what were glial in origin, and more specifically, if neuronal, a system where we could return to exactly the same neuron from animal to animal. Aplysia seemed to offer this possibility. As you will see the first job was to find out what could and what

could not be done with the preparation in pursuit of the rather grandiose type of problems mentioned initially. This, in part, determined the direction and goals of subsequent work.

I would like to review for you first some of the basic data I have collected on RNA metabolism in a single particular neuron of the abdominal ganglion of Aplysia, and at the same time introduce the preparation to you. Then I will move on to the quantitative changes we have observed during various types of stimulation, and for various durations. By "we" I mean myself and my two neurophysiological collaborators, Daniel Kernell of the Nobel Institute for Neurophysiology in Stockholm, and Sol Erulkar of the University of Pennsylvania. Finally I will present some evidence which I believe indicates a genomic transcriptional response to synaptic stimulation.

Aplysia, or the sea hare, is a large marine mollusc somewhat reminiscent of a very large garden slug (see Figure 1). They exist in most of the oceans of the world but since they are most readily available off the coast of Southern California, *Aplysia californica* is the species of choice in this country. When the animal is pinned out, ventral or foot side up, and an incision made the length of the body, the ganglia composing the nervous

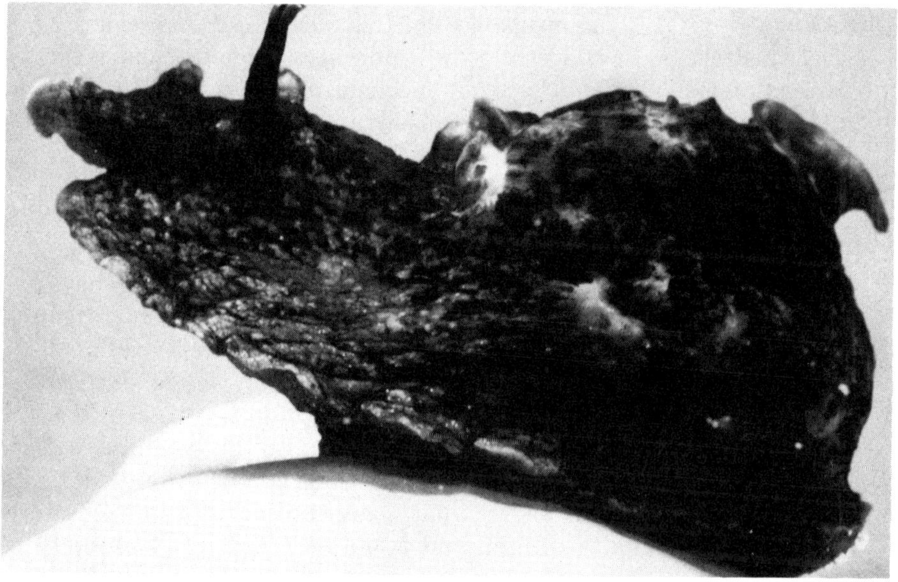

**Figure 1. Photograph of a 15 cm long *Aplysia californica*.**

system are easily identified (see Figure 2). The abdominal ganglion which has both the least connective tissue and the largest neurons may then be excised and placed in saline or organ culture medium and kept for many days. Of the 1,000 or so neurons in the ganglion, about 30 have been identified by Kandel and his group (Frazier et al., 1967). The giant cell, or R2, will be the neuron of particular interest in this talk. This is one of a pair of cells. The homologous neuron is in the left pleural ganglion and referred to as the left pleural giant. Both neurons are large monopolar cells, .5-.7 mm in diameter, whose major synaptic input appears to terminate on its axon and axonal branches. In

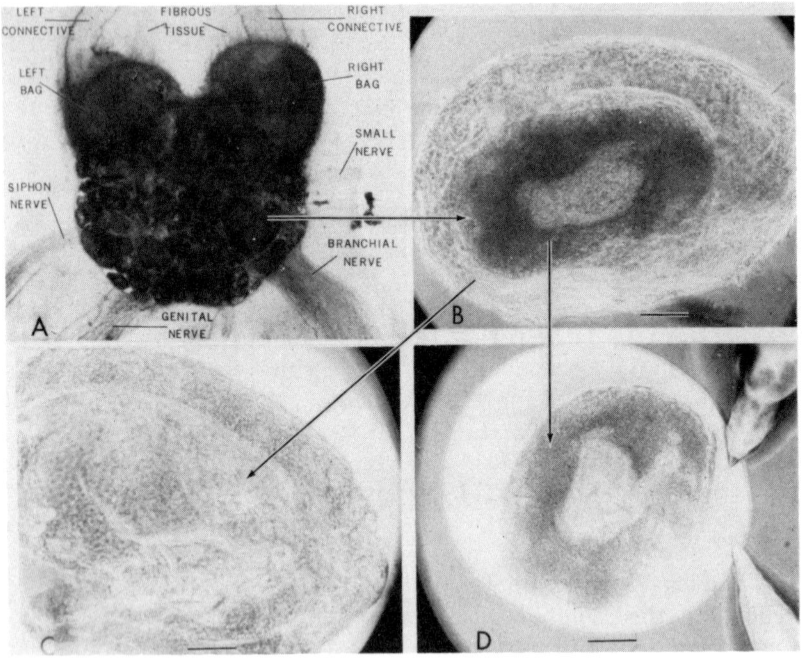


**Figure 2. (A) Photograph of a living abdominal ganglion (from Coggeshall, 1967, by courtesy of the author). The connectives contain axons running to and from other ganglia of the nervous system, and the nerves distribute to the periphery. Cell R2, whose main axon runs out in the right connective, is the large cell at the blunt end of the arrow. (B) Cell R2 photographed with phase-contrast optics after fixation and removal from the ganglion. Cell R2 was subsequently dissected into cytoplasm (C) and nucleus (D) with the aid of dissecting needles (shown in D) attached to a micro-manipulator. In B, C. and D the unmarked bar is 100 $\mu$m.**

the case of R2, its main axon leaves the ganglion via the right connective, and, although normally quiescent in vitro, all 5 ganglionic nerves carry synaptic input to the cell. This means that essentially all neural activity of the cell may be controlled by the experimenter. The neuron is hyperpolarized by acetylcholine and the excitatory transmitter is unknown. The only known neuron which plays onto R2 is a weakly inhibitory interneuron. Table 1 shows some of what is known biochemically about this nerve cell.

**Table 1**

Characteristics of neuron R2 from a 200 gm. *Aplysia californica*

| | Volume | Diameter | Source |
|---|---|---|---|
| Size: | .075 mm$^3$ | 510 $\mu$ | Lasek, 1970 |
| | .050 mm$^3$ | 460 $\mu$ (*A. fasciata*) | Hughes, 1967 |
| | .050 mm$^3$ | 460 $\mu$ (*A. depilans*) | Arvanitaki and Tchou Si-Ho 1942 |

Dry Weight (after TCA extraction): 9.2 ± 1.5 (n = 8) (Peterson, 1971)
Protein: 7.4 ± 0.6 $\mu$g (Giller and Schwartz, 1968)
RNA (O. D.$_{260}$): 1.6 ± 0.3 $\mu$g (n = 13) and 3.2 ± 0.3 $\mu$g (n = 4) (Peterson, 1971)
DNA: 0.67 $\mu$g and 0.135 $\mu$g (Lasek, 1970)
Choline Acetyltransferase (E. C. 2.3.1.6): 2.26 ± .18 x $10^{-9}$ moles/hr. (Giller and Schwartz, 1971a)
Acetylcholinesterase (E. C. 3.1.1.8): 18.5 ± 8.1 x $10^{-9}$ moles/hr. (Giller and Schwartz, 1971b)

The basic experimental methodology I will discuss involves incubating the isolated ganglion in a buffered sea water solution containing high specific activity — $^3$H uridine, or other labeled components, at 17°C either in a test tube or in a special neurophysiological chamber. When done in the latter, electrical activity, either spontaneous or induced, is monitored with a citrate-filled intracellular recording micropipette. At the end of the incubation, the ganglion is pinned out and "fixed" with a cold ethanolic acetate buffer solution which is subsequently replaced with a solution containing glycerol to facilitate microdissection.

The neuron is then removed from the ganglion and RNA is extracted either from the whole cell or after dissection into

nucleus and cytoplasm. RNA extraction is done on a microscale by pronase-SDS digestion and ethanol precipitation (Peterson, 1970). That the material labeled is RNA has been checked by actinomycin-D inhibition and RNAase digestion.

Initial efforts were directed at determining what were the molecular weights of the major classes of RNA in R2, and what were their kinetics of labeling (Peterson, 1970).

Figure 3 shows the pattern of cytoplasmic RNA labeled for 24 hours from five pooled neurons. It is clear that the resulting electrophoretic pattern is strongly dependent upon the method of RNA extraction; the sold line shows the method most often used of pronase-SDS extraction followed by ethanol precipitation. Two rRNA peaks are apparent at approximately 18 and 27S as well as the large 4S peak. Phenol extraction results in, for the most part, unacceptable losses with a minimal increase in purity.

Figure 4 shows the pattern of labeling of nuclear and cytoplasmic RNA in neuron R2 following 30, 60, and 80 minutes of incubation. A major nuclear peak may be seen to emerge at about 38S. In order to resolve this more clearly by obtaining more counts, several nuclei were pooled following an 80 minute incubation, as were several whole neurons. Figure 5 shows that by 80 minutes the 38S peak is well developed and two small peaks are apparent at about 31 and 21S. At the same time, as shown in Figure 6, the whole neurons show these peaks as well as those of the finished rRNA. Thus we can estimate that about 80 minutes are required for this neuron to transcribe and process rRNA. Figure 7 shows the complete pattern obtained at 24 hours for neuron R2. In this figure I would call your attention not only to the previously mentioned RNA species, but also a small but consistent peak at about 14S. This peak has been found in RNA from R2 but not from R14 or R15. In connection with this, Dr. David Wilson of Caltech (Wilson, 1971) has reported that the dominant labeled protein in R2 is about 60,000 M.W., whereas very little of this material is found in labeled protein from R15 and R14. Calculations estimate that to code for a 60,000 M.W. protein an mRNA of about 14S would be required.

In order to explore further the processing of rRNA, ganglia were incubated for either 1 or 2 hours in labeled uridine, followed by incubation in Actinomycin-D for 2 or 6 hours. During the incubation in labeled uridine the 38S nuclear RNA became labeled,

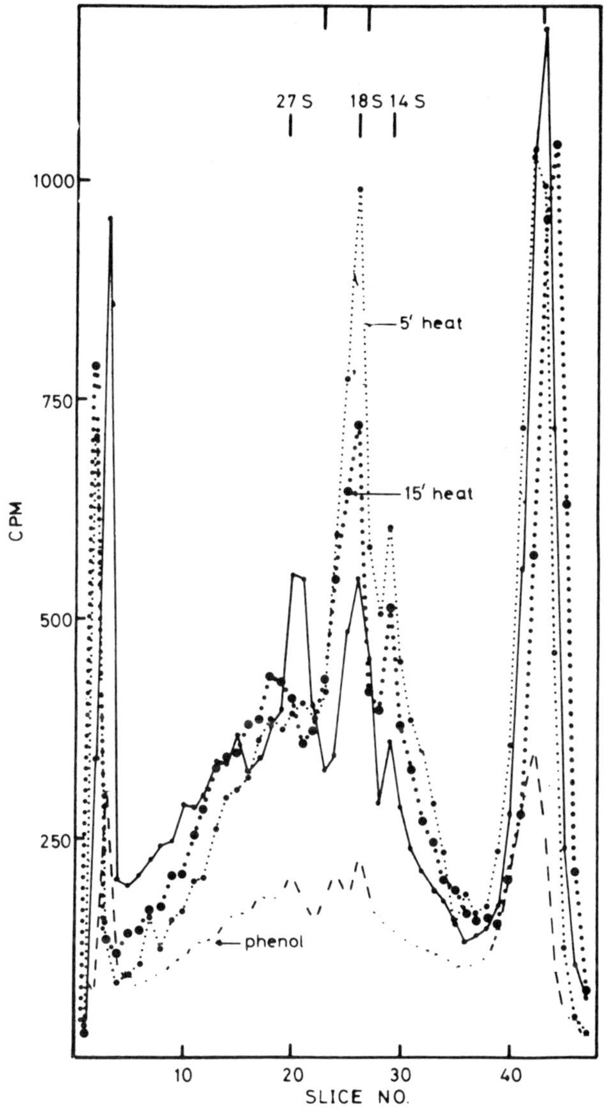


Figure 3. Five cytoplasms were pooled from a ganglion incubated 24 hr in U + C. After pronase-SDS incubation and dilution the sample was divided into four equal parts. One was immediately precipitated (solid line) and one was further extracted with cold phenol (open squares). Of the remaining two, one was heat-treated for 5 minutes (small solid circles) and the other heated for 15 minutes (large solid circles). All were precipitated twice with NaCl-ethanol prior to electrophoresis (from Peterson, 1970).

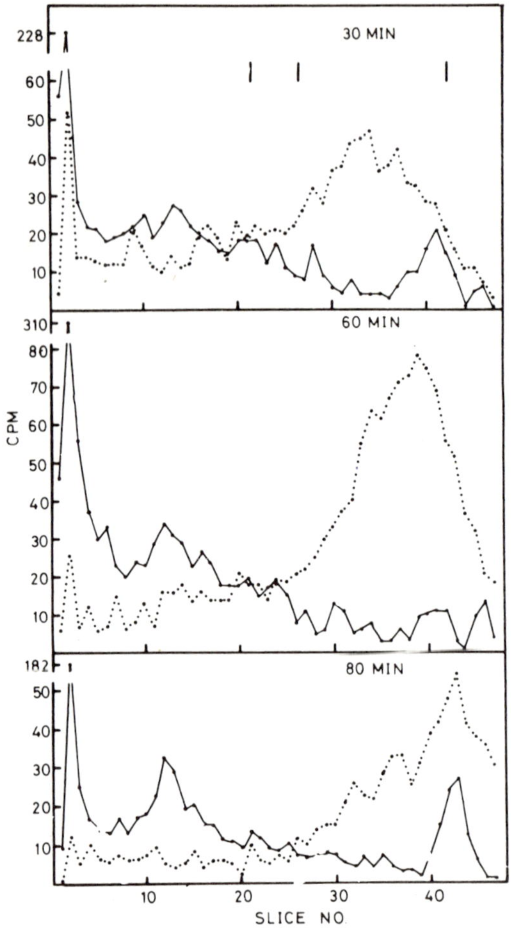


Figure 4. Electrophoretic separations of RNA extracted from the nucleus (solid line) and cytoplasm (dotted line) of R2 cells. The origin is at the left and the unmarked lines at top indicate the positions of 23 S, 16 S and 4 S *E. coli* RNA (left to right). The R2 cells were removed from abdominal ganglia which had been incubated in 0.4 ml of sea water containing 200 μc of ($^3$H) cytidine and 200 μc of ($^3$H) uridine for the times indicated. RNA was extracted from the 30 and 60 minute incubated cells by heat treatment and from the 80 minute incubated cell by cold phenol extraction. The nuclear and cytoplasmic samples were processed identically and simultaneously. Since the nuclear samples show little material between slices 30 and 40 it is unlikely that the large amount of cytoplasmic label in this region is a degradation artifact. CPM = counts/min. (from Peterson, 1970).

and during the subsequent Actinomycin-D block, this label was shifted to the 31, 21 and 18S peaks. Although nuclear-cytoplasmic transfer of rRNA was severely inhibited under these conditions, this did provide some evidence for the processing of rRNA from a 38S transcribed precursor.

In summary, this revealed that the cell contained rRNA's of 18 and 27S which were transcribed as a 38S precursor with at least two intermediates at 31 and 21S. In this there was nothing particularly remarkable. The only real peculiarity was the fairly consistent precursor of a 14S peak. Originally this was thought to be a by-product of the non-conservative rRNA processing, however another possibility has already been mentioned. More important for future work was the fact that, although RNA turnover was rather slow (of the order of .05%/hr.), there was ade-

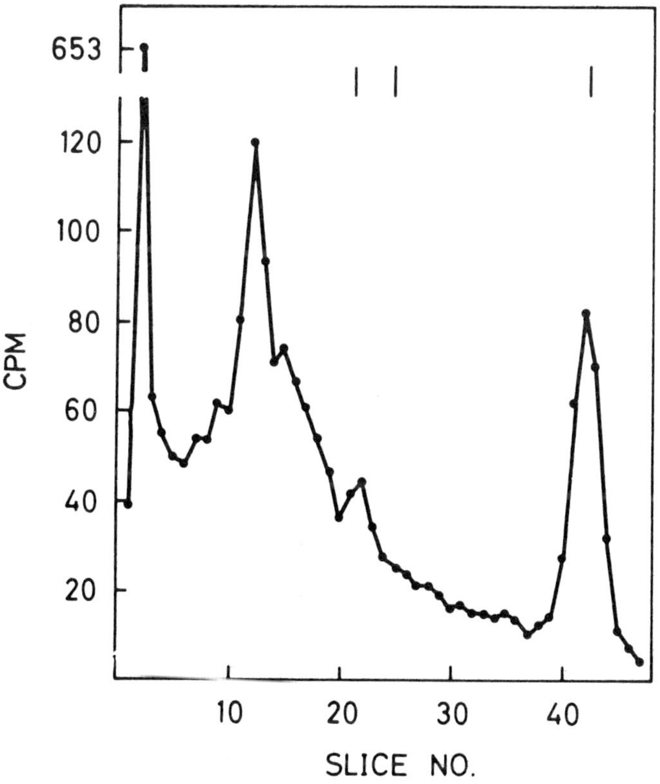


**Figure 5. Agarose electropherogram of nuclear RNA from 5 pooled neurons incubated 80 minutes in buffered sea-water containing ¾ uridine. Origin at left.**

quate incorporation of label to allow qualitative analysis in conjunction with experimental approaches. This latter statement must be qualified however. We have received some batches of Aplysia where the uptake is uniformly about 1/10 of normal. Since this occurs primarily in the fall we feel it is probably some sort of seasonal variation.

Thus armed with these basic data we felt justified in moving to combined biochemical-neurophysiological experiments to see what effect, if any, synaptic bombardment would have upon DNA transcription (Peterson & Kernell, 1970). These experiments involved incubations and analysis of $^3$H uridine uptake in R2 using either R2 from another animal or the LPG from the same animal as controls. Initial control experiments indicated that penetration of R2 by the recording microelectrode *per se*

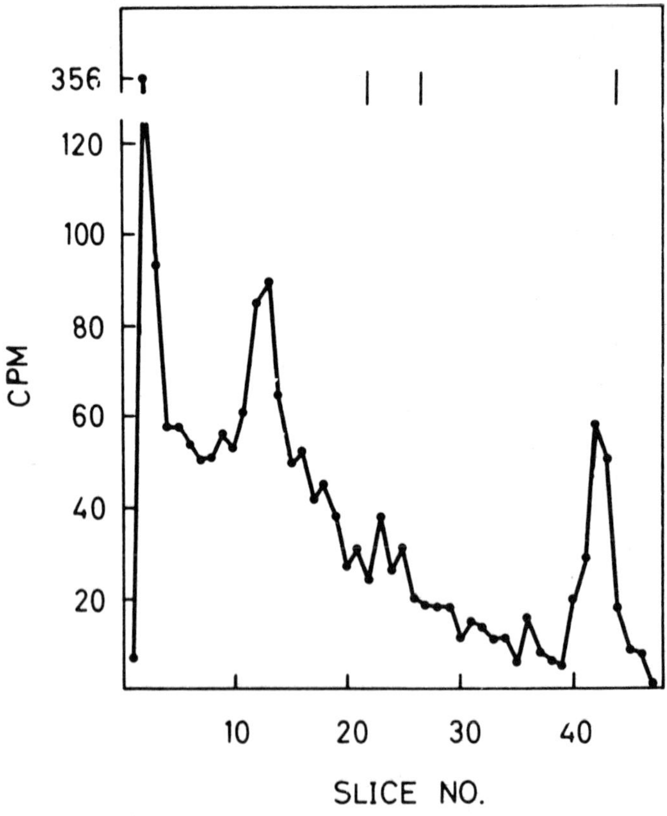


**Figure 6. Agarose electropherogram of whole cell RNA from 5 pooled neurons incubated 80 minutes in buffered sea-water containing ¾ uridine. Origin at left.**

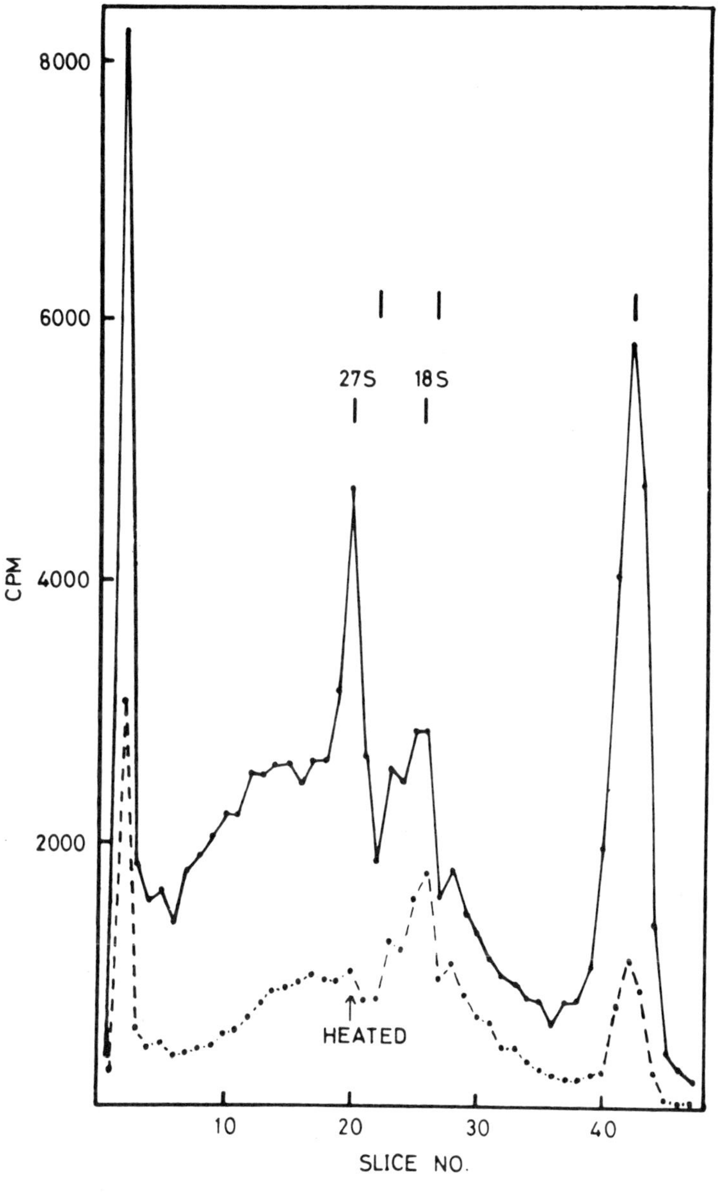


Figure 7. Agarose electropherogram of RNA extracted from neuron R2 after 24 hours of incubation in labelled uridine. The dashed line represents part of the sample which was heated 5 minutes at 55° C prior to electrophoresis and shows a transition to lower molecular weights.

had no apparent effect on the rate of incorporation of labeled U into R2. This was checked by measuring uptake in relation to controls in penetrated but unstimulated R2 cells, and by checking that *changes* in uptake which accompanied stimulation were observed with both intra and extracellular recording.

Next we began measuring uptake of $^{3}H$ uridine into R2 during stimulation of R2 through the various nerves of the ganglion. The rate of stimulation was about 1/1.5 sec. delivered to either branchial, genito-pericardial, siphon or left connective nerves. At this frequency synaptic conduction generally began to fail within 10-20 minutes (presumably due to transmitter exhaustion) and so stimulation was switched to another nerve. In this way the stimulus was rotated among the inputs, returning to previously depleted nerves after synaptic conduction had been restored.

The result of this regimen of suprathreshold stimulation was that after 90 minutes there was a marked increase in labeled uridine incorporated into RNA of the stimulated R2 neuron as compared to its unstimulated control (see Fig 7). That this reflected nuclear DNA transcription was indicated by the fact that at 60 minutes the increase was in nRNA, whereas by 90 minutes this was in cRNA. It should be mentioned here that the earlier ground work had shown that about 80 minutes at this temperature are required for label in rRNA to appear in the cytoplasm. I would also point out here that increased incorporation resulting solely from increased pool specific activity cannot be ruled out. Although we have been unable to measure pool S.A. directly, Berry (1969), who has obtained similar results, has found that under similar circumstances no increase in TCA soluble nucleotide pool radioactivity occurs. Other experiments we have done indicate that the nucleotide pool is apparently exceedingly small, so that its specific activity is probably very similar to that of the external medium which is common to both stimulated and unstimulated neurons.

With these reservations in mind there are two obvious questions which arise from the results. What is the adequate stimulus for increased transcription — EPSP, postsynaptic spike, or both? The second question is, what kind of RNA is being transcribed as a result of stimulation? To the first question we have a rather clear answer. To the second we have several partial answers.

In order to determine whether it is the postsynaptic spike or the subthreshold EPSP which is responsible for the increased

incorporation, we first analyzed the incorporation as a function of the number of spikes produced. In our hands, clearly, there was no correlation. We then drove R2 by passing current directly through the recording microelectrode, producing spikes but no EPSP's (see Figure 8). This produced no significant increase in uridine incorporation and suggested that the previously observed increase was due to EPSP's. Finally we did the converse experiment and stimulated at an intensity which was adjusted to produce EPSP's subthreshold to the spike firing level. This stimulus, of predominantly EPSP's, produced that previously observed increase in incorporation into RNA (Kernell & Peterson, 1970) (see Figure 9). At this point we could say that the augmented

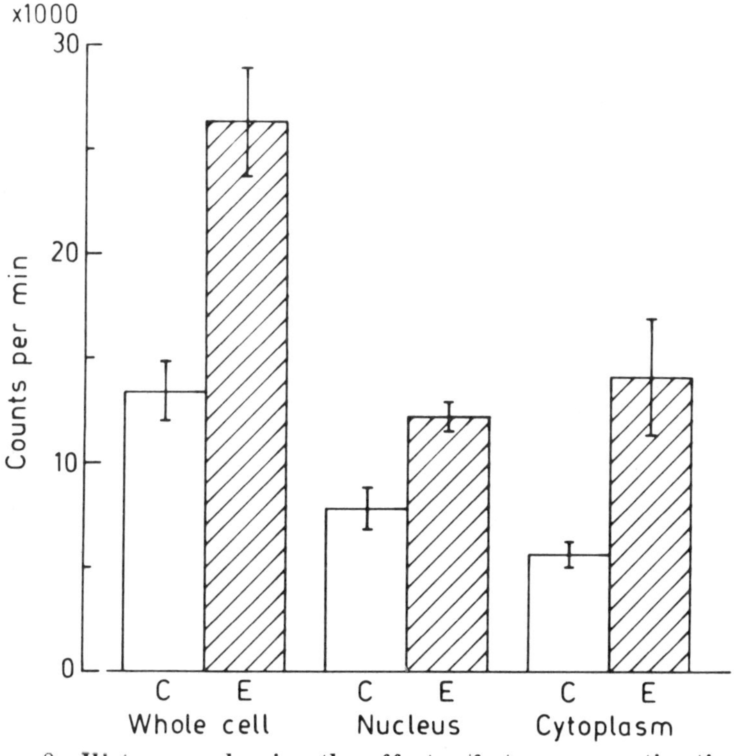


Figure 8. Histogram showing the effects of strong synaptic stimulation on the incorporation of labelled nucleosides (counts per min.) into the RNA of the whole cell R2, the nucleus, and the cytoplasm respectively. Separate mean values are shown for experimental cells (E) and the corresponding control cells (C). Vertical bars give the S. E. of the respective mean values. Each column represents the mean value of samples from eleven cells (from Peterson and Kernell, 1970).

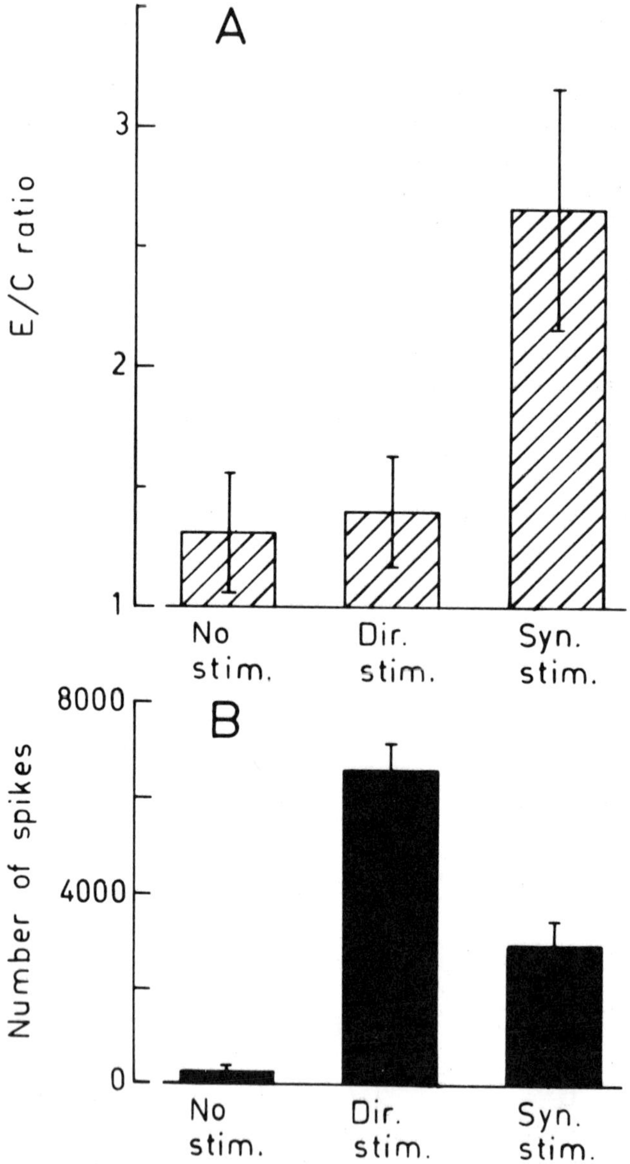


Figure 9. Histograms showing (A) the relative amount of RNA labelling (E/C ratio) and (B) the total number of spikes in experimental cells penetrated by an intracellular microelectrode and subjected to strong synaptic stimulation (Syn. Stim.; n = 4), direct electrical stimulation (Dir. stim.; n = 6), and no stimulation at all (No stim.; n = 5). Columns show the mean values for the respective groups, and vertical bars give the S. E. of the respective mean values (from Peterson and Kernell, 1970).

local synaptic activity appears to enhance DNA transcription. Furthermore the absence of a response to direct electrical activation would argue against the notion that the effect of EPSP's is mediated by the non-propagated electrical activity associated with it, and is likely a result of the release of transmitters and/or trophic factors released from the synaptic terminal. However we have no direct evidence to this point.

One approach to an indirect line of evidence is as follows. If the effect is chemically mediated, since the synapses are located several hundred microns away from the cell body, one would expect a delay between the onset of synaptic stimulation and the nucleic acid response. This delay would be necessary for intracellular messengers to get from subsynaptic locations to the nucleus either by diffusion or active transportation processes. On the other hand a response to the electrical event which invades the body might be more immediate.

This point was checked in collaboration with Dr. Sol Erulkar

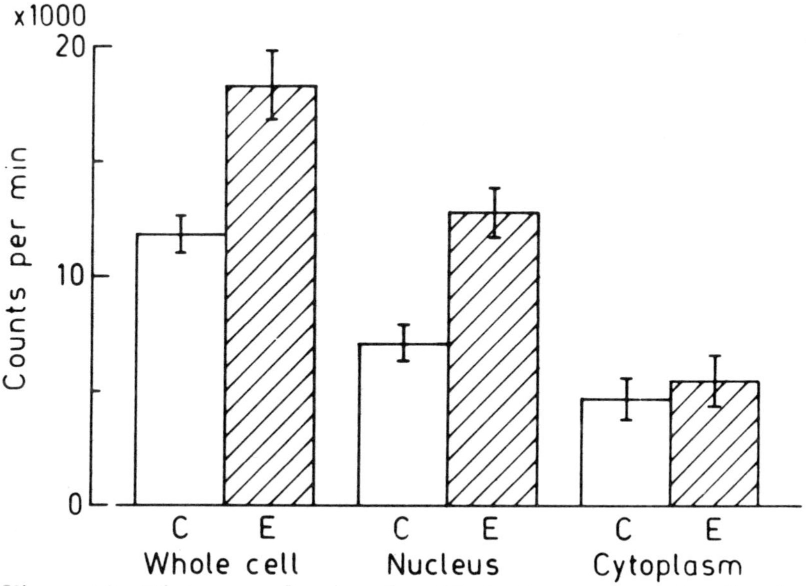


Figure 10. Histogram showing the effect of weak synaptic stimulation on the incorporation of labelled nucleosides (counts per min.) into the RNA of the whole cell R2, the nucleus, and the cytoplasm respectively. Separate mean values are shown for the experimental cells (E) and the corresponding control cells (C). Vertical bars give the S. E. of the respective mean values. Each column represents the mean value of samples from five cells (from Kernell and Peterson, 1970).

by taking a closer look at the time course of the increased incorporation accompanying synaptic stimulation. To our surprise, this showed a significant *decrease* in incorporation during the initial 10 minutes of stimulation, followed by a return to normal at about 20 minutes and then a linear increase out to 90 minutes (see Figure 10). This would indicate a delay in the increased incorporation, and again indirectly argue for a chemical effector.

In summary, the adequate stimulus for enhanced incorporation appears to be the local EPSP, and is probably chemically mediated.

The second question I posed earlier was what type of RNA incorporated the increased label during stimulation? Is this a general metabolic increase going into all species of RNA equally, or are certain selected portions of the genome being transcribed at an increased rate?

The first approach to this was to compare labeling patterns of RNA from stimulated and unstimulated neurons on the basis of molecular weight by gel electrophoresis (Peterson & Kernell, 1970). One of these is shown in Figure 11. Inspection of a number of them suggested that rather than an across the board increase, the main enhancement of labeling was occurring in the middle to higher M.W. regions. However, since this is the area of pre-ribosomal and rRNA, it was unclear whether the increase was in rRNA or non-rRNA. If the increase was in non-rRNA this might be of significance because of the possibility that stimulation was activating transcription in parts of the genome not normally transcribed. Although it was clear by this time that there were no large "stimulation-RNA" peaks it was felt that the only strategy which might answer this question would be competitive RNA-DNA hybridization. However before undertaking what seemed to be a rather foreboding course, i.e., competitive hybridization with all the necessary controls and microscaling, it seemed reasonable to try one other experiment which might cast some light on the ribosomal or non-ribosomal nature of the increase. This involved measuring the degree of methylation of the RNA synthesized during stimulation. The logic of this experiment was based on our previous finding (Peterson, 1970) that, as in most other cells, the vast bulk of labeled methyl groups transferred from labeled methyl-methionine into RNA went into rRNA and tRNA. Therefore if the degree of methylation during stimulation equaled or exceeded the degree of in-

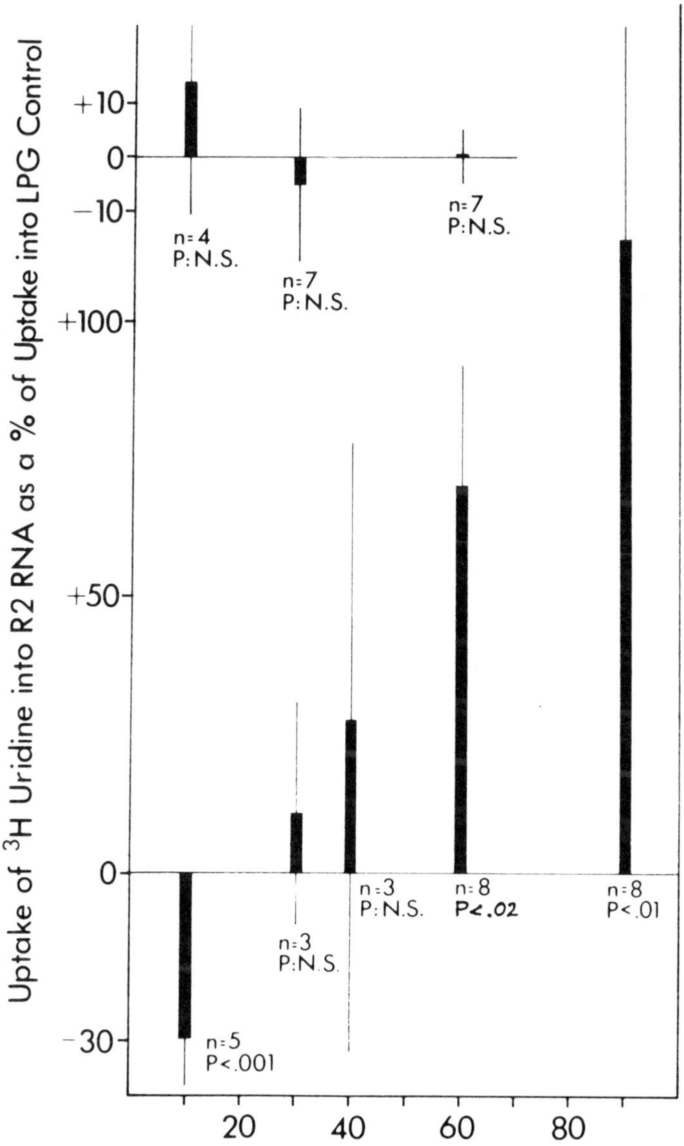


Figure 11. Abdominal and left pleural ganglia were incubated in sea-water containing $^3$H-uridine for the times indicated. R2 was stimulated with single pulses (1/1.5 sec.) for the same times. Acid precipitable counts incorporated into R2 RNA are presented as a percentage of those incorporated into the LPG control neuron. The top graph indicates results obtained when R2 was penetrated by the recording microelectrode but *not* stimulated. The bars indicate one standard error of the mean; N. S. indicates not significant.

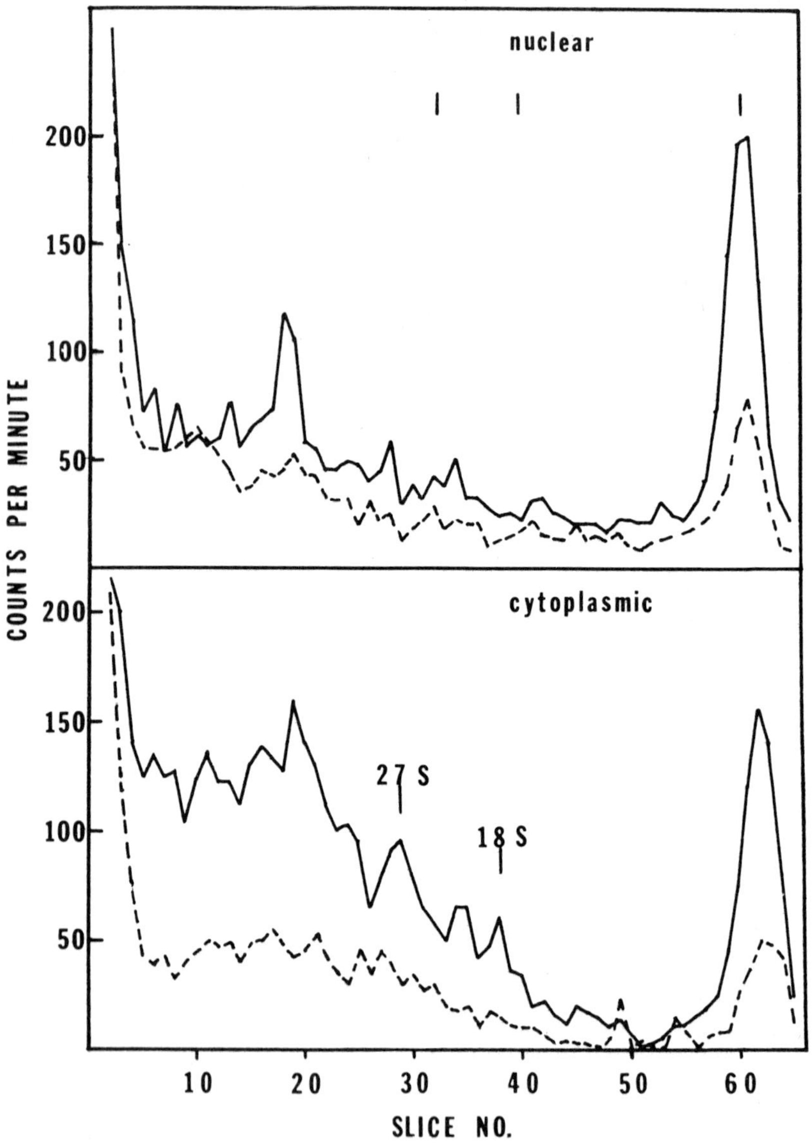


Figure 12. Diagram showing the electrophoretic mobilities of nuclear and cytoplasmic labelled RNA from a synaptically stimulated R2 cell (solid lines) and its control (interrupted lines). Incubation time: 86 minutes from the onset of synaptic stimulation. Unmarked vertical lines in upper graph indicate, from left to right, the positions of 23 S, 16 S and 4 S *E. coli* carrier RNA. The positions of ribosomal RNA of cell R2 in Aplysia (27 S and 18 S) are indicated in lower graph (from Peterson and Kernell, 1970).

creased incorporation, there would be little reason to believe that the increased incorporation was related to non-rRNA. On the other hand, if the methylation was less than the increased uridine incorporation, possibly the increased uridine label was going into non-rRNA. Thus, R2 and its control LPG were incubated for 60 minutes in $C^{14}$ uridine and $^3H$ methyl-methionine, during which time R2 was synaptically driven. Following this the RNA and protein were extracted by a variation of the Hess and Thalheimer PCA procedure, RNA estimated spectrophotometrically, and the RNA and protein counted for both labels. Table 2 shows the results of this experiment.

These results indicate that the RNA made during stimulation was significantly less methylated than when unstimulated. In fact, if one estimates the methylation occurring in the additional RNA synthesized, this represents a decrease of 30%. This experiment also showed a marked increase in incorporation of methionine into protein — however with no data on pool specific activities, I interpret this with great caution.

**Table 2**

| | *E* | *C* |
|---|---|---|
| RNA (μg/cell) | 1.59 ± .08<br>(13) | 1.40 ± .05<br>(8)<br>3.2 ± .1<br>(4) |
| Moles methionine (x $10^{-15}$) incorporated into protein/cell | 415 ± 57<br>(13) | 89 ± 17<br>(13) |
| Moles uridine (x $10^{-15}$) incorporated into RNA/μg RNA | 33 ± 5<br>(11) | 13 ± .9<br>(11) |
| | P < .00001 | |
| Methyl incorporation into RNA as a percentage of uridine incorporation | 22.6 ± 2.5<br>(10) | 29.6 ± 4.0<br>(10) |
| | P < .02 | |

± = S.E.M.
( ) = n
60 minutes incubation plus stimulation (37/min) in 4.5 x $10^{-5}$ M methyl-T-methionine and 1.6 x $10^{-5}$ M $^{14}C$ uridine.

## Conclusions

If I may I would now like to return to my opening remarks. There I asked, does the stimulus, to induce a plastic neuronal response, intervene in that neuron's metabolism at transcriptional, translational or purely membrane levels?

The experiments I have described strongly suggest a response at the transcriptional level. Unless we assume that the neuron transcribes its genome merely for exercise, this will almost surely result in a response at the translational level, and shifted patterns of protein synthesis. Could these changes provide the complete response? Undoubtedly not, because if these processes are to express themselves by modifying synaptic efficacy (synaptic pathway) then they must modify the synaptic membrane itself. Therefore it seems likely that the synaptic stimulus will intervene at all these levels, probably making investigation of the problem at least three times as complicated.

## Acknowledgments

I would like to acknowledge the kind hospitality of Professor Jan-Erik Edström in whose laboratory this work began and where it was supported in part by the Swedish Medical Research Council and the U.S.PHS grant no. NB 1871. It has also been supported by PHS grants MH 15767, NS 08759, and NS 02941.

## Literature Cited

Arvanitaki, A., and S.-H. Tchou. 1942. Les Rois de la croissance relative individuelle des cellules nerveuses chez l'*Aplysia*. Bull. Histol. Tech. Microscop. 19:244-256.

Arvanitaki, A., and N. Chalazonitas. 1949. Reactions bioelectriques neuroniques à la photoactivation specifiques d'une heme-protein et d'une corotene-protein. Arch. Sci. Physiol. 3:27-44.

Berry, R.W. 1969. Ribonucleic acid metabolism of a single neuron: Correlation with electrical activity. Science 166:1021.

Frazier, W.T., E.R. Kandel, I. Kupfermann, R. Waziri, and R.E. Coggeshall. 1967. Morphological and functional properties of identified neurons in the abdominal ganglion of *Aplysia californica*. J. Neurophysiol. 30:1288.

Giller, E.G., and J.H. Schwartz. 1968. Choline acetyltransferase: Regional distribution in the abdominal ganglion of *Aplysia*. Science 161:908-911.

Giller, E., and J.H. Schwartz. 1971a. Choline acetyltransferase in identified neurons of abdominal ganglion of *Aplysia californica*. J. Neurophysiol. 34:93.

Giller, E., and J.H. Schwartz. 1971b. Acetylcholinesterase in identified neurons of abdominal ganglion of *Aplysia californica*. J. Neurophysiol. 34:108.

Hughes, G.M. 1967. The left and right giant neurons of *Aplysia*, p. 423. *In* J. Salaniki [ed.] Symposium on neurobiology of invertebrates. Plenum, New York.

Kandel, E.R., and L. Tauc. 1965. Heterosynaptic facilitation in neurons of the abdominal ganglion of *Aplysia depilans*. J. Physiol. 181:1-27.

Kernell, D., and R.P. Peterson. 1970. The effect of spike activity versus synaptic activation on the metabolism of ribonucleic acid in a molluscan giant neurone. J. Neurochem. 17:1087-1094.

Lasek [personal communication, 1970]

Peterson, R.P. 1970. RNA in single identified neurons of *Aplysia*. J. Neurochem. 17:325-338.

Peterson, R.P. [unpublished observations, 1971]

Peterson, R.P., and D. Kernell. 1970. Effects of nerve stimulation on the metabolism of ribonucleic acid on a giant molluscan neurone. J. Neurochem. 17:1075-1085.

Strumwasser, F. 1965. The demonstration and manipulation of a circadian rhythm in a single neuron, p. 442-462. *In* J. Aschoff [ed.] Circadian clocks. North-Holland, Amsterdam.

Wilson, D.L. 1971. Molecular weight distribution of proteins synthesized in single identified neurons of *Aplysia*. J. Gen. Physiol. 57:26-40.

## Discussion

*Dr. Kates:* Well, I think this is a paper that we will be discussing here for some time to come. I thought it was a very clear presentation of very elegant experiments, and here is why I think they are elegant. First of all, they are experiments that were done on single cells whose stimulation is entirely controlled. Chemical experiments on these cells are thereby not only experiments on the single cell but involve sub-cellular fractionation. They are the kind of experiments that, when I came to this symposium, I thought could only be done if you had neurons growing in tissue culture, which is the subject of my talk tomorrow. It modified my viewpoint somewhat. So I think they are really very interesting and novel studies in this respect. Now, some things that were of immediate interest to me: There are people who study RNA synthesis in whole sections of brain, involving many cells. After stimulation, after learning experiences, they find the same order of increases that you observed in a single cell. There has always been some doubt about whether the reason the in-

creases were on the order of a hundred percent was just because of an inability to focus on the relevant cell. It seems that maybe it is not — maybe in a mammalian brain all cells, or many cells are participating by increasing a relatively small amount. I don't mean to say that your differences are insignificant.

*Dr. Peterson:* There is one point there. There are papers dealing with changes in RNA concentration, for something — per microgram protein, or per milligram wet weight — and there are whole series of experiments dealing on changes in uptake of radioactivity. But the interesting thing is that in some papers they were reporting doubling of RNA concentrations, whereas in these experiments we are getting an increase of at most a tenth of a percent of the actual amount of RNA, so I have never been really able to reconcile these.

*Dr. Kates:* Well, I don't see that the cell has enough . . . I don't know about these experiments where the RNA doubles in actual amount. It would seem to me that this would take quite a bit of time.

*Dr. Peterson:* It would to me too. Fast doubling of the RNA would be distressful.

*Dr. Kates:* Another point I want to make next: These studies of the single cells I think are going to be influencing the viewpoint of these other studies done with whole sections of tissues. In looking at the differences, the qualitative differences, in your RNA made after stimulation you used several techniques. One was a size fractionation. You couldn't pick up any difference in any of the hybridization procedures. There are a couple of reasons why you might not expect to see much difference, which I think you are probably aware of. One is that you are looking primarily at a class of redundant RNAs, RNAs that occur very often, and one might not necessarily expect these RNAs to differ much from one cell to another. I interpreted your data as not showing much of a difference, qualitatively.

*Dr. Peterson:* The hybridization, you mean?

*Dr. Kates:* Yes. You interpreted it as showing a qualitative difference. The difference in the two plateaus with all these experiments is significant. You are looking at raw data except for the counting error, which, I think, works out to be twice the width of the lines. You made the point about this being predominantly highly redundant material. It doesn't really say that you could not turn on a highly redundant gene.

*Dr. Peterson:* Alright, there is another way. Since nobody knows really what redundant RNA represents, it may represent non-information RNA; this is a possibility that is still open. If it didn't, then the informational RNA would be buried in there as a small percentage of this non-informational redundant RNA, and you might in fact be looking at the protein that is synthesized. Then you are looking just at the product of the informational RNA.

*Dr. Kates:* The question is, might you not magnify the differences by going now to the protein? That is just a suggestion.

*Dr. Peterson:* This is what Dave Wilson at Caltech has done. He does find that you can, almost. He takes proteins from these single neurons and puts them on an SDS gel which separates them out pretty much by molecular weight. But you can just about fingerprint these neurons on the basis of the protein distribution. You can shuffle these up and show him one and he'll tell you which neuron it is.

*Dr. Kates:* You said the proteins which they synthesize?

*Dr. Peterson:* Yes.

*Dr. Kates:* That is fantastic.

*Dr. Peterson:* It's phenomenal. You could guess that neurons that do different things would have different proteins, but it is the first time I know of that anybody has demonstrated it. Pacemakers have one protein. R15, which is a pacemaker cell, has one. R2 has another one, and R14 has another. They are wonderful. The other comment I would make on the question of the RNA, and frankly I haven't figured out if it makes the situation worse or better, is this: In hybridization . . . remember, there are in that one cell 137,000 copies of the genome, with the complete genome to start with . . . so if I turn on one gene . . . if I turn it on all at once, I get . . .

*Dr. Kates:* Yes. You have the equivalent of $10^5$ HeLa cells, or mammalian neurons, in that one cell. That's certainly, in terms of mass . . . I can see why you can do these types of experiments.

*Dr. Peterson:* Perhaps the only really genetic component of this talk is this wild polyploidy in this cell. If someone can figure out why this neuron needs to be so polyploid, I would love to know. It might be selective polyploidy. In other words, only some genes might be. There is only about one-half percent of

satellite DNA visible by cesium chloride centrifugation. There is not a satellite type of magnification?

*Dr. Kates:* Right. But you might not be able to actually fractionate them on the basis of density. The genes have the same density as the bulk of the DNA.

*Dr. Peterson:* Well, if they did it would be peculiar because on the basis of melting curves, the GC content is very peculiar for these cells. It is about 31%.

*Dr. Kates:* It is different from other cells from the same organism?

*Dr. Peterson:* No. It is very low.

*Dr. Kates:* Maybe we should have questions.

*Dr. Gamow:* I believe Dave Wilson suggests that he would like to look at different protein amounts as the circadian rhythm goes. Now do the cells that you are working with have this circadian rhythm?

*Dr. Peterson:* No. But I'd love to see him do that. All the cells that have pacemaker activities — particularly R15 which has the circadian rhythm — have the 12,000 molecular weight protein. Cells that are not pacemakers don't. But it would be fascinating to see if that one peak showed a circadian fluctuation. I mentioned in the paper that Tom Berry had done similar work at the same time we were doing this. He had stimulated R2 and also found an increase in RNA incorporation. I didn't know until Dr. Kates just told me that he had also found the same thing, when he separated it into EPSP *versus* spike. In his hands apparently also the EPSP does it and the spike does not. You mentioned Dave Wilson. He has been unable to confirm these resuits. But I think I know why, and if I might take just a minute, it is an interesting point. I know you are supposed to be the discussant.

*Dr. Kates:* No, go ahead.

*Dr. Peterson:* Again, another mistake we made was when we decided to get very scientific about this thing. If this is the ganglion, it's got these nerves coming off and R2 is sitting here. Previously, what we had done was to stimulate this nerve until it pooped out and then we went to this one. After 5-10 minutes it would drop out, and then this one. Then we came back and just stimulated. We were just trying to keep the response going.

We decided that this was very sloppy. What we should do is to stimulate just one nerve in this experiment. Maybe then we could compare the effect of this input with this input, etc. When we did that, we got absolutely no change in RNA synthesis. This varied though. We felt something terrible must be happening. Well, about this time a couple of interesting things happened. One was perhaps questionable. A paper by Kirka came out describing RNA changes, changes in RNA synthesis, in the cockroach with learning. It was the leg dipping in the water routine. There he emphasized that the animal which could learn, the cockroach that could learn, had a much greater incorporation than the cockroach that wasn't learning, the yoked control. But the one that wasn't learning also had an increase over the one that was just resting. He made the point that it appeared to be the timed input, the changing pattern of input, which seemed to be important. The next day Frank Morrell, who has worked for many years with these epileptiform foci, came by and described these fantastic experiments. If you implant an electrode in the hippocampus of a cat and put your high frequency stimulus through this stimulating electrode for 2 seconds every 24 hours — just 2 seconds every day — on about the 10th day the animal starts to display seizure activity on the EEG. By the 15th day, he is showing full-blown seizure responses to the stimulus. Then you can drop the stimulus or eliminate it or just give it a very small stimulus for maybe a second and he will still give a full-blown seizure. The interesting thing was that if you stimulated the cat all day long, he never developed the seizure response. Or if you stimulated the cat for half an hour he would never develop the seizure response. It had to be this very short time period. But all of these began arguing . . . it almost reminds me of the discussion about the rabbit who pushes on the random button when he is trying to get away from the fox . . . but there is something about the quality of the stimulus for this response that has to be varying — you just can't keep pounding away on the same thing. This kind of sloppy thing we were doing kept presenting new and different products of stimuli in the cell. Wherever we have tried this since, just keeping a nice scientific input from any of these nerves, we get no response. To go back to your question which started this, along with Dave Wilson . . . He had the same approach to this that we did, on second thought. You know he was going to do it scientifically, clearly. I think that

is why he has been unable to prove it thus far. I think he is going to go back to doing it sloppily like we did.

*Dr. Best:* When you say, stimulating the cat all day long, you mean in the hippocampus?

*Dr. Peterson:* Yes.

*Dr. Best:* Well, hippocampus isn't just any old neuron. It's part of that limbic system.

*Dr. Kates:* I was wondering if there is an alternative interpretation. First of all, what is the likelihood that you are doing a monosynaptic reflex trick here, where there is no other neuron interposed between where you are stimulating for the orthodromic input and the giant cell. Is there any likelihood that there is another neuron interposed?

*Dr. Peterson:* I have no idea.

*Dr. Kates:* It seems to me, if I recall correctly, that these are probably direct synaptic inputs. I wonder if some of the inputs don't have greater efficacy than others, rather than being random, because it also occurs to me that some inputs to a neuron are far more hard-wired, if you will, than others. If you tried to play games with certain synaptic interconnections, some of them are really, really fixed, but if you play games with others they habituate or show decrements or increments in the response, I wonder if in this you haven't located one that is doing this trick.

*Dr. Peterson:* I would have thought that we would have picked it up, when we did each one individually.

*Dr. Kates:* You did each one individually and that didn't work?

*Dr. Peterson:* Yes.

*Dr. Kates:* Then the next problem is, it would appear that the sign of electrical response might not be of any significance to this cell, so that one of the things I would like to ask if you have on your agenda, would be the following: Either stimulate directly one of these cells — I don't remember which one it is, it has a weak inhibitory input — and then hyperpolarize that cell to a reverse potential, and then also leave it at normal potential so you can get an IPSP instead of EPSP. Then conversely, depolarize this cell so that these inputs now become IPSP's. In other words, I would like to bet that the sign has nothing to do with this. That it is the gates.

*Dr. Peterson:* I would say that too. Joe Schwartz, in Eric Kandel's lab, has spent a good deal of time incubating the ganglia in labeled uridine with acetylcholine which hyperpolarizes R2, and then analyzing it. They were not recording simultaneously, so they really didn't have a monitor on the effects of acetylcholine, but what they found after doing a number of these was mainly aggravation. Joe said that they had the impression that they were getting an increase in incorporation in RNA, but the results were simply too variable.

*Dr. Kates:* Do you bathe the cells in acetylcholine?

*Dr. Peterson:* No, all the gates are just wide open. The way you do it, there is work; the gates close and open, open and close, and things are happening, as opposed to a steady state phenomenon, so I wouldn't be surprised at that result.

*Dr. Sharpless:* Do I understand that this ganglion is in a bath when you do it? The axon of the R2 cell as well?

*Dr. Peterson:* The axon of both the control and the experimental cell is cut. This is known to induce changes in RNA synthesis. This part of chromatolysis is very old.

*Dr. Sharpless:* Suppose it were intact and you were stimulating that cell electrically, rather than synaptically. Would you expect to get a different result?

*Dr. Peterson:* I would not and I'll tell you why. As you may know, there is quite a bit of literature by people who have stimulated and looked for changes in RNA, going back to the early 1940's. We started going through this literature and fortunately after about the fifth or sixth paper we found a review of this literature by a Russian named Pevsnor. We have reviewed all of these papers and scored those that found an increase in RNA and those that found a decrease and those that found no change. I think he had some 35 references on this, and it was very interesting. All those that increased, as near as you could tell from the original paper, were synaptic activation. All those with no change were antidromic activation. People had felt, well I can do it cleaner if I stimulated antidromically. So to get at the motoneurons — increases in motoneurons — in a classical way, they stimulated the dorsal root and found increases in RNA and motoneurons. Then a few years later to try and confirm this, Lew and Chambers came along and they thought they would do it a little better. They would stimulate antidromically. . . .

*Dr. Sharpless:* But did they stimulate the cut ventral root?

*Dr. Peterson:* I think the intact ventral root — in fact, I think they did it both ways, come to think of it. They found no change. So that seems to be the history of the thing. Once we had this clue to go on, one could explain all of these seeming paradoxes. The only problem one really can't explain is described in a couple of papers where with stimulation you get a decrease in RNA. But all of these, that I know of at least, are brain slices — macrobrain type of experiments, with brain slices mounted on electrodes, where the stimulation is a little bit more subtle than sticking it into a 100 volt house current. But invariably both protein and RNA go down in most cases.

*Dr. Sharpless:* Do you think the change in RNA might be due to or correlated with the amount of change produced in synaptic activation? That is to say, you stimulate one nerve and you get some form of fatigue, and adaptation stops working. You stimulate another, and you get some. Is it possible that the more change you produce in the synaptic endings converging on that cell or the sub-synaptic regions, the more RNA would be found? Do you understand my point?

*Dr. Peterson:* The epitome might be to just flood the cell with the transmitter. Then you should get the maximum . . .

*Dr. Sharpless:* No, I am asking whether the incorporation of RNA is correlated with the change of synaptic function rather than simply with stimulation.

*Dr Peterson:* Oh. I think in a way that is what this business of rotating around among nerves says, and that is what I think that wild experiment of Frank Morrell's says — that it is two seconds a day.

*Dr. Sharpless:* I might have another explanation.

*Question:* What happens to the RNA, when you stimulate it with the electrical stimulation, after you remove the impulse?

*Dr. Peterson:* You mean where does it go?

*Question:* Are you taking total counts?

*Dr. Peterson:* In other words, incubate and stimulate for an hour and then turn off the stimulation and let it incubate for 4 or 5 more hours?

*Question:* Yes, and you remove the label or add actinomycin D or . . .

*Dr. Peterson:* Yes. We haven't done it. I'm sorry.

*Dr. Kates:* I'd like to ask you a question about the major question that you posed — does the genome participate in plasticity — and how your studies might relate to that. What is the fastest possible plasticity that one can get a neuron to undergo in this system?

*Dr. Best:* Can I ask you a question that may allow you to answer them both at the same time? Is there anything that you can do that significantly changes the number of synapses?

*Dr. Peterson:* I know of nothing to answer your question first. To answer your question it depends on what you call plasticity. That's the hooker.

*Dr. Kates:* The one that happens the fastest.

*Dr. Peterson:* Probably the one that happens fastest is what Kandel has described as heterosynaptic facilitation.

*Dr. Kates:* How long does that take about?

*Dr. Peterson:* It takes about 10 seconds.

*Dr. Kates:* That's not RNA and protein synthesis?

*Dr. Peterson:* Right. For the plastic effects, which probably most of us right now are interested in, such as this heterosynaptic facilitation, it can be described, I gather from psychologist friends, as a classic Pavlovian conditioning of this neuron. But it only takes 10 seconds to get it, and it lasts for about an hour or an hour and a half, so it's got a long duration. But as you say, it's hard to imagine RNA involved in that.

*Dr. Best:* Now do you mean 10 seconds of treatment to get it, or does it have a development time?

*Dr. Peterson:* Ten seconds of pairing a conditioned and unconditioned stimulus.

*Dr. Best:* But then if you test later, does it have a development time? That is, is there anything like a long term and short term component to it?

*Dr. Peterson:* No. Not that I know of.

*Dr. Sharpless:* This experiment seems to me to concern the other kind of plasticity that Kandel talks about. Habituate one nerve, habituate the other, and then habituate the third. My question is, could the amount of RNA incorporation — labeled

uridine into RNA — be a consequence of the amount of habituation of the cell that the total number of synapses on that cell has gone through rather than the amount of synaptic action?

*Dr. Peterson:* That's an interesting idea. In a sense, habituation would be the stimulus for the long term. . . .

*Dr. Sharpless:* It may be that the sum of the things that the cell is manufacturing is responsible for it. Habituation — some form of desensitization of some sort.

*Dr. Peterson:* I never thought of that; it's an excellent idea. We have assumed, with no honestly good reason, that this is transmitter depletion. As any neurophysiologist will tell you, there are at least 59 reasons why a synapse drops out — a whole variety of them. One thing that has made us think it is not post-synaptic, whatever it is, is that other nerves will immediately activate at the original level.

*Dr. Sharpless:* It could be pre-synaptic but still be a consequence of something produced by the post-synaptic cell, couldn't it?

*Dr. Peterson:* It's an interesting idea. One of the things that we are hoping to do is to try to find the excitatory transmitter. We've got some hunches, and a little bit of evidence that it may well be dopamine. If it is, then we can identify this ion by electrophoresis, and we may have a more reliable preparation than the synaptic stimulation. On the other hand, we may lose the whole effect. In a way, I hope we do because, as I said, I am kind of off on this tropic kick at the moment and maybe it will be something like that.

# The Use of Spider Web Measurements in the Study of Plasticity of Behavior*

PETER N. WITT
*North Carolina Department of Mental Health Division of Research, Raleigh*

THE ADULT female of the spider *Araneus diadematus* Cl., after mating, usually lays in autumn of each year between 100 and 200 fertilized eggs, surrounds the eggs with a yellowish cocoon spun with a special silk from six cylindrical glands, and dies several days to weeks later. Depending mainly on environmental conditions, but usually in the spring of the following year, about 100 spiderlings, small images of the adults (Figures 1 and 2), hatch. They soon emerge from the cocoon, letting out threads as they walk along; they form a silken sheet on which they can be watched climbing up and down; the sheet looks similar to the communal web of social spiders (Shear, 1970). A few weeks later, animals draw apart and build individual orb webs, small but perfect in design. From that day until the end of their life, about six months later, females construct an orb web nearly every morning in 20 to 30 minutes, thus producing altogether between 100 and 200 webs per individual.

* Much of the work reported here was carried out with the support of a grant from the National Science Foundation to Peter N. Witt.

Figure 1. This young female *Araneus diadematus* Cl. spider sits in in the hub of its orb web in front of a scale indicating 20 mm. The animal weighed 20.2 mg at the time the picture was taken and had emerged from the egg cocoon 2 months and 6 days before. Note abdomen on top and cephalothorax below, and 8 legs, holding the spider in its normal head-down waiting position.

Figure 2. This adult female *Araneus diadematus* Cl. spider sits in the same position as the animal in Figure 1. It weighed 369.8 mg at the time the picture was taken and had hatched 10 months and 8 days before. Scale as in Figure 1. Note similarity of the two animals in all but size.

It appears (Peters, 1969) that spiderlings, when taken out of the cocoon before the time at which they normally leave, build incomplete, irregular thread patterns. A gradual transition to the geometric orb could be traced by Peters through several stages. On the other hand, crowding of normally hatched spiderlings could postpone appearance of the first orb web for several days in our laboratory; whenever it appears, however, the early orb web built by a normally hatched young animal is, though the animal has had no contact with adult individuals of its species nor has it "seen" another orb, a miniature image of the adult web (Figures 3 and 4).

The characteristic pattern, size, regularity, and shape of the web are the results of a rapid sequence of movements with which the spider pulls the silk out of the spinnerets and fastens it to other threads or to the surface of surrounding structures, while

**Figure 3. This baby web was built by the same animal whose web appears in Table 1, line 1, on May 8, 1968, when it weighed 27.4 mg. Observe the relatively high number of radii, small mesh size and even spacing of threads.**

at the same time it continuously probes, turns, stretches and bends, with clock-like precision. The web structure — its shape, size, regularity and detail — is the subject of the following discussion. It will be tried as far as data permit to establish correlations between web measurements and the age, size, sex and other characteristics of the animals. One can assess how well the functions of the web (i.e. as a trap for food, a transmitter of information, and a support for the moving animal) are preserved under the influence of adverse inner and outer circumstances. From this it should be possible to tentatively speculate about the coding of the pattern in the central nervous system of the individual, how much of the code is transmitted from parents to offspring, and how different species preserve their different codes.

**Figure 4. On September 4, 1968, the same female *Araneus diadematus* Cl. as the one in Figure 3 weighed 158.2 mg and built this "adult web." Note the similarity in details of frame structure in webs built by the same animal many months apart.**

*The Pattern:* The basic orb web pattern is uniform throughout life. A dense little platform, the hub, forms the central part. From the hub a varying number of radii run — more or less evenly spaced — toward the periphery. Radii terminate on an irregular frame, which serves as the anchoring device of the web to surrounding structures. The spiral is laid after all scaffolding has been finished; it may, in some instances, come from another set of silk glands than the scaffolding thread. While the spiral produces the dense meshwork which traps the prey, the scaffolding thread serves as support for the running spider, as well as the transmitter of vibrations.

**Figure 5. The same spider as in Figures 3 and 4 built this "old-age web" on January 7 of the following year; it weighed 250.0 mg at that time. Note the relatively large, irregular meshes in a full sized web, an observation confirmed by the high figure for standard error of median mesh size, sample South in Table 1, on the right. Scale in the upper left corner is the same in all three pictures.**

When the baby spider leaves the dense communal silk sheet and builds its first individual web, the orb contains all components. The same is true for the large, fine-spun web of the adult female *Araneus diadematus*, and for the wide-meshed, erratic, last web of a spider at the end of its life (Figure 5).

There are many species of orb weaving spiders, and it is doubtful whether they should all be placed within the single family *Argiopidae* (Gertsch, 1949). Different species build the same pattern; some webs can be identified through additional species-specific structures like the stabilimentum or hatched band of *Argiope aurantia*, *argentata*, etc. (Figure 6), and others by reductions, like the free sector of *Zygiella*. Still others, however,

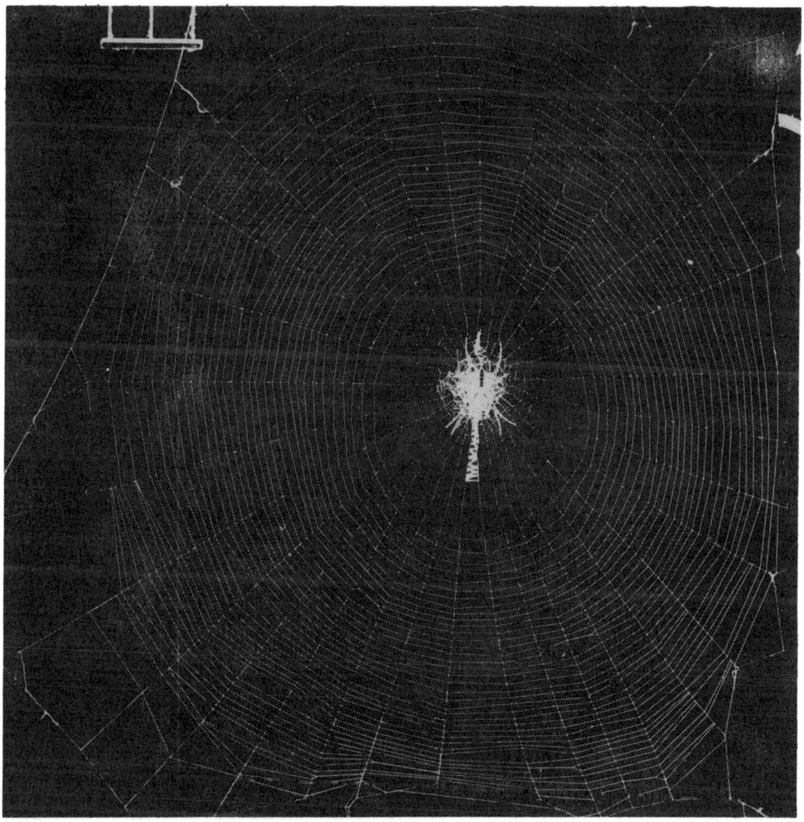

**Figure 6. This web was built by an adult female *Argiope aurantia* Lucas 10 months after hatching, weighing 165.2 mg at that time. Compare this web to the one in Figure 4; note the species' specific "hatched band" or "stabilimentum" built by *Argiope*, and the similarity of the pattern in webs built by different species.**

build webs which do not permit easy identification of the species of the builder. The pattern is always the same, even if the direction of the two-dimensional structure varies between vertical (i.e., *Araneus*) and horizontal (i.e., *Uloborus*, see Eberhard, 1969). The spiral can be sticky (*Araneus*, *Argiope*) or dry (*Uloborus*), but it consists always of circular and pendulum turns laid across radii.

Individual as well as species and age characteristics of a web lie not in the orb pattern. Severe disturbances of the builder through brain lesions, missing legs, or drugged states may distort webs, but do not affect the pattern. Given the anatomy and physiology of the builder, and its food requirements, no alternate solutions to construction seem to exist. There are sheet webs, triangular webs, and a number of other thread patterns; they are built by different spider families, and they deserve separate discussion. In my opinion there are no patterns of existent webs which can be regarded as a transition between these and the orb, or as the orb's immediate predecessor.

*The Size:* Web photography permits identification of size measures for any of the web components. Selected web measures are fed into the computer which, under the present program (see Witt et al., 1968) prints out the number of radii, the number of spiral turns in any of the four cardinal directions, the size of the three concentric areas (the center area, surrounded by the innermost spiral turn, the catching area, bounded by the innermost and outermost spiral turn, and the frame area, comprising the open spaces between outer spiral and irregular frame), and the total thread length.

Every one of the size measures can vary extensively, partially in correlation with the species, age, body size, and drug pretreatment of the animal, and partially in what we call "unpredictable" fashion, for lack of insight into possible causes. For example, the catching area of webs of one individual adult female comprised 542 $cm^2$ on August 13, 771 $cm^2$ on August 25, and 229 $cm^2$ on August 31, a decrease of 67% in one animal's web in one week.

There is, however, a definite pattern of web growth in each individual and species from early to adult web, followed by a slowing of the growth (*Neoscona vertebrata* McCook, Witt and Baum, 1960) or a steep decline in catching area (*Argiope aurantia* Lucas, Reed et al., 1969) (Figure 7). Witt and Baum (1960) have speculated that increase in web size during part of an indi-

vidual's life may be correlated to growth in leg length. It is interesting to note in Figure 7 that two groups of animals of one species reached significantly different web sizes at peak growth. This may be related to the phenomenon described in the appendix to this paper.

Given an individual adult *Araneus diadematus* spider, it has been shown that a 30% change in body weight is correlated with a statistically significant change in thread length. Christiansen

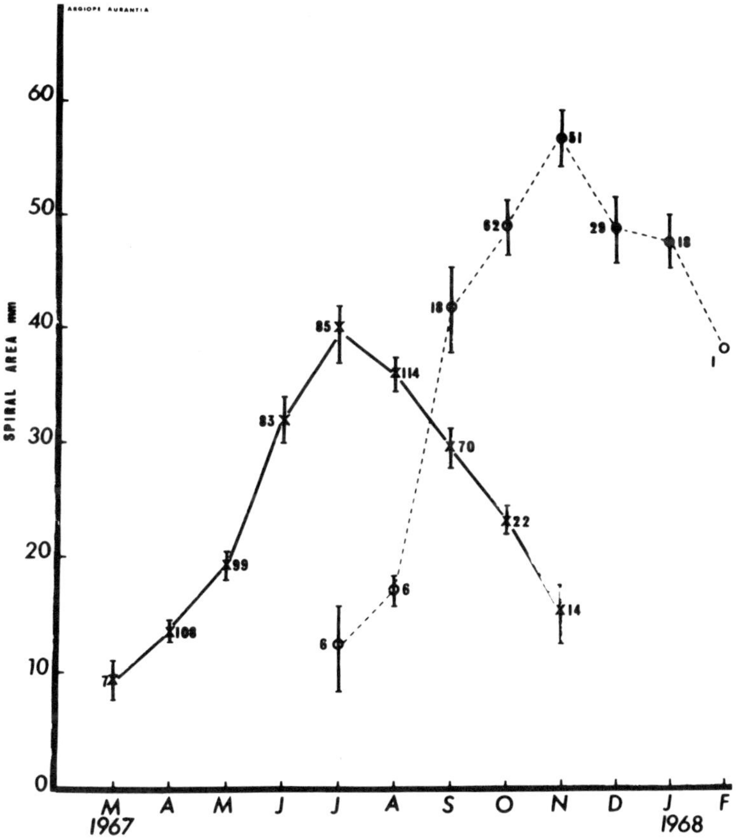


Figure 7. In this figure from Reed et al. (1949) spiral area means are plotted for all webs built by 8 littermate female *Argiope aurantia* Lucas raised in the laboratory (left) and animals caught outdoors (right) for each month. Vertical lines indicate standard errors. Small figures indicate the number of webs in each group. Peak sizes of catching zone coincide roughly to times of last moult and sexual maturation in each group; peak size is significantly higher for the "wild" group, even though they had been in the laboratory for more than 4 months.

et al. (1962) demonstrated in 15 spiders, which had built webs with a mean thread length of 15.95 meters (standard error ± 16.9 m) on one day, that increasing the animal's body weight by 30% resulted in 15 webs built the following day with 10.76 m average thread length (standard error ± 0.95 m). Control animals which had been treated with glue alone showed no significant change in thread length. And after the weight had been removed from the test animals, their webs were built again with longer threads.

The same authors hypothesized that the heavier spiders had used thicker thread to hold themselves on the web without danger of breakage. If no more silk than before had been available, the thicker thread had to be shorter. Actual measurements of thread thickness, calculated as $\mu$g protein in the web per meter thread, showed doubling of the quotient from 11.5 ± 2.1 $\mu$g/m in controls to 22.6 ± 5.4$\mu$g/m under additional weight. Such observation indicates that a spider empties its silk supply during web construction and adjusts size measures to available length of thread.

The latter assumption is supported by experiments which explore the effects of drugs which speed up and therefore probably increase (physostigmine, carbachol) or slow down (atropine, diisopropyl fluorophosphate = DFP) silk-protein synthesis in the ampullate glands (Peakall, 1964a). A relatively large silk supply 36 hours after 1 mg/kg physostigmine caused animals to build relatively large webs (Witt, 1963), while a supply which was small, 16 hours after 4 mg/kg atropine, resulted in small webs (Witt, 1962). Exceptions to such a relationship between silk supply and web size have been found when webs were built by spiders after 100 mg/kg of a drug like the tranquilizer diazepam (Valium): silk supply appeared normal, but the thread used for building was relatively short and thin. The animals had not emptied the ampullate glands at the end of web-building, possibly as a consequence of the drug's effect on the central nervous system (Reed and Witt, 1968).

Peters (1953) has shown that when one compares orb webs built by adult animals of different species, the size of the catching area and thread length and other web measures are characteristic for the species rather than for the size (or leg length) of the builder, e.g., the relatively small adult female of *Gasteracantha cancriformis* L. (Emerit, 1969), with a leg length of about 5 mm, builds an orb pattern nearly indistinguishable in size and detail from that of the adult female *Argiope argentata* Fabr. with a

leg length of more than 20 mm. Under these circumstances size measures appear to be specific for a species. All data on web-size indicate that the amount of silk in the gland, available at the time of construction, is a limiting factor. The web of *Araneus* contains at least three separate silk-components which come from three different sets of glands: 1. the scaffolding thread, which contributes most to the weight of the web, comes from the large ampullate glands; 2. the viscid thread of the sticky spiral is supplied totally or in part by the small aggregate glands; 3. the attachment discs come from the piriform glands (Peakall in Witt et al., 1968). The scaffolding thread has been analyzed most thoroughly, and the ampullate glands have been observed and their function has been measured *in vitro*. They have been found by Peakall (1964b) to be empty at the end of a web-construction period, to fill up a few hours afterwards, and to show signs of an increasingly rapid discharge of protein from the epithelium into the lumen followed by a high rate of uptake of amino acids into the epithelium after cholinergic treatment. Any analysis of web-size determinants has to take gland function into consideration.

*The Fine Structure:* A size measure of a different kind is the average mesh size of the web. It is a simple measure of a complex process which comprises the spacing of central angles during radius construction as well as the laying of the sticky spiral across radii. The relatively even spacing of radii requires animals to determine one part of this measure early in a radius construction period. Laying of the sticky spiral across radii is seen as a rhythmically executed sequence of movements, which is continuously monitored for proper spacing and successful execution. Experimental interferences with web construction through burning of threads during building (Reed, 1969) or exchanging spiders from one web to another (Peters, 1970) show that a spider, in laying thread, tries to realize a certain distance, guided by cues from existing structures.

Average mesh size is an expression for the density of the structure, and determines the minimum prey size which can be caught. Our computer print-out contains one figure for the mesh width of a web (equals: catching area divided by number of radii and spiral turns), and another figure for a median mesh size sample, separately taken for the upper and lower part of the web. A large web, for example, can be narrow-meshed when it is constructed with a relatively long thread, or the same catch-

ing area can be wide-meshed because a shorter thread divides it into larger rectangles. Average mesh size increases in a nearly linear fashion throughout the lifetime of a female spider, e.g., for *Argiope aurantia* Lucas from 20 $mm^2$ in March to 150 $mm^2$ in November (see Reed et al., 1969). It appears to be the most predictable of our measures, provided that species, age, etc. of the builder are known.

*The Shape of the Orb:* The web is neither a perfect circle, nor does the hub lie in the center. It is most frequently oval in the direction of gravity. But even horizontal webs are not circular (Eberhard, 1969). A spider which usually builds oval webs, like *Araneus diadematus*, when forced into a horizontal frame, builds a more circular web than before. This indicates gravity as one determinant of web shape (Mayer, 1953). Older animals (e.g., *Nephila*) have more catching area below than above the hub; this may be related to speed and ease of downward motion of the old animal in contrast to upward motion when catching prey. There appears to be no good analysis of these measures in the literature. The building plan changes according to age, a fact which could be explained as a consequence of changes in the central nervous system and/or the mobility of the legs in the aging animal..

Some drugs like strychnine and adrenochrome appear to have an effect on shape, causing rounder or longer webs (Witt, 1971). Nothing is known about an individual animal's preference for one shape over another.

*Regularity of Thread Spacing:* There are three different and partially independent ways in which we can measure in the web the ability of a spider to perceive and process cues from existing threads and to execute the appropriate movement for laying a new thread at the right place: 1. regularity of central angles between radii; 2. regularity of spiral spacing; 3. regularity of mesh size. The present computer program counts oversized angles, determines the median angle, calculates the standard deviation of central angles, determines relative (to web size) deviations of spiral turns for each of the four cardinal directions, and prints out the standard error of median mesh size separately for the upper and lower part of the web.

All these measures have been found to be sensitive to drug influence, and can be modified either toward increased or decreased regularity. Such change measures revert to control values as soon as the drug has left the body (see Witt, 1971). First

figures have just been obtained, and two examples are shown in Table 1, which indicate that regularity of thread spacing changes with the age of the animal. No evidence has as yet been found which would indicate the cause for such a change.

Other regularity measures might indicate behavioral mutants; they are not expressed in the present web measures. Witt described (1956) a spider, the webs of which were photographed

**Table 1**

| | 2 Month | 4 Month | 6 Month | 8 Month | 10 Month | 12 Month after hatching |
|---|---|---|---|---|---|---|
| Spider #7 | 1.40 | 2.37 | 5.00 | 2.61 | 4.00 | 12.40 |
| (N) se | (10) ±0.5 | (8) ±1.07 | (8) ±2.13 | (18) ±2.35 | (14) ±1.23 | (5) ±3.36 |
| Spider #26 | 0.87 | 3.20 | 3.52 | 5.90 | 6.25 | 14.33 |
| (N) se | (8) ±0.36 | (5) ±0.83 | (17) ±3.07 | (10) ±2.28 | (12) ±4.33 | (3) ±4.05 |

Table 1: Regularity of mesh size in the bottom part of webs of 2 *Araneus diadematus* Cl. Both animals hatched on March 20, 1968, from one cocoon and were raised in aluminum cages in the laboratory; new webs were photographed 2-3 times per week. Mean figures are calculated for all observed webs during alternate months of their lives; below means, in brackets, is the number of webs measured, followed by the standard error of the mean. Note the initially slow and later rapid increase in the irregularity of webs.

for more than three months, that showed an unusual and characteristic thread-free ring in the middle of the catching zone. Two other individuals, *Zygiella-x-notata*, built webs with different characteristic frame structures which were preserved even when the spiders were forced to build in mirror image. These and other observations indicate that a search for mutants expressed in web regularity measures is promising.

*In Summary:* We have found that construction of the elaborate, geometric orb web is achieved by each spider without direct instruction from its parents or other web-experienced individuals. The pattern is complete and perfectly executed with the first effort during an individual's life. The basic pattern never varies. Components of the pattern like size, shape, regularity, and fine structure, and added extras like the hatched band, permit, singly or in combination, identification of species, age, and sex of the builder. The human eye is in this respect superior to any of our measuring devices, but the computer gives some hope of combining measures in the proper way for complete evaluation of the

interaction of the components. A search for objective documentation of the individual or family web characteristic is underway, but results are only preliminary.

If we assume that the central nervous system of spiders is constructed and functions in a similar way to that of crayfish and insects, we can transfer some of the findings in the central nervous system from those animals to spiders. According to Wilson as interpreted by Kennedy, 1971, "locomotor and other behavioral rhythms are established by central connections, and do not depend upon feedback from peripheral sense organs for their timing. All the details of the temporal program are stored in connections between drivers or oscillators and motor neurons. Command fibers may select such blocks of output in different proportions, but primarily their role is permissive rather than instructive. It is thus appropriate to refer to such output as a "motor score." The genetic determinants of central nervous connections fix it, as surely and as precisely as the composer and arranger of a piece of music fix its structure. The question arises whether the existence of such specific command fibers would be sufficient to explain genetic components in web building behavior?

Though web-construction can be broken down into smaller patterns of movements which are repeated over and over, and may be guided by command fibers, the data on web-size suggest the existence of an overall plan. While the smaller patterns would account for construction of radius after radius and section after section of the spiral, the overall plan would determine the exact spacing, assuring completion of the structure with the available thread. Such a plan could be computed in and directed from a central nervous structure like the central body (Witt, 1969) based on data from genetically transmitted information (i.e. species-specific features) plus the amount of silk available and the feed-back from probing legs. Properties of the silk gland and its capacity to synthesize and store silk may play an essential part in the construction plan. Size and fine-structure of the web would be the end result of an interaction of all components.

The complexity of the pattern as a record of behavior — its regularity and predictability on the one hand, and its variability on the other — is the attraction as well as the shortcoming of web studies. The orb-web offers many opportunities for evaluation, but it presents such an abundance of measurable detail that it is not an easily analyzed model for the study of the genetics of behavior.

# Appendix: Preliminary Data on a Possible Genetic Component in Web Building*

*JOHN O. RAWLINGS, North Carolina State University, Raleigh,* and PETER N. WITT

A REWARDING approach to the study of behavior has been the utilization of the webs of orb-weaving spiders as a permanent record of a significant part of their behavior pattern. The preceding paper provides a review of some of this work as well as a general description of web structure and spider behavior with respect to web building. It is the purpose of this note to report on a preliminary effort to determine the degree of genetic control of the web building behavior of the spider, *Araneus diadematus* Cl.

The information utilized in this analysis is a sampling of web measurements for individual spiders from eight different families available in Dr. Witt's laboratory; all members of each family originated from the same cocoon and are presumably full sibs. The relationships among the eight families are unknown although the eight cocoons were obtained from a common source reputed to be a natural population of *Araneus diadematus* in the vicinity of the supplier's home. Due to the lack of mating

---

* Contribution from the Department of Statistics, North Carolina State University, and the Division of Research, North Carolina Department of Mental Health, Raleigh, North Carolina. Paper No. 3465 of the Journal Series of the North Carolina State University Agricultural Experiment Station, Raleigh, North Carolina. This work was supported in part by Public Health Service Grant GM 11546 from the Division of General Medical Sciences.

control in the production of the families used in this study and due to the fact that all members of the same family had of necessity the same cocoon environment, no attempt is made to relate the results to genetic components of variance; rather, the results are presented in terms of repeatability and the intra-class correlation of full sibs. The correlation of full sibs, in this case, is measuring the combined effects of common genetic components and the common environmental effects of being reared in the same cocoon. If certain assumptions were met, namely, 1) that all parents are unrelated and random members of a population, and 2) that there is no environmentl correlation among the full sibs of the same family, the correlation of full sibs would be equal to one half of the heritability (measured in the broad sense in that it includes nonadditive types of genetic variation).

The number of offspring per family varied from 1 to 29, the total number of spiders being 91. For each spider two webs spun approximately one week apart were measured; the webs being chosen, insofar as possible, from those spun at comparable physiological maturities of the spiders as measured by their body weights. (Body weight has been observed to have a marked effect on some web characteristics, see Witt 1971, and there appear to be very different rates of development even for sibs from the same cocoon.) In addition to this initial approximate standardization to a common body weight, all data were adjusted by covariance for the remaining spider weight differences if the tests showed a significant ($\alpha = .10$) association of the variation in spider weight with that of the web characteristic being analyzed.

The following definitions were utilized:

$\hat{\sigma}^2_f$ = the estimate of the component of variance among families.

$\hat{\sigma}^2_{s/f}$ = the estimate of the component of variance among full sibs within families.

$\hat{\sigma}^2_d$ = the estimate of the component of variance between duplicate webs from the same spider.

The repeatability of the web measurement is estimated as

$$r = 1 - (\text{proportion of total variance due to } \sigma^2_d)$$

$$= 1 - \frac{\hat{\sigma}^2_d}{\hat{\sigma}^2_d + \hat{\sigma}^2_{s/f} + \hat{\sigma}^2_f}$$

Repeatability measures to what degree a given spider at the chosen level of maturity could be expected to duplicate its web. Since the webs chosen for a particular spider were built only one week apart, this measure of repeatability cannot be expected to apply to random webs in the spiders' lifetime.

The full-sib correlation, assuming one web per spider, is to be estimated as

$$t = \frac{\hat{\sigma}_f^2}{\hat{\sigma}_d^2 + \hat{\sigma}_{s/f}^2 + \hat{\sigma}_f^2}$$

The limiting value of the full-sib correlation, approached as the number of webs measured per spider, gets large, and is given by

$$t^* = \frac{\hat{\sigma}_f^2}{\hat{\sigma}_{s/f}^2 + \hat{\sigma}_f^2}$$

As previously stated, if the necessary assumptions hold, twice the full-sib correlation is an estimate of heritability in the broad sense. Since the theoretical upper limit on heritability is unity, any tendency for the estimates of the full-sib correlation to exceed 0.5, other than that expected from sampling variation, suggests that there is some consanguinity between the mates and/or that there are positive environmentally induced correlations among full sibs from the same cocoon. The term "Environment" is used here in a broad sense to include any maternal influence transmitted through the cocoon, as well as the usual environmental effects.

It should be kept in mind that estimates of variance components are subject to large errors of estimation. This is particularly so for the component of variance among families. Consequently, the estimates obtained from this preliminary study should be regarded only as indications of the order of magnitude of the components or correlations.

Appendix Table 2 presents the estimates of the variance components, repeatability, and the full-sib correlations for twenty-seven web characteristics. For a detailed discussion of these web measures see Reed, Witt and Jones (1965) and the preceding paper. The repeatability measure reflects the consistency of an individual spider in the construction of its webs. A web characteristic for which a given spider produces exactly the same result in different webs would show a repeatability of 1.0. On

## Appendix Table 2

Estimates of Variance Components, Repeatability and Full-sib Correlation for Web Building Characteristics of *Araneus diadematus* Cl.

| Web Characteristics | Components of Variance | | | Repeatability | Full-sib correlation | |
|---|---|---|---|---|---|---|
| | $\hat{\sigma}^2_d$ | $\hat{\sigma}^2_{s/f}$ | $\hat{\sigma}^2_f$ | r | t | t* |
| Number of radii | 7.71 | 6.63** | 3.59** | .57 | .20 | .35 |
| Number of oversized angles | 2.42 | 0.19 | 0.31** | .17 | .11 | .62 |
| Median angle (°) | 1.91 | 1.82** | 0.31* | .53 | .08 | .14 |
| Angle regularity (°) | 1.78 | 1.00** | 0.69** | .49 | .20 | .41 |
| Width/length | 0.011 | 0.006* | 0.001* | .39 | .07 | .18 |
| Radius ratio N/S | 0.033 | 0.010* | 0.007** | .34 | .14 | .42 |
| Radius ratio E/W | 0.018 | 0.003 | 0.004** | .29 | .18 | .62 |
| Number of spirals W | 29.3 | 26.0** | 29.6** | .65 | .35 | .54 |
| Number of spirals N | 31.7 | 23.5** | 21.9** | .59 | .28 | .47 |
| Number of spirals E | 24.9 | 36.0** | 23.2** | .70 | .28 | .40 |
| Number of spirals S | 37.1 | 39.8** | 31.6** | .66 | .29 | .44 |
| St. dev. of center angles (°) | 1.55 | 0.84** | 0.62** | .48 | .20 | .43 |
| Frame area ($mm^2$) | $19.3 \times 10^6$ | $4.8 \times 10^6$* | $4.9 \times 10^6$** | .33 | .17 | .51 |
| Spiral area ($mm^2$) | $18.7 \times 10^7$ | $13.5 \times 10^7$** | $7.8 \times 10^7$** | .53 | .20 | .37 |
| Center area ($mm^2$) | $44.8 \times 10^4$ | $28.7 \times 10^4$** | $25.0 \times 10^4$** | .55 | .25 | .46 |
| Frame area/Spiral area | 0.014 | 0.003* | 0.00 | .17 | .00 | .00 |
| Center area/Spiral area | 0.0004 | 0.0003** | 0.0002** | .56 | .22 | .39 |
| Thread length (m) | $23.0 \times 10^6$ | $18.1 \times 10^6$** | $15.3 \times 10^6$** | .59 | .27 | .46 |
| Mesh width ($mm^2$) | 159.9 | 59.5** | 38.2** | .38 | .15 | .39 |
| Rel. dev. of spiral turns W | 0.0095 | 0.0032** | 0.0006 | .29 | .05 | .17 |
| Rel. dev. of spiral turns N | 0.0187 | 0.0056* | 0.0005 | .25 | .02 | .08 |
| Rel. dev. of spiral turns E | 0.0134 | 0.0008 | 0.0013** | .14 | .09 | .61 |
| Rel. dev. of spiral turns S | 0.0109 | 0.0041** | 0.00 | .27 | .00 | .00 |
| Median mesh size N ($mm^2$) | 285.4 | 59.4 | 51.6** | .28 | .13 | .46 |
| Median mesh size S ($mm^2$) | 184.0 | 87.6** | 43.2** | .42 | .14 | .33 |
| St. error of med. mesh size N ($mm^2$) | 0.0114 | 0.0026* | 0.0006 | .22 | .04 | .18 |
| St. error of med. mesh size S ($mm^2$) | 0.0005 | 0.0001 | 0.0002** | .37 | .25 | .68 |

*, **Component significantly different from zero at the .05 and .01 levels of probability, respectively.

the other hand a web characteristic for which a particular spider is no more likely to produce the same result on two trials than are two different spiders would show a repeatability of 0. A repeatability of zero would imply that there is no individual, and hence no genetic or family control, of the variation expressed for the particular trait.

The repeatabilities from this particular set of data varied from a low of 0.14 for "relative deviations of spiral turns on the east axis" to a high of 0.70 for the "number of spirals on the east axis of the web." As a class, all size measures — radii number, number of spirals in the four cardinal directions, frame area, spiral area, center area, and thread length — showed reasonably high repeatabilities, on the order of 0.5 or greater, with the "number of spiral" measurements being the largest.

The significance levels of the components of variance among families also apply to the full-sib intraclass correlations. Thus, a full-sib correlation of the order of 0.07 to 0.10 appears to be significantly different from zero. All of the traits which are primarily measures of size were significant, the full-sib correlations varying from $t = 0.17$ for "frame area" to $t = 0.28 - 0.35$ for "number of spirals" in the four cardinal directions. The measures of mesh size — mesh width and median mesh size sampled from the upper, N, and lower, S, part of the web — while having significant full-sib correlations of $t = 0.13$ to $t = 0.15$, were only about half as large as those for characteristics measuring web size. In terms of predictability for the individual spider (repeatability) and predictability of the family characteristics (full-sib correlation), the number of spiral turns in each of the four cardinal directions appears to be the most predictable characteristic.

The web characteristics for which there was little or no evidence of genetic control ($t < 0.1$) are "median angle," "width to length" ratio, "frame area to spiral area" ratio, "standard error of the median mesh size" and "relative deviations of the spiral turns" in the four directions. Several of these are measures of regularity of the web construction. There is some indication, however, that part of the regularity, or lack of regularity, of the spider behavior is under "family" control to some degree. The "number of oversized angles" shows a low but significant correlation of $t = 0.11$, the "standard deviation of central angles" and "angle regularity" both have significant correlations of $t = 0.19$, and the "standard error of the median mesh size" sampled from

the lower part of the web shows a reasonably high correlation of $t = .26$.

There is a strong suggestion from these results that there is some kind of individual and family control or "preference" for many of the web characteristics. It is impossible from these data to determine how much of this similarity in behavior of the members of the same family is due to their genetic relationship in contrast to that due to the commonness of the environments in which they were reared.

The correlations listed as $t^*$ are the values the full-sib correlations would take if the web characteristics for each individual spider were precisely evaluated in such a way that the $\sigma^2_d$ component of variance could be ignored in the variance of spider means. The proximity of many of these values to 0.5, and some to greater than 0.5, the theoretical upper limit on the full-sib correlation with the previously stated assumptions, suggests that in these cases either the differences among individual spider performance are almost entirely a reflection of genetic differences or that there is a sizeable "learning" component induced via the cocoon in which the full sibs were reared.

While it was never intended that these data would provide the necessary information for making definitive conclusions regarding the genetic control of web building characteristics, the suggestions of some degree of genetic control were deemed sufficient to warrant the presentation of this preliminary report and to provide encouragement for controlled genetic studies.

## Literature Cited

Christiansen, A., R. Baum, and P.N. Witt. 1962. Changes in spider webs brought about by mescaline, psilocybin and an increase in body weight. J. Pharmacol. 136:31-37.

Eberhard, W. 1969. Computer simulation of orb-web construction. Amer. Zool. 9:229-238.

Emerit, M. 1969. Contribution a l'étude des gasteracantes de Madagascar et des iles voisines. Thesis, Montpellier.

Gertsch, W.J. 1949. American spiders. D. van Nostrand, New York.

Kennedy, D. 1971. Nerve cells and behavior. Amer. Sci. 59:36-42.

Mayer, G. 1953. Untersuchungen über Hertstellung und Struktur des Radnetzes von *Aranae diadema* und *Zilla-x-notata* mit besonderer Berücksichtigung des Unterschieds von Jugend- und Altersnetzen. Z. Tierpsychol. 9:337-362.

Peakall, D.B. 1964a. Effects of cholinergic and anticholinergic drugs

on the synthesis of silk fibroins in spiders. Compar. Biochem. Physiol. 12:465-470.

Peakall, D.B. 1964b. Composition, function, and glandular origin of the silk fibroins of the spider *Araneus diadematus* Cl. J. Exp. Zool. 156:345-350.

Peters, H.M. 1953. Weitere Untersuchungen über den strukturellen Aufbau des Radnetzes der Spinnen. Z. Naturforsch. 8b:355-370.

Peters, N.M. 1969. Maturing and coordination of web-building activity. Amer. Zool. 9:223-228.

Peters, P.J. 1970. Orb web construction: Interaction of spider and thread configuration. Anim. Behav. 18:478-484.

Reed, C.F. 1969. Cues in the web-building process. Amer. Zool. 9:211-222.

Reed, C.F., P.N. Witt, and R.L. Jones. 1965. The measuring function of the first legs of *Araneus diadematus* Cl. Behaviour 25:89-119.

Reed, C.F., and P.N. Witt. 1968. Progressive disturbance of spider web geometry caused by two sedative drugs. Physiol. Behav. 3:119-124.

Reed, C.F., P.N. Witt, and M.B. Scarboro. 1969. The orb web during the life of *Argiope aurantia* (Lucas). Develop. Pyschobiol. 2:120-129.

Shear, W.A. 1970. The evolution of social phenomena in spiders. Brit. Arachnol. Soc. 1:65-80.

Witt, P.N. 1965. Die Wirkung von Substanzen auf den Netzbau der Spinne als biologischer Test. Springer, Berlin-Göttingen-Heidelberg.

Witt, P.N. 1962. Effects of antropine on spiders' web building behavior and thread production. Fed. Proc. 21:80.

Witt, P.N. 1963. Effects of physostigmine on spiders' web building behavior and thread production. Pharmacologist, Fall.

Witt, P.N. 1969. Behavioral consequences of laser lesions in the central nervous system of *Araneus diadematus* Cl. Amer. Zool. 2:121-132.

Witt, P.N. 1971. Drugs alter web-building of spiders. Behav. Sci. 16:98-113.

Witt, P.N., and R. Baum. 1960. Changes in orb webs of spiders during growth. Behaviour 16:309-319.

Witt, P.N., C.F. Reed, and D.B. Peakall. 1968. A spider's web. Springer Verlag, Heidelberg.

## Discussion

*Dr. Gamow:* I enjoyed it very much. Lovely, beautiful. I wish Jerry Feldman weren't here, because the first thing I want to ask you is about circadian rhythms. You mentioned it makes one every morning. How does it know it's morning — I mean, do you have a light program of some sort?

*Dr. Witt:* Our knowledge is quite sketchy on that. We know two factors which are overwhelmingly important for what I would call the "release" of web building. Whether they build or not depends on other things — the filling of the gland, its saturation, and all this type of thing. However, at the time that the sun rises, it has been shown that throughout the summer as this time changes, web building changes with the time of sunrise. In the laboratory, where we have a click from dark to light and at the same time a slight rise in temperature (through a temperature programmer), all the spiders take off on web building. If you have 100 spiders, then the span between their starts is about 2 minutes, not more than that. So the change in temperature and the change from dark to light seem to be the main stimuli. In consonance with other findings on circadian rhythms, we can start to change that kind of thing. We can do two things. One we did was that we kept the light even and the temperature even — whether you have a low temperature or a high temperature, or all light or all dark doesn't really matter. If these two things stay completely even, your web building frequency goes down from 98 out of 100 spiders to 2 or 3 out of 100 — they just don't build. If you vary one and not the other, you get something in between. This was a very crude experiment. The other thing I have tried is to reset the time — get the light-dark click and the rise in temperature at, instead of 5:00 in the morning, 9:00 in the morning — which is more convenient anyway. What happens (and we haven't done this in a very protracted experiment) is that you get a complete mess. You get something that you never see normally — spiders building all over the day. One builds early, one builds late, and all day long there is some web building going on. So you mess it up in a way. Now what would happen if you do that for a very long time, or if you do it from hatching on or even earlier, we do not know. It has not been tried.

*Dr. Gamow:* I guess it is a similar kind of question, but you mentioned that they can build it without sight. You can blind them. There is some kind of preset program of how many loops and how many rings and how many ellipses they are going to make. What if you carefully, while they're building in another part of the web, unwind another part of the web? Do they come back and realize that there is something kaput there and that they have to go back and patch it up?

*Dr. Witt:* Yes. Actually the question you're asking really is

where do they get their cues from. You could assume three different things. You could assume that they have a clock that runs off, and if you could interrupt them for a while, for instance, they will start as if they had built all the time in between and just go on. The other possibility if you interrupt them is that 10 minutes later they will start where they stopped and just go on from there. Or you can transfer them onto another web, either their own web brought to an earlier stage or a web of another spider in a different stage of building. This has been done by a fellow called Paul Peters, and the results just came out in *Animal Behaviour* last week. He was a student of mine. What it actually tells you, very briefly, is that the overriding determinant is the cue from the existing structure. Which means that the spider builds on from the point where building was interrupted. If you bring its own web back to an earlier stage, by taking radii out and things like that, it will replace those taken out radii, up to a point. But basically it tests the structure on which you put it. What this fellow did was, he took the spiders out of the webs and kept them for 10 minutes or so in a little vial and then put them back, either on their own web or the web of another animal, or their own web changed. They took the cue from what they found after the 10 minutes. Now there are restrictions to this. If the web is at a very much earlier stage, it was not finished in full size — and naturally not, because they didn't have all the silk any more. That is obviously a very important input into the pattern. But basically, the cue was taken from the existing structure.

*Dr. Gamow:* I remember in *Green Mansions*, Rheba was supposed to have a sarong made out of spider web. Is the silk very different from that of a silk worm?

*Dr. Witt:* It's very similar, but not exactly the same. Nobody has ever been able to use it commercially.

*Dr. Gamow:* Can you say something about the engineering? One of the pictures we have is that, if you want to be a good engineer and you can't figure out how to do something, look how nature does it. But I still maintain, I hope correctly, that we can't now build a spider web. For gun sights and microscopic things, we still have to use the genuine spider.

*Dr. Witt:* That's right. We can't make a similar kind of thread at this point. In the spider, the ordinary orb web builder makes six different ones. Because, for instance, for wrapping the prey,

thread with a slightly different amino acid composition and slightly different mechanical properties is used than the thread that makes the ground structure of the web. Obviously it does serve a different function. It comes from a different set of glands. For making the sticky spiral, which is a very thin gluey thing, it uses a different thread, and yet a different one for the dry threads that transmit the vibrations. We can't make a single one, and they can make six. I think that one of the most interesting questions is how the whole thing is formed. You know, if you look into the gland, you have a fluid mass, and at some point it becomes a thread. Where does that happen and how does that happen? Nobody knows. There are in spinning, I understand, different processes, depending on the jets and pressure and all this kind of thing. No, nobody has figured out how the spider really does it.

*Dr. Gamow:* What are these things that predetermine the diameter of the thread?

*Dr. Witt:* It seems that they can. Now, there are two ways you could think about it. Wilson, in England, has found a control valve which can narrow and widen the duct. Whether that determines the thickness of the thread is not certain. It could also be used just to hold the thread, because the spider can hold itself up on the thread without using the legs — just through the spinnerette. That could be with the control valve.

*Dr. Best:* Do they squirt some sort of catalyst into this liquid material that polymerizes it as they run it out?

*Dr. Witt:* Nobody knows. Another possibility is that the speed of pulling determines the thickness. That is a possibility.

*Dr. Best:* Now, is there an enzyme or something that is put out just as it comes out of the spinnerette?

*Dr. Witt:* That is a good possibility. The ampullate gland which produces the structural thread comes in three parts. One is a layer of synthesizing cells, and you can show this *in vitro.* You can label them and measure everything. Actually what happens is that amino acids flow in and the flow comes out into a kind of collecting bag which has a very thin epithelium which does not synthesize. This can be shown with labeled substances. You find something which is an unbelievably long duct, with a very thin epithelium, going all the way to the spinnerette. Now if the function of the duct was nothing but connecting the fluid

source to the spinnerette and getting it out of there, it could be a fraction of the size. So the length has interesting implications. It could be used in the formation of the thread.

*Dr. Best:* It looks like a loop of Henle in there.

*Dr. Witt:* Right. Another thing, it could be used either in reabsorption of certain solvents, maybe, or in adding certain things to it. All this is too speculative at this point.

*Dr. Peterson:* To get back for a minute to your drug studies, I take it that when you gave the depressants, the animal made essentially a complete web but it just wasn't as big as normal and it got tired, or something, and quit ahead of time. I'm thinking of this in relation to your quotation.

*Dr. Witt:* It was an incomplete web.

*Dr. Peterson:* Yes, that's what I mean.

*Dr. Witt:* You see, if it had built the angle already, it would have to be a little wider than in the other web, and the spiral which starts at the periphery has to start closer to the center than in the control web in order to make a smaller web. There is no way of making size adjustments later in web building.

*Dr. Peterson:* What I am trying to get back to is your quotation from Kennedy. Your interest, and I think all of ours, is in these command neurons and command functions. In your introduction you pointed out how there are several quite different parts of the web, that is, the frame is one thing, the radials are another, the central part another, and so on. What I am thinking is that it would be very nice, in his command neuron concept, if you had a command function for a frame, and then a second command function for spirals, and a third command function. But it doesn't seem to work that way.

*Dr. Witt:* I don't think so. If you try to break it down, then you see that in building a radius the same thing is repeated over and over again, and you could imagine that there is one motor pattern — could be one command neuron or three or four, with one super command or so. You know, probing the size of the angle, running out and fastening the radius, and going back to the center, and so on.

*Dr. Peterson:* You have been unable to fractionate the behavior such that the animal does everything but put in the spirals?

*Dr. Witt:* I think you can fractionate the behavior very well. Naturally, it does have to change at a certain point in time when it goes from the dry structure to the sticky one. You could think of another whole system now coming into function.

*Dr. Peterson:* That's all right, but I mean that you've been unable to fractionate — to turn one on and off.

*Dr. Witt:* No, no. You see, you still have to have this overall adjustment in size, which is done at a very early point in the web and which has nothing to do with a detailed movement. It means that every time the spider pulls for the last spiral, it has to space the next one a little bit further away. The whole thing has to be reset at a different level. You know, it is the same detail but now every piece of the detail is a little bit bigger than before That can be reset at the beginning of web building — at the moment you put the weight on, at the moment you give your tranquilizers, at the moment you destroy part of the web and it has to restart. At that same moment, the click-clack-click-clack-click is set at a wider and a wider spacing. That's the thing that mystifies me.

*Dr. Peterson:* The ultimate command is really nucleus disparity. You've got so much silk and so much weight and you're stuck.

*Dr. Witt:* There may be one computer figure here which says how much do I have in my gland, how much frame have I already laid out, and so on. How thick do I have to make the thread today, because how much do I weigh — have I over-eaten yet, or something like that. Then ticker-tick-tick, you know, and you get an exact thread length, and then you produce a pattern that corresponds to that length, because the end of web building and the emptying of the gland coincide.

*Dr. Best:* On this triggering of web building, since this obviously has a very important role connected with food getting, I've not heard you comment on what happens to the web building activity if you fast the spider or, if when he starts to build a web, you glut him — really start just feeding him until he is stuffed? Does he say, oh the hell with building the web.

*Dr. Witt:* We've only done two sorts of experiments. One is starving the spider. The most interesting thing actually was to read what had been said up to then. Nobody had ever measured it, but everybody had an idea about it. There were basically two schools — and I called them the psychologists and the physiolo-

gists, or biochemists. The psychologists say that if you starve a spider, the web gets bigger and bigger because it gets hungrier and hungrier. The biochemist says that if you starve a spider, the web gets smaller and smaller because it doesn't have material any more. You could really subject spiders to the test and measure exactly who was right and who was wrong. We starved them up to six weeks. For the first two to three weeks the webs stayed exactly the same size, with the same amount of detail, the same amount of silk — everything stays the same. Now, you wonder where it comes from. Our animals lost over 50% of their body weight. Which means that web building is high in importance in the animal's life. You begin to sacrifice body constituents, obviously proteins and amino acids.

*Dr. Best:* What happens in the converse situation?

*Dr. Witt:* Let me follow this story for a little while. The next thing that happens, after about two weeks, is that the web is still full size and becomes wider spaced — which means the meshes grow bigger, there are less radii and less spiral turns — which means a full-sized trap but sacrifice of the smallest flies. They still get all the medium and big sized ones. After six weeks, the web really became very small and very miserable, and the spider was very small and miserable at that time. We did try the opposite. But we didn't try over-feeding. We tried one other thing, which I think is very interesting and which shows how simple-minded one can be. This involves two populations. One had grown up in the laboratory and the other one we had caught outside and then grown in the laboratory. We said, if there would be very much larger webs, maybe, the ones that had been outside were really rewarded for building a large web, because they caught a lot of food. In the laboratory, it didn't matter. We fed them their regular diet. So they were kind of reinforced in web building, and that made them build larger webs. So that again was open to direct experimentation. We had four groups of spiders, all littermates. With a litter of 100, you can easily make four groups of 25. 50% of the animals we raised in tubes so they never built a web, and 50% built a web every day. We then subdivided those groups. Of the tube-raised, 50% of them got their fly into their mouth every day, or twice a week, and the other 50% were put on a web to hunt for flies every day. But they never built a web. 50% of the web builders were always fed in a tube, and 50% had to hunt for their flies. Now, there was absolutely no difference between the hunting for the

fly and the feeding by mouth. This means it had nothing to do with success or failure of the web, or whether the web was large or small. I guess there are a lot of things that are simpler in nature to do than to adjust to all contingencies. For example, think of the taking down of the web every day; the web is neatly taken down about an hour before a new web is built. Sector by sector, the web is rolled up, redigested, and over 90% of today's web — silk, amino acids — appears on the day after tomorrow's web — with radioactive tracers. This is done every day. Now, you could think, how stupid is that — because they could assess the web and say, well, how much is left from the last day, did I catch a lot of flies, and according to that either build a new one or repair the old one. They don't. Then, what I say is it is probably easier to take a web down and build a new one every day than to make a decision every day. So, to adjust the web size every day to anything that you want to do doesn't seem to be worth doing.

*Dr. Peterson:* This means that if they can eat and utilize their own web, they can really go for a remarkably long time without food.

*Dr. Witt:* That's right. During the deprivation experiment we took the web away every day. We took it out of the cage.

*Dr. Best:* In some ways in a deprivation experiment the problem is that, what else can the spider do, but go ahead and make this biochemical investment.

*Dr. Witt:* If you can't build a web, that's the end of everything. So you give up very late.

*Question:* The interesting aspect of silk structure is that it is, depending upon what class of silk, very high in certain amino acids.

*Dr. Witt:* That's right.

*Question:* In particular, there are some glycine silks that are 70% glycine. If an animal can eat its web every day, the old web, then it need not synthesize large amount of these amino acids. In the starving experiments, it is amazing that a spider losing 50% of its weight can still put out that much of almost a pure polymer amino acid, as they did.

*Dr. Kater:* You don't know that they did. They may have changed the amino acid composition, isn't that possible?

*Dr. Witt:* Right. Who knows. I don't know how you could do that.

*Question:* Take a particular type of silk. All the silks are in certain classes — they are either very high in glycine or glumatic acid.

*Dr. Witt:* If you assume a 100 milligram body weight for an adult female, and a .1 milligram web, then the web constitutes about .1% of the body weight of the animal: Now, you have to think about water and all these kinds of things — it's a substantial amount.

*Question:* Well, yes . . . except if occasionally you starve them . . .

*Dr. Witt:* Again, we don't know how much water was involved there.

*Question:* Another thing that was interesting was this duct gland structure. It would seem that the purpose of that long, looped structure would be to align polypeptide chains in a laminar flow situation because any long polypeptide chain, when it is forced through a tube with sufficiently high velocity, will align very well to the flow of the liquid. That is a requirement for forming this kind of material. So it would be easy to check — just by looking with cross polarizers in the . . .

*Dr. Witt:* Exactly. I'm trying to get some. There are a few other things that are happening, and if you look at web building, it goes in leaps and bounds — which means that one radius is laid at a very rapid speed, then the animal sits in the middle and what I call "dawdles" for several seconds. Imagine that it has 20 minutes to make 1,500 connections, then why dawdle after every thread for a few seconds? Now, that period can have a number of meanings. It can have something to do with the silk formation. We calculated that the volume of the duct would about correspond to one double radius, or it can have something to do with information processing or something like that. I don't know. But you know, it would really need the short time to form the silk before it is rolled out.

*Dr. Gamow:* One thing I was very curious about in connection with Dr. Best's amoeba: It seems like the pseudopods when it is building its glass house become more sticky at certain points. To get to the children's question: How come the spider doesn't get stuck to its own web?

*Dr. Witt:* Actually, nobody knows. The thing is that it always runs along the dry structure, and a spiral is not even able to hold the spider — it's a very thin thing. It always runs along the dry structure — the radii and so on — and carefully avoids the others. But what you can do — and I have tried — you can swing that animal into the web, and it certainly must touch the sticky stuff. It does not get stuck. Now, if you watch a spider, which you can do in your free moments, you can see that it continuously brushes itself and licks itself. You could hypothesize that it puts some little wax or repellant on in order not to stick to the glue. It's a possibility. I don't know.

*Dr. Wilson:* Is it possible to cross breed some of these species and obtain viable hybrids, or even fertile hybrids?

*Dr. Witt:* Whether you can cross breed species, I have no idea.

*Dr. Winston:* I wondered about the size of the frame and so on. Do you always put them in the same size frame?

*Dr. Witt:* Yes, we do.

*Dr. Winston:* Does the same spider stay in about the same place and build his web (in the wild) in the same place every day?

*Dr. Witt:* I have a graduate student at present who works on the problem of web location, and it turns into many tough problems. Let us assume that the first web is built. Let's not think about the distribution of young ones and all this type of thing. Let us say that the first web is built. What happens now if you withdraw food or create a gradient of humidity, or whatever you can think of which makes the site unpleasant? The spider absolutely does not move. For weeks and weeks it takes its web down every day and builds it in the same place. If we give it no food and a lot of buzzing and things going on, it always builds in the same place.

*Dr. Gamow:* Now I understand why a spider builds its web in my shower.

*Dr. Witt:* When you begin to think about it, if these animals have 100-200 offspring, all they need for preserving the species is two. Why should they have a mechanism to search around for a good place? They can spread out their 100 children and 2 fall on a fat place and 98 fall on not such a fat place — which is probably what happens. Anyway, all we can say at present is,

if the site is once selected — and we don't know how that comes about — it sticks to the site very, very stubbornly.

*Question:* You say that it begins at the top?

*Dr. Witt:* You know, the strongest is the bridge on the top of the web — which sometimes can go across a river or from one tree to another — and that contains an enormous amount of silk. That's preserved, and the basic structure — which is three or four radii at the hub — and it is taken down in kind of pie-shaped pieces. Then, from this basic structure the new radii and frame are built, and the spiral is put in.

*Question:* You said you could tell individual differences when the spiders . . .

*Dr. Witt:* The story is that the methods we developed were all for statistical purposes. As you do the drug experiments or the aging thing, you do want to extract the law and not the exceptions. So, we developed the tradition of not looking at the individuals, and I think it probably needs a new approach. Particularly the frame structure seems to be very characteristic for individuals. But we have described individuals that had strange webs which were characteristic for that one individual all throughout the lifetime. They could be a mutant.

*Dr. Murphey:* I just wanted to make a brief suggestion with regard to that. A very nice index of symmetry is to take the perimeter squared over the area. It will give you a very nice index of the symmetry of the shape.

*Dr. Witt:* I wanted to do a vector analysis, where you get two measures for the symmetry of the web — the direction of asymmetry and the amount of asymmetry.

*Dr. Murphey:* This is a very simple thing and easy to do.

# Tissue Culture and Its Application to Neurobiology

MOLLY ARSENAULT, JOSEPH KATES, and KURT SCHLESINGER
*University of Colorado, Boulder*

TISSUE culture is becoming an increasingly important technique in neurobiology. There are several reasons for this. One can grow large quantities of a homogenous cell type in tissue culture for biochemical and physiological studies. *In vivo*, on the other hand, neural tissue is surrounded by supportive glial or Schwann cells. It can be difficult to attribute certain kinds of experimental results specifically to nerve cells in such a mixture, since one does not know the function of the supportive cells. The reproducibility of experiments increases when one uses cells continually growing in culture, rather than explanting new cells from an organism. One can manipulate the environment of cultured cells; i.e., one can control cellular growth rate by adjusting cell density, pH, or serum concentration. Thus, cells grown in tissue culture have a relatively controlled environment.

Three general types of cells can be grown in tissue culture: (1) non-proliferating cells, (2) non-cancerous cells which reproduce in culture, and (3) cancer cells. Only nerve cells of the first and third types have been maintained in culture.

Non-proliferating differentiated cells can be maintained in culture for several months. Levi-Montalcini and Angeletti (1963) have reported that dissociated embryonic chick sympathetic neurons can be maintained in tissue culture and undergo axonal

regeneration in the presence of nerve growth factor. Sympathetic ganglia from 10-day chick embryos have been explanted and maintained in culture for 68 days (England and Goldstein, 1969; Lodin, Booher and Kasten, 1970). Lodin et al. (1970) have done time-lapse microcinematography of explanted chick spinal ganglia cells. They pointed out that the cellular changes they observed agree with Cajal's (1959) description of nerve cell differentiation during embryological development. Lodin et al. (1970), during the course of *in vitro* differentiation, first observed multipolar, stellate cells, with short processes. Then, following the disappearance of many of the short processes, the neurons produced long processes and displayed bipolar or unipolar morphology. Since many cells did not become fully unipolar, the authors concluded that supportive glial cells might be necessary for neural maturation.

The above type of experiment requires going back to an organism and explanting cells repeatedly. Consequently, a new cell sampling might not be identical to the previous one. Another disadvantage of this approach is the difficulty of separating neurons from other kinds of cells. This makes it difficult to attribute neural differentiation to the neuron alone and to rule out the need for other cells in the process.

Proliferating non-cancerous cells are another cell type which can be studied in tissue culture. Cell lines can be established in culture from mouse and human fibroblasts which proliferate for limited periods of time when first explanted.

In 1963, Todaro and Green disaggregated mouse embryo tissues and established cell lines from fibroblasts. Their experiments gave relatively precise criteria for the establishment of a cell line. When fibroblasts were first cultured, they underwent a progressive decline in growth rate and massive cell mortality upon successive transfer. By 30 to 70 days in culture, the remaining cells could maintain a rising growth rate. Cell lines which were established by these criteria (i.e., 3T3 cells) survive repeated passage in culture. According to Hayflick and Moorhead's (1961) criteria for distinguishing between established lines and normal cell strains, these 3T3 cells represent cell lines because they showed the essential capacity for unlimited growth *in vitro*.

Established cells lines also showed the remarkable ability to grow at low inoculation densities. Todaro and Green (1963) suggested that established cells showed an increased ability to function as independent organisms. They attributed this increased

autonomy to changes in the synthetic processes of established cells, i.e., established cell lines could compensate for the leakage of cell substances into the medium, whereas freshly cultured cells could not. Thus, the established cell lines were not representative of the freshly cultured cells. This implied that to establish a cell line, one inevitably selects for cells which will grow continually in an abnormal environment.

Seeds (1971a, 1971b) developed a system in which massive mortality of freshly explanted brain cells was obviated. Pre- and post-natal mouse brain cells were grown as aggregates in suspension. This reduced cellular division, and it was observed that the specific activities of acetylcholinesterase (ACE), cholineacetyltransferase (CAT), and glutamic acid decarboxylase (GAD) increased as a function of time in culture. The activities of the enzymes of the phosphogluconate pathway (PGP) decreased relative to the specific activities in newly explanted cells. In mouse brain *in situ* the activities of ACE, CAT, and GAD increased while PGP enzymes decreased during the same period of time. According to these biochemical criteria, the *in vitro* system resembled the *in vivo* development of mouse brain.

Cancer cells are able to adapt to tissue culture and subsequently replicate more easily than the two types of cells discussed above. Freshly cultured cancer cells in general do not undergo the initial growth rate decline or massive mortality exhibited by freshly cultured cells. Many types of cancer cell lines have been established successfully. In the presence of ACTH, adrenal cortex tumors will synthesize steroid hormones. Buonassisi, Sato and Cohen (1962) have shown that alternate successive passages of steroid-secreting mouse tumor cells between culture and animal (the alternative animal-culture passage technique) give rise to cells with increased growth capacity in culture and enhanced hormonal activity. Benda, Lightbody, Sato, Levine and Sweet (1968) have isolated a mouse brain glioma which retains the ability to synthesize S-100 protein specific to brain. Pierce (1967) has shown that teratoma stem cells in culture generate embryonic-like tissue. This tissue can differentiate into brain, nerve, tooth or skin cells, both *in vivo* and *in vitro*. Thus, tumor cells grown in tissue culture can retain their differentiated properties and ability to differentiate.

A feasible method of getting neurons into tissue culture might be to transform them with an oncogenic virus. One could explant neural tissue (i.e., mouse brain) and grow it in tissue

culture. Since nerve cells do not divide, one could treat the entire cell population with BUdR and specifically kill only growing non-neural cells. One could try to transform the remaining nerve cells by treating them with an oncogenic virus (i.e., SV 40). The capacity for neural differentiation of the resultant cancer cells would, of course, have to be characterized. If a clone contained neural-specific morphological, biochemical and electrophysiological properties, it could be studied as a putative nerve cell system.

Using Buonassisi's alternative animal-culture passage technique, Augusti-Tocco and Sato (1969) obtained a mouse neuroblastoma cell line in tissue culture from the spontaneous sympathetic ganglia tumor C 1300. They compared levels of tyrosine hydroxylase (TAT), CAT, and ACE in cultured neuroblastoma cells and mouse brain, and found the levels comparable. This showed that the neuroblastoma cells retained some biochemical functions characteristic of nerve cells. They justified the presence of both adrenergic and cholinergic enzymes in neuroblastoma cells with Burn's (1971) theory of multiple neural transmitters. According to this theory, when a cell is stimulated, acetylcholine (ACh) is released from the perikaryon, causing an evoked potential. In response to the nerve impulse, the boutons release norepinephrine (NE).

In our laboratory, studies have been performed with cultured neuroblastoma cells from tumor C 1300. When grown in 10 percent calf serum, these cells exhibit the epithelial-like morphology and high saturation density characteristic of cancer cells. When calf serum is removed from the medium, the cells stop dividing and undergo morphological and biochemical changes. The cells lose their rounded morphology, and extend "processes" morphologically similar to axons (Kates, Winterton and Schlesinger, 1971; Seeds, Gilman, Amano and Nirenberg, 1970).

Time-lapse cinematography shows that during the first 48 hours after removal of calf serum, the cells exhibit rapid movement, extend and retract their "processes," and have ruffled membranes along their leading edges. In some cases the neurite extensions are formed by cytoplasm left behind as the cell moves away. Cells retract their "processes" completely just before undergoing mitotic division. After 48 hours in serum-free medium, cell movement has diminished considerably and by 96 hours, the cells are largely immobile and become either unipolar or bipolar, forming extensive networks of "processes" with other cells.

Shubert and Jacob (1970) have shown that BUdR induces differentiation of neuroblastoma cells in the presence of calf serum. They find that when DNA synthesis is inhibited by CAR, BUdR still induces differentiation in 60 percent of the cells. They conclude that BUdR does not induce differentiation through its incorporation into DNA. Rather, they present evidence that BUdR increases the affinity between the cell and the surface of the culture dish, which in turn induces the synthesis of specialized cell products required for differentiation.

It is of interest to look at changes in DNA, RNA, and protein synthesis concomitant with the morphological differentiation of neuroblastoma cells. Experiments showing incorporation of radioactive thymidine, uridine, and leucine into neuroblastoma DNA, RNA, and protein, respectively, indicate that DNA synthesis stops 48 hours after removal of calf serum, while RNA and protein syntheses continue for at least 120 hours (see Figure 1).

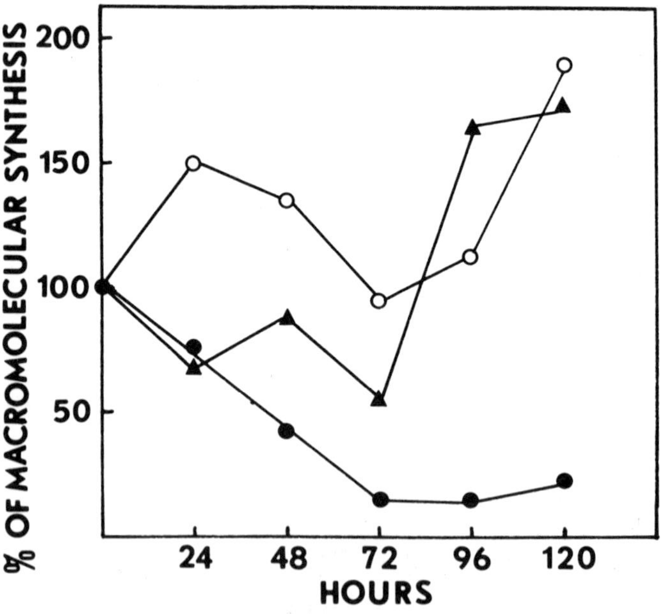


Figure 1. Neuroblastoma cells were labeled with $^{14}$C-thymidine for 1 hour, allowed to grow for 2 days, and then transferred to serum-free medium. Cells were then labeled with $^{3}$H-thymidine (● —— ●), $^{3}$H-leucine (▲ —— ▲), $^{3}$H-uridine (0 —— 0) 1 hour prior to harvesting. The ordinate is the $^{3}$H/$^{14}$C ratio expressed as percent of 0-hour incorporation of label. The $^{3}$H/$^{14}$C ratio was employed to control for variation in the amount of cells harvested in each sample.

This continued synthesis of RNA and protein may be required for the elaboration of the differentiated functions. In fact, it is shown below that ACE induction is dependent upon continued RNA and protein synthesis.

Concomitant with cellular differentiation, the activity of ACE in neuroblastoma cells is increased twenty-fold (Kates et al., 1971; Seeds et al., 1970). This is shown in Figure 2. The induction of ACE activity is inhibited by actinomycin-D and cyclohexamide, suggesting that *de novo* protein and RNA syntheses are required (see Figure 3). If ACE is solubilized from cell membranes with Triton X 100 and run through a 5-20 percent sucrose gradient, two peaks of enzyme activity are recovered, while only one peak is recovered from eel ACE (see Figure 4).

The data on induction of other enzymes from differentiated

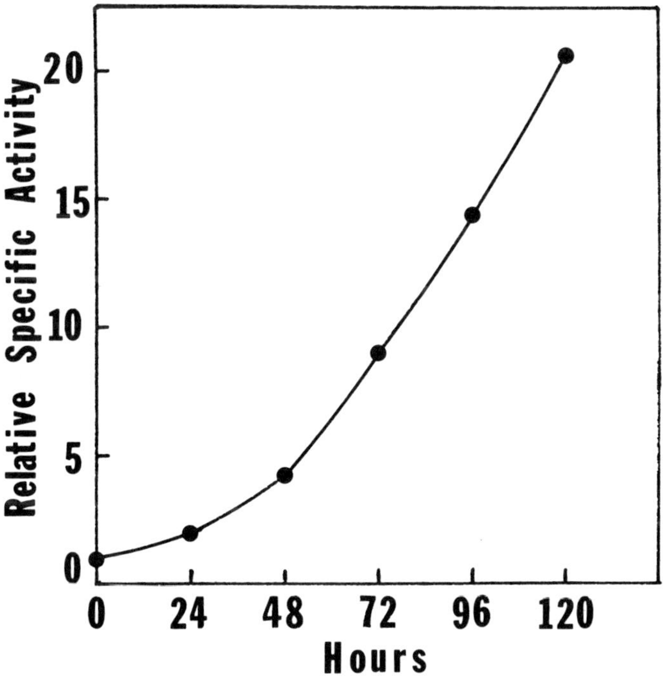


**Figure 2. Increase in specific activity of acetylcholinesterase after removal of serum from the culture medium of mouse neuroblastoma cells. Cell cultures in logarithmic growth were transferred to serum-free medium at time "0" on the abscissa. Generally about 0 nmol of acetylcholine were hydrolyzed per min/mg protein in non-induced cells.**

neuroblastoma cells are not as clear. Blume, Gilbert, Wilson, Farber, Rosenberg and Nirenberg (1971) reported no increase in catechol-O-methyltransferase (COMT). Tyrosine hydroxylase was not induced (Kates et al., 1971). This suggests that the adrenergic neural transmitters are not induced. Cholineacetyltransferase (CAT) was induced in some neuroblastoma clones and not in others (Kates et al., 1971; Amano, Richelson and Nirenberg, 1971; Rosenberg, Nandeventer, De Francesco and Friedkin,

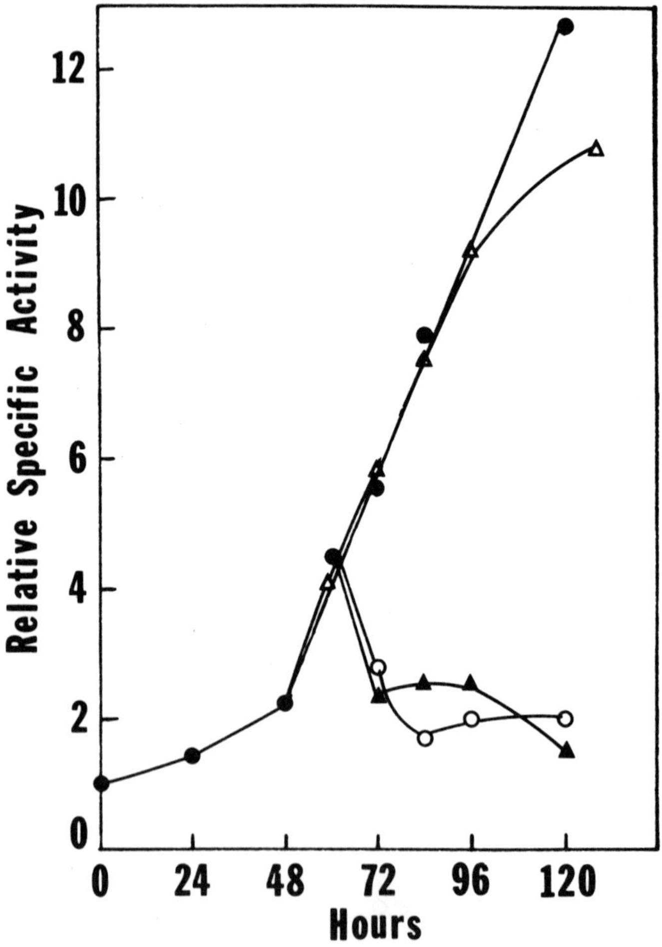


**Figure 3. Increase in specific activity of acetylcholinesterase (see Figure 1 for details): (• —— •) = no drugs; (0 —— 0) = 0.5 μg/ml of actinomycin D added; (▲ —— ▲) = 5μg/ml cyclohexamide; (Δ —— Δ) = $10^{-6}$M colchicine.**

1971). Neuroblastoma cells from our laboratory did not show CAT induction.

Amano et al. (1971) have isolated clonal lines from neuroblastoma tumor C 1300 which show biochemical characteristics specific for each clone. One clone shows induction of ACE and CAT activities ("cholinergic"). Another clone shows induction of ACE and TAT activities ("adrenergic"). Several clones have ACE activity yet are neither "cholinergic" nor "adrenergic." These data suggest that neurological characteristics of neuroblastoma tumor C 1300 are clone-dependent, which may explain the varying biochemical results discussed above.

Experiments from our laboratory show that there is competition between neuroblastoma cell division and cell differentiation. If one inhibits cell division with FUdR or CAR, ACE (an index of differentiation) is induced (see Table 1). The reverse is also true. If serum is added back to neuroblastoma cells after 48 hours in serum-free medium, the cells lose their neural morphology, show high saturation density growth, and the specific activity of ACE returns to the original level.

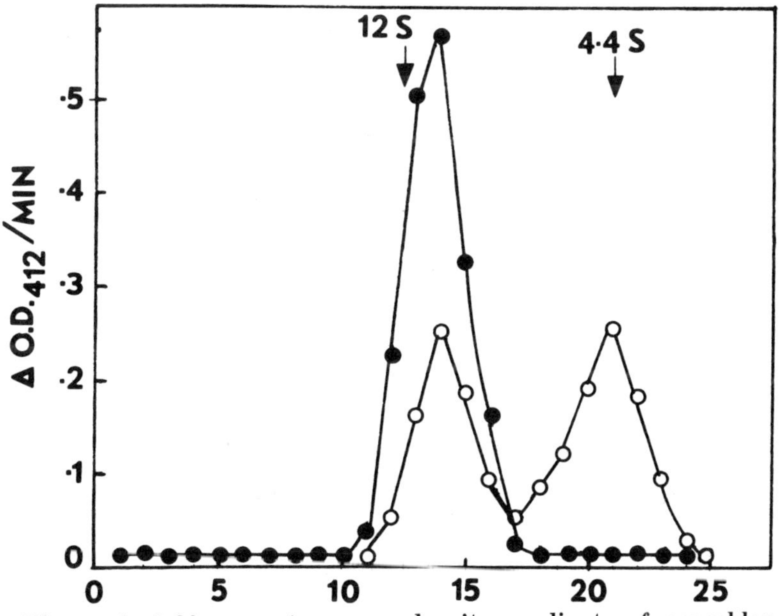


**Figure 4.** 5-20 percent sucrose density gradients of neuroblastoma cell extract (0 —— 0) and purified eel acetylcholinesterase ( • —— • ). Specific activity of ACE measured by the method of Ellman at $O.D._{412}$. (Ellman, Courtney, Andres and Featherstone, 1961).

**Table 1**

Induction of Acetylcholinesterase by Inhibitors of Cellular DNA Synthesis

| Medium | Relative Specific Activity Acetylcholinesterase |
|---|---|
| Complete | 1.0 |
| Complete + FUdR | 5.1 |
| Complete + CAR | 4.8 |
| Serum-free | 5.7 |

We have noted a correlation between the morphological changes of neuroblastoma cells and the induction of ACE. Clones which exhibit bipolar and unipolar morphology upon differentiation show a concomitant rise in ACE activity. Clones which remain multipolar with short "processes" after 120 hours in serum-free medium have low enzyme inductions (see Figure 5). These clones which remain multipolar after four days in serum-free medium also continue to synthesize DNA and to undergo cell

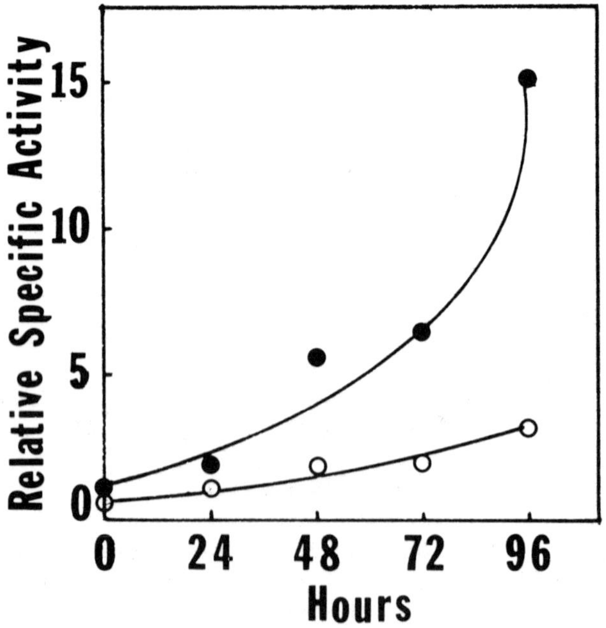


**Figure 5. The increase in specific activity of ACE in two clones of neuroblastoma cells: (• —— •) = control cell clone; (0 —— 0) = clone which undergoes little morphological differentiation.**

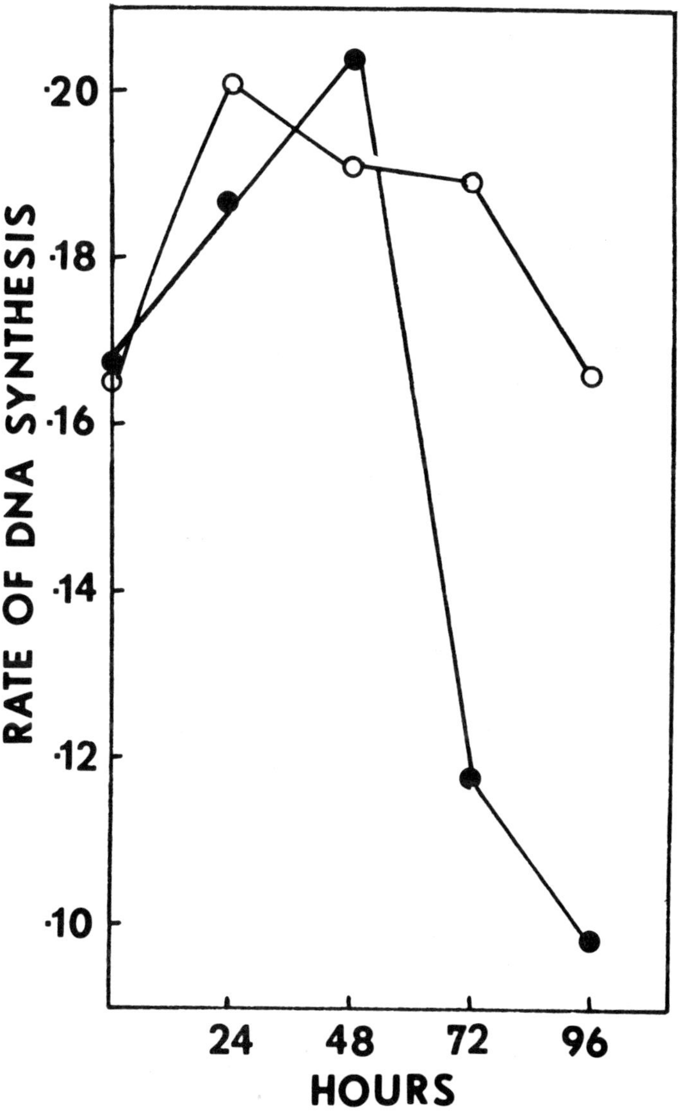


Figure 6. Neuroblastoma cells were labeled with $^{14}C$-thymidine before transfer to serum-free medium. Cells which undergo little differentiation (0 —— 0) and control cells (• —— •) were labeled with $^{3}H$-thymidine 1 hour before harvesting. The $^{3}H/^{14}C$ ratio is expressed as the rate of DNA synthesis.

division, while clones which exhibit bipolar and unipolar morphology do not continue to synthesize DNA at appreciable rates (see Figure 6). This is further evidence that there is competition between division and differentiation in neuroblastoma cells.

Colchicine, which disrupts microtubules and causes axonal retraction, does not prevent ACE induction when added after 60 hours. Morphologically the differentiated neuroblastoma cells lose all "processes" and appear epithelial. This suggests that once these "processes" have been formed, their disruption does not affect ACE synthesis. Other kinds of experiments will be required to determine the necessity of "process" development for ACE induction.

Previously, Olmsted, Carlson, Klebe, Ruddle and Rosenbaum (1970) had shown the presence of microtubule protein in neuroblastoma. They used vinblastin to disrupt microtubules into filaments and large cytoplasmic crystals, and specifically characterized the vinblastin precipitate as disrupted microtubule protein.

We have preliminary evidence that if cell growth is arrested by serum withdrawal, levels of microtubule protein initially do not increase. This is consistent with the observation that protein synthesis is not required for initial "process" formation (Amano et al., 1971). Apparently pre-existing microtubule protein can be utilized (Kates et al., 1971).

Thus, neuroblastoma cells grown *in vitro* can exhibit morphological and biochemical characteristics unique to nerve cells. It is important to show that they can also exhibit excitable membranes. Nelson, Ruffner and Nirenberg (1969) have shown that 65 percent of the cells have "active" membranes, which depolarize upon application of an electrical stimulus. Nine percent of the cells generate action potentials.

Harris and Dennis (1970), using a clone of cultured neuroblastoma cells, have found that each cell shows an electrogenic response to current applied through a bridge circuit. They can elicit action potentials by iontophoretically applying ACh to the neuroblastoma cells. Each of the cells is insensitive along one of its "processes," which suggests (a) the presence of specific sites for the cholinergic receptors, and (b) that the "processes" could be subdivided into axons and dendrites.

Harris, Heineman, Shubert and Tarakis (1971) have recently shown that neuroblastoma cells can form contact with muscle, producing in that muscle a selective sensitivity to exogenous ACh at the site of contact. This parallels the early stages of

synapse formation *in vivo*. In synapse formation, the post-synaptic cell is uniformly sensitive to neural transmitters before it makes contact with a nerve (Katz and Miledi, 1964). When contact is made with an ingrowing nerve, the post-synaptic cell develops local sensitivity at the point(s) of neural contact, and the remaining cell surface becomes insensitive to neural transmitters. Additional synapses cannot be made on the insensitive regions but a functional synapse(s) develops on the sensitive area(s). Harris et al. (1971) grew myoblasts in tissue culture for seven days and showed them uniformly sensitive to exogenous ACh. They then plated cultured neuroblastoma cells on top of the muscle cells. Within several days, the neuroblastoma cells formed contacts with the muscle cells. Using iontophoretic application of ACh to test muscle sensitization, the authors showed that muscle sensitivity to exogenous ACh was inversely proportional to the distance from neural contact at which ACh was applied. They found that muscle cells showed highest sensitivity to ionophoretically applied ACh at the point of neural contact. As the ACh was applied at increasing distances from the neural contact, the muscle cells showed correspondingly larger insensitivity to the ACh. Thus, in this respect, the interaction between muscle cells and neuroblastoma cells paralleled synapse formation *in situ*. The authors attempted to show functional synaptic transmission, but were unsuccessful.

Thus, differentiated neuroblastoma cells resemble nerve cells morphologically, biochemically and electrophysiologically, except that synaptic transmission remains to be shown. Although one must be cautious in his interpretation of results obtained from cultured cells, the neuroblastoma system might be developed into a powerful system to use in answering important neurobiological questions.

## Literature Cited

Amano, T., E. Richelson, and M. Nirenberg. 1971. Neurohormone synthesis by different clones of neuroblastoma. Fed. Proc., Abs. #193, p. 1085.

Augusti Tocco, G., and G. Sato. 1969. Establishment of functional clonal lines of neurons from mouse neuroblastoma. Nat. Acad. Sci., Proc. 64:311-315.

Benda, P., J. Lightbody, G. Sato, L. Levine, and W. Sweet. 1968. Differentiated rat glial cell strain in tissue culture. Science 161:370.

Blume, A., F. Gilbert, S. Wilson, J. Farber, R. Rosenberg, and M.

Nirenberg. 1970. Regulation of acetylcholinesterase in neuroblastoma cells. Nat. Acad. Sci., Proc. 67:786-792.

Buonassisi, V., G. Sato, and A.I. Cohen. 1962. Hormone-producing cultures of adrenal and pituitary tumor origin. Nat. Acad. Sci., Proc. 48:1184-1190.

Burn, J.H. 1971. Release of noradrenaline from sympathetic endings. Nature 231:237-240.

Cajal, S.R.Y. 1959. Studies in vertebrate neurogenesis. [Transl. by L. Guth] C.C. Thomas, Springfield, Ill.

Ellman, J.L., K.D. Courtney, V. Andres, and R.M. Featherstone. 1961. A new and rapid colorimetric determination of acetylcholinesterase activity. Biochem. Pharmacol. 7:88-95.

England, J.M., and M.N. Goldstein. 1969. The uptake and localization of catecholamines in chick embryo sympathetic neurons in tissue culture. J. Cell Sci. 4:677-691.

Harris, J., and M.J. Dennis. 1970. Acetylcholine sensitivity and distribution on mouse neuroblastoma cells. Science 167:1253-1255.

Harris, J., S. Heineman, D. Shubert, and H. Tarakis. 1971. Trophic heteraction between cloned tissue culture lines of nerve and muscle. Nature 231:296-301.

Hayflick, L., and P.S. Moorhead. 1961. The serial cultivation of human diploid cell strains. Exp. Cell Res. 25:585-621.

Kates, J., R. Winterton, and K. Schlesinger. 1971. Induction of acetylcholinesterase activity in mouse neuroblastoma cells. Nature 229:345-347.

Katz, B., and R. Miledi. 1964. The development of acetylcholine sensitivity in nerve-free segments of skeletal muscle. J. Physiol. 170:389-396.

Levi-Montalcini, R., and P.U. Angeletti. 1963. Essential role of the nerve growth factor in the survival and maintenance of dissociated sensory and sympathetic embryonic nerve cells *in vitro*. Develop. Biol. 7:653-659.

Lodin, Z., J. Booher, and F.H. Kasten. 1970. Phase-contrast cinematographic study of dissociated neurons from embryonic chick dorsal root ganglia cultured in the Rose chamber. Exp. Cell Res. 60:27-39.

Nelson, P., W. Ruffner, and M. Nirenberg. 1969. Neuronal tumor cells with excitable membranes grown *in vitro*. Nat. Acad. Sci., Proc. 64:1004-1010.

Olmsted, J.B., K. Carlson, R. Klebe, F. Ruddle, and J. Rosenbaum. 1970. Isolation of microtubule protein from cultured mouse neuroblastoma cells. Nat. Acad. Sci., Proc. 65:129-136.

Pierce, G.B. 1967. Teratomacarcinoma: Model for a developmental concept of cancer. Current Topics in Develop. Biol. 2:223-246.

Rosenberg, R.N., L. Nandeventer, L. DeFrancesco, and M.E. Friedkin. 1971. Regulation of the synthesis of choline-O-acetyltransferase

and thymidylate synthetase in mouse neuroblastoma in cell culture. Nat. Acad. Sci., Proc. 68:1436-1440.

Seeds, N.W. 1971. Biochemical development of brain cell cultures. Fed. Proc., Abs. #431, p. 1127.

Seeds, N.W. 1971. Biochemical differentiation in reaggregating brain cell culture. Nat. Acad. Sci., Proc. 68:1851-1861.

Seeds, N.W., A.G. Gilman, T. Amano, and M. Nirenberg. 1970. Regulation of axon formation by clonal lines of a neural tumor. Nat. Acad. Sci., Proc. 66:160-167.

Shubert, D., and F. Jacob. 1970. 5-bromodeoxyuridine-induced differentiation of a neuroblastoma. Nat. Acad. Sci., Proc. 67:247-254.

Todaro, G.J., and H. Green. 1963. Quantitative studies of the growth of mouse embryo cells in culture and their development into established lines. J. Cell Biol. 17:299-313.

## Discussion

*Dr. Best:* As to neoblasts in planaria. I have already noted that there is a sort of stem cell that's totipotent in the planarians that can turn into any of the some 16 or 17 different types of cells that they have. Now, when one cuts the head off, one of the first things that one sees is a mobilization of these neoblasts up to the region of the cut surface and particularly an aggregation around the cut ventral nerve cords. Usually, one of the first structures to differentiate in this formation of the new head is a kind of brain primordium. In the formation of this brain primordium, what appears as sort of the primary event — or in tissue culture of neoblasts — is this sending out of a process from one end that's a long one, and then from the other end ejection of a sort of shorter one that is at the opposite end of the cell. So, the first thing that happens is that it kind of forms this bipolar sort of object, and there is a great deal of probing with these feet, in which they test one another out and sometimes they stick and will contract and pull one another in and form a kind of connecting mass. And the other thing is that, with planarian neoblasts, if one — I haven't done this experiment, but it was done by Ansedin and Buchsbaum back about 8 or 9 years ago — if one takes a bunch of these neoblasts and puts them in tissue culture, part of the time they will form sheets and part of the time they will glob up into little worms. And these little worms are viable and proceed to crawl off. So that I think there are some very interesting possibilities here for sort of crystallizing out a little brain.

*Dr. Kates:* Is the idea to get an embryonic stem cell that gives rise to a whole variety of neurons?

*Dr. Best:* There may be a lot of the same kind of considerations involved in crystallizing a brain out as are involved in crystallizing out physical crystals. You know, I think there is a lot of . . . it depends on the order and temperature and maintenance of the conditions as to whether you are going to grow a large crystal or get a bunch of little imperfect sub-crystals. This whole business of organ reconstitution, or tissue assembly, is probably a new enough game at this point that we don't know nearly as much about crystallizing out an organ as the crystallographer types know about growing big crystals.

*Dr. Kates:* May I say something to that? I think that it's here that tissue culture systems may be useful. Let's say that you have a variety, five types of neurons, some of which you know interact with each other in the nervous system *in vivo*, and some of which you know don't. And let's say, for a moment, that the fact that they're cancer cells does not modify surface specific determinants that might cause these cells to interact specifically. Then you can test, just in a test tube, by mixing these neurons to see if there is selective adhesion of the ones that would normally interact with each other *in vivo* as opposed to the ones that wouldn't.

*Dr. Best:* Apropos of this business of crystallizing out a brain, one of the things that one sees in the natural product all the time is this presence of glia or glia-like cells that . . . in the flatworm brain there is a sort of cell that we call "neural accessory" cells but which we really believe to be a kind of archetypical glial cell. And I was wondering if you would care to comment on what you might think the role, or importance, of the glial cells would be to the business of crystallizing out an ensemble of neurons that look like a brain.

*Dr. Kates:* I will just say that it is going to be interesting to add other kinds of cells to these neuroblastoma cultures to see how they interact with neurons — if they change the developmental sequence that we see. The nice thing here is that if you know you're looking at just one type of cell, then you can add any other selected cell type. You can look at neuroblastoma with and without [the added cells] to see if it causes any effects. When you have a cell in culture you can control its environment,

and that is one of the things that leads to an understanding of things at the molecular level.

One point I think that should be announced is that Montelcini had in fact a very small number of embryonic neuroblasts and provided the conditions under which the neuron produced what appeared to be morphologically the most complete nervous system within the adult cockroach, which I think was rather interesting. The other point, I would like to inquire as to what morphological things might be worth looking at, and I would like to propose a possibility. We've all, I think, throughout the whole time, agreed that the surface properties of these cells could be very important for many aspects of what's going on. In 1967, a technique for handling large numbers of cells was developed by Collins whereby he was able to electrophorese and isolate cells on the basis of charge. This could provide you with a method for determining differences in cell surface properties and then try to correlate these with other properties, such as biochemical activity and so on.

*Dr. Peterson:* I hate to be one of these guys who asks, have you done this yet . . . but I will, of course. One thing that struck me was that this cell, after you withdraw the serum, immediately expresses itself — from what you said — as a post-ganglionic sympathetic neuron. It's got cholinesterase, it is cholinoceptive, and it's got the enzymes to produce catecholamines, which would imply that this thing is wired, prior to that last division — in fact, probably from way back — which in the thinking of the developmental biologists is not to be considered. Well, let me put it another way. What would be interesting would be to see if this cell also has cholinacetyltransferase.

*Dr. Kates:* Yes, it does.

*Dr. Peterson:* And monoamine oxidase?

*Dr. Kates:* Not tested yet.

*Dr. Peterson:* Well, what I'm getting at then is, if that is true and if, for instance it synthesizes monoamine oxidase or something . . . then this is not a differentiated neuron. This is a pretty all-purpose, general duty neuron which actually makes it more interesting, as you say, because then you can start adding contaminants and see what is repressed.

*Dr. Kates:* Are you suggesting that neuroblastoma might be an early developmental form of a neuron?

*Dr. Peterson*: If it's got cholinesterase and transferase and the enzymes for catecholamine synthesis, it's really general duty. You can put this in any place in the body and it's going to do the job of the neuron, if you repress the right activities. So it's not really acting like a post-ganglionic neuron in this situation. A post-ganglionic neuron shouldn't have the transferase.

*Dr. Kates:* There are people who think that there is some basis for believing that a sympathetic neuron should have both transmitters. Neuroblastoma is used as an example of this idea. I think it is a weak argument because you really have to prove that it is a sympathetic neuron and that is where the whole thing sort of falls down a little bit. It's a tumor that was propagated in mice and never looked at for a long, long time and then, suddenly, people started studying it.

*Dr. Best:* How about using something like susceptibility to the Levi-Montelcini growth factor?

*Dr. Kates:* We tried some experiments with nerve growth factor; it didn't seem to have any effect when the cells were in serum, and when we removed the serum it didn't enhance anything.

*Dr. Peterson:* Can I get in a second question? Have you tried the effect of added acetylcholine on the cholinesterase activity?

*Dr. Kates:* That's on our list. We may be doing that next week.

*Joyce:* Some culture lines undergo changes in their chromosome number and, in any one culture, they are highly variable. Has that happened?

*Dr. Kates:* We should look at them.

*Dr. Best:* On this business of the retraction of the cell processes at the time of mitotic division you sort of suggest that maybe there is something like . . . some kind of principle operating here like a conservation of microtubules and a conservation of the rate at which microtubules can be assembled. What do you think is the . . .

*Dr. Kates:* Well, please correct me. My feeling is this has not been completely studied. It would be nice to do an electron microscopy study of cells and different stages as they approached mitosis, and really see what's happening. It might be possible if there were some way of injecting labeled neurotubule protein and watching it redistribute from one structure to the next. But that gets pretty fancy.

*Dr. Feldman:* Nirenberg has this result where you make a hybrid between a neuroblastoma and an L-cell and the thing is very much like a neuroblastoma, implying that the differentiated state is dominant to the undifferentiated state which is maybe something that you may not have expected. I was wondering whether you have any ideas — whether that gives you any clues as to what the nature of the differentiation process is.

*Dr. Kates:* Well, first you've got two nuclei in the same cytoplasm — and evidently the other cell nucleus is not repressing. At least some of the genes are causing the membrane to be stimulatable. Electrical studies indicated that the hybrid had an active membrane — had a potential across the membranes, and could generate an action potential. So, I think it shows that there is nothing in an L-cell that would repress those nerve cell genes. It may also suggest that in the membrane of a normal cell there is room, so to speak, to stick in neurologically active molecules. It would be interesting to know the topology of this. Are these neural function units stuck in areas of otherwise normal non-neural membranes? It would be interesting to see if you could create a kind of mosaic or whether it is all intermingled on the membrane. I think that these are interesting questions that require identifying some surface component, being able to label it with an antibody, or something, and then looking at the surface.

*Dr. Crumpacker:* At first, when you were talking about one technique — taking living cells out and hoping that a few of them might reproduce. These are taken out of some mature sort of tissue?

*Dr. Kates:* Even when they're taken out of embryonic tissue one has difficulties establishing a cell line.

*Dr. Crumpacker:* Can we assume that they are all the same genotype? Do they have . . . I don't know, but let's assume they're all the same genotype and have the same chromosome number. What are you selecting for?

*Dr. Kates:* It's really hard to say.

*Dr. Best:* Well, while he was talking away, I was thinking some more about this business of the receptor sites and things binding on to one another, and I was wondering if you couldn't use these systems to do the cell counterpart to hybridization experiments, where you could take out already differentiated cells, and maybe one way of asking how differentiated this cell is, would be to add

back in either cells of a given type that had differentiated and see if you could get them to sort out with a given class of cells here, or maybe cell membranes from this given type of cell. You know, using some kind of labeling of these cells and then feeding them back into the tissue culture and see if they'll sort out . . . sort of like some of the tissue reconstitution studies, those like Steinberg has done. It would give you a way of exploring the similarity or dissimilarity of these surface groupings, wouldn't it?

*Mrs. Galag:* When you return the cell to the medium, then do the cells revert to neuroblast stage? Do they de-differentiate?

*Dr. Kates:* Yes. They de-differentiate and that's how you can tell if it's still O.K. We haven't done careful studies on quantitatively how many of the cells recover and so forth, but you do get them going again.

# Genetic and Biochemical Approaches to Biological Clocks: Implications for the Study of Behavior

JERRY F. FELDMAN
*State University of New York at Albany*

## I. Introduction

THE term "biological clock" has been used to describe nearly every type of biological phenomenon in which a series of events occurs with a fixed temporal relationship. The term clock, however, implies a specific time-keeping function for the organism and the daily, or circadian, clocks which will be discussed here seem to satisfy this requirement, as will be shown below. Although we know essentially nothing about the physical or chemical components of a circadian clock, we can recognize its existence by its control over some directly observable parameter in the organism. Such a parameter — biochemical, physiological, or behavioral — would express a 24-hour rhythm either in the level of the response or by an event occurring at a particular time of day.

Although there are many known cases of daily rhythms under natural conditions, only a unique class of these rhythms persists in the laboratory in so-called constant conditions — continuous darkness and constant temperature — with a "free-running" period which closely approximates, but is not exactly

equal to, the 24-hour period of the earth's rotation. This slight, but significant, deviation from 24 hours of the free-running period of these circadian (*circa* = about, *diem* = a day) rhythms is the strongest single piece of evidence that the rhythm is under the control of an endogenous timing mechanism referred to as a biological clock.

In addition to persistence in constant conditions circadian rhythms typically show two other properties which support the concept that they represent adaptive timing devices. First, they can be synchronized to, or entrained by, environmental light and temperature cycles. In this way the biological rhythm can assume a unique, and presumably adaptive, phase relationship with the daily environmental cycle. Second, in contrast to most chemical and metabolic systems, the rate of operation of a circadian clock, as measured by its period length or frequency, is relatively constant over a wide range of physiological temperatures — i.e., the $Q_{10}$ of the period length has a value close to 1. This property of a temperature-compensated period length is exactly what is demanded if the internal oscillation is to be used as a timing device, whose rate should not change with fluctuations in ambient temperature.

Circadian rhythms have now been reported in organisms representing all of the major taxonomic categories except bacteria and blue-green algae; several reports of circadian rhythms in procaryotes (Halberg and Conner, 1961; Guillot, 1971) remain unconfirmed. In eucaryotic microorganisms a wide spectrum of responses are under the control of a circadian clock, including motility (Brinkmann, 1966), phototactic response (Bruce and Pittendrigh, 1956), sexual reactivity (Barnett, 1966), cell division (Sweeney and Hastings, 1958; Edmunds, 1966), photosynthesis (Hastings et al., 1960), and luminescence (Hastings and Sweeney, 1958). Among invertebrates and vertebrates, there are examples of rhythms in locomotory and feeding activity (Roberts, 1960; DeCoursey, 1960), hormonal levels, neural activity (Strumwasser, 1965), and drug sensitivity (Halberg et al., 1960), for example, and in plants rhythms in leaf movement (Bunning, 1960), petal movement (Bunsow, 1953), and $CO_2$ emission (Wilkins, 1959) have been described. These are only a few examples of what has become a vast array of descriptions of circadian rhythms, and the interested reader can fully indulge his curiosity by consulting a number of books and symposia (e.g., Bunning, 1964; Sweeney, 1970; Aschoff, ed., 1965; Menaker, ed., 1971).

To the behavioral geneticist circadian clocks present two important features: First, quantitative measurements of any behavioral parameter must take into account these daily fluctuations. Even if the particular parameter being studied is not itself rhythmic, experiments attempting to alter the behavioral response must take into account oscillations in the response of the organism to any external treatment or manipulation — such as drug administration, sleep deprivation, or learning ability. Second, since the biological clock represents at least one of the many inputs which affect the behavioral response, a thorough understanding of clock mechanisms and coupling between the clock and the behavior would, of necessity, shed some light on the cellular or physiological basis of the response.

## II. Biochemical Approaches

Our laboratory is primarily engaged in attempting to elucidate the cellular and biochemical nature of the clock. Despite many attempts, little has so far been learned about the biochemical nature of the endogenous oscillation involved in circadian rhythms. Biochemical approaches have fallen into two classes: measurement of various biochemical oscillations and determination of the effects of metabolic inhibitors on rhythmicity.

A number of biochemical oscillations with circadian periodicity have been observed in a wide variety of organisms. For example there are rhythms of photosynthesis (Sweeney, 1960), luciferin activity (Bode et al., 1963), and protein synthesis (Feldman, 1968) in microorganisms; $CO_2$ emission (Wilkins, 1960) in higher plants; and levels of coenzyme A, glycogen, and succinic dehydrogenase (Halberg, 1960; Glick et al., 1961) in mitotically synchronized mammalian liver and adrenal cells. However, in no case has it been shown that the oscillation under study is the clock itself or part of the clock rather than some other periodicity caused by the clock.

In studies with inhibitors several generalizations appear to be emerging. Despite negative results of early studies with respiratory inhibitors (Bühnemann, 1955), more recent data in plants, animals, and some microorganisms suggest a requirement for oxygen, presumably for oxidative phosphorylation, for at least part of the circadian cycle (Wilkins, 1967; Bünning et al., 1965; Sargent, personal communication). In addition the involvement of macromolecular synthesis in clock operation is suggested from experiments with actinomycin D, an inhibitor of DNA-dependent

RNA synthesis (Karakashian and Hastings, 1962; Strumwasser, 1965; MacDowell, 1964; Vanden Driessche, 1966). Nevertheless, the nature of the requirement for oxygen or RNA synthesis remains a mystery.

The involvement of RNA synthesis in clock operation is consistent with a number of proposed models based on periodic gene activation. Monod and Jacob (1961) pointed out that appropriate coupling of two of their regulatory systems would generate an oscillation, while Goodwin (1963) showed from theoretical calculations that the components of a single genetic regulatory system of DNA-RNA-enzyme-metabolite-DNA would oscillate if one assumed classical absorption kinetics for the binding of the co-repressor (metabolite) to the repressor and the repressor to the DNA. Ehret and Trucco (1967) have proposed a "chronon" model of clock timing based on sequential transcription of adjacent cistrons in DNA analogous to the models for control of DNA replication in bacteria (e.g., Helmstetter, 1969). Other attempts to relate clock cycles to cell cycles (e.g., Halberg, 1960; Bruce, 1965; Edmunds, 1966; Willie and Ehret, 1968) remain speculative.

## III. The *Euglena* System

*Euglena gracilis* is a unicellular green flagellate which exhibits a circadian rhythm of phototaxis — cells are strongly responsive (positive phototaxis) during the subjective day and unresponsive or weakly responsive during the subjective night (Bruce and Pittendrigh, 1956). The period length of the rhythm in constant conditions is about 24 hours and can be synchronized to light and temperature cycles (Bruce and Pittendrigh, 1958) and the $Q_{10}$ of the period lengh is very close to 1.

Several years ago we began a series of experiments to determine whether protein synthesis was involved in the *Euglena* clock. Although previous experiments with actinomycin had shown that this drug causes loss of rhythmicity in *Gonyaulax* (Karakashian and Hastings, 1962), this criterian of abolishing a rhythm is not sufficient for demonstrating that the inhibited reactions are required for the oscillation. This is dramatically demonstrated in experiments with the circadian rhythm of photosynthesis in *Gonyaulax* (see Hastings et al., 1961) using the drug dichlorophenyldimethyl urea (DCMU), a specific inhibitor of photosynthesis. If DCMU was added at appropriate concentrations, photosynthesis was completely inhibited, but the inhibition was reversible since photosynthesis resumed its circadian oscillations at normal levels after the drug was washed out. Most

important, however, was the fact that the phase of the oscillation after recovery exactly coincided with that of a control culture which had never been exposed to the drug. This correspondence of phase between experimental and control cultures occurred irrespective of the time or duration of the drug pulse and strongly suggested that, although the process of photosynthesis itself had been inhibited, the underlying circadian clock had continued to oscillate normally. Thus when photosynthesis resumed, after the pulse, it simply became recoupled to the underlying "driving" oscillation. Since this type of experiment suggests that the reactions of photosynthesis themselves are not part of the driving oscillation, it led to the important distinction between the clock and the "hands" of the clock which required that all biochemical experiments — whether measuring a biochemical oscillation or using an inhibitor — distinguish between these alternatives (Hastings, 1960).

In our experiments with *Euglena*, therefore, we set the following requirement for demonstrating involvement of protein synthesis in the clock mechanism itself: either an alteration in *phase* after a pulse of an inhibitor or a change in *period length* in the presence of sub-lethal concentrations of the inhibitor. We also concentrated our efforts on cycloheximide, because this inhibitor fulfilled three important conditions: its mechanism of action was known, it was effective at low concentrations so that additional side effects might be minimized, and its action was reversible (Noll, 1966; Schweet et al., 1966). With cycloheximide, then, we were able to demonstrate changes in both period length and phase (Feldman, 1967a). At concentrations of 0.2 to 4.0 $\mu$g/ml cycloheximide caused an increase in period length whose magnitude was a function of the concentration of the drug — i.e., the higher the cycloheximide concentration, the greater inhibition of protein synthesis was observed and the longer the period of the rhythm. Phase shifts were also demonstrated by pulsing the drug for various lengths of time (24-72 hours) and observing the phase of the rhythm after the inhibitor was removed. The steady-state lengthening of the period and the phase shift strongly suggest that interference with protein synthesis also interferes with the oscillation mechanism and that protein synthesis is somehow necessary for normal clock operation.

This conclusion, however, is subject to several important qualifications. First, it is not certain that the only primary effect of cycloheximide is to inhibit protein synthesis and its

effect on the rhythm may have been caused by inhibiting other processes. To overcome this objection, two other inhibitors — ethionine and 5-fluorouracil — were tested. Although these two drugs are structurally different from cycloheximide and from each other, and their mechanism of interference with protein synthesis is similarly different, nevertheless they also caused a lengthening of the period of the rhythm (Feldman, 1967b). It seems reasonable to conclude, therefore, that all three drugs inhibit the clock because of their inhibition of protein synthesis. On the other hand streptomycin, which specifically inhibits chloroplast and mitochondrial protein synthesis but not cytoplasmic protein synthesis (Kirk, 1962; Clark-Walker and Linnane, 1966; Linnane and Stewart, 1967), had no effect on the rhythm at concentrations up to 2 mg/ml. It seems likely, therefore, that the protein synthesis necessary for the clock is cytoplasmic rather than organnellar. This suggestion is supported by the recent observation (Schnabel, 1968) that bleached *Euglena* totally devoid of chloroplasts can express a normal circadian rhythm. A second qualification is that alteration of pools of metabolic intermediates and/or the levels of energy pools might result as secondary effects due to inhibition of protein synthesis and that the observed effects on the rhythm might result from any of these or other secondary effects. Another type of experimental approach will be necessary to overcome this objection.

An additional observation with cycloheximide is the fact that a given concentration of inhibitor produced a much greater % inhibition of protein synthesis than % lengthening of the period. For example, at 50% inhibition of protein synthesis there was only 10% lengthening of the period. By assuming that as a chemical reaction protein synthesis has a $Q_{10}$ of about 2, it is possible to show a correlation between the effects of temperature on the clock and the effects of cycloheximide on the clock. This correlation led to a model (Feldman and Stevens, 1971) in which the temperature-compensation mechanism of the clock includes a device to compensate for changes in the rates of protein synthesis. Further support for this model, which represents an attempt to identify the chemical basis of the clock's temperature-compensation mechanism, has come from experiments with a mutant of *Neurospora crassa*, to be described below.

## IV. The *Neurospora* System

*Neurospora crassa* exhibits two distinct types of morphological rhythms as well as a rhythm of $CO_2$ production. One of the

morphological rhythms, periodic production of conidia (asexual spores), is typically circadian — the period length of the rhythm is about 24 hours and is temperature-compensated and the rhythm is entrained by light-dark cycles (Pittendrigh et al., 1959; Sargent et al., 1966). The other morphological rhythm, alternation between monopodial and dichotomous branching of the hyphae (Sussman et al., 1964), is not entrained by light-dark cycles and the $Q_{10}$ of its period length varies from 1.2 to 2.0 depending on the medium on which it is grown (Berliner and Neurath, 1965; Feldman and Stevens, 1971). Most previous work has been done on mutants which express one or the other rhythm; the *patch* and *band* strains exhibit a conidiation rhythm while the *clock* mutant expresses the hyphal-branching rhythm.

A. *Inhibition of the Rhythm by Cycloheximide*

We have used the *clock* mutant to obtain further evidence to support our model of temperature compensation based originally on the data from *Euglena* discussed above. Since this model depends on the similarities of the effects of temperature and cycloheximide on the rhythm, the model predicts that a correlation should exist between the temperature coefficient of a rhythm and the extent to which its period length can be affected by inhibitors of protein synthesis. If one could utilize a circadian system with a $Q_{10}$ value different from that found in *Euglena*, one would predict that its response to cycloheximide should also be different. Better still would be the use of a system whose $Q_{10}$ can be varied so that one could directly determine whether a correlation exists between $Q_{10}$ and period lengthening by inhibitors.

As mentioned in the section above, the mutants of *Neurospora* present a unique opportunity to test our model, since the *band* mutant has a $Q_{10}$ between 0.95 and 1.2 on all media (Sargent et al., 1966) while the $Q_{10}$ of the rhythm in the *clock* mutant depends on the medium on which it is grown. For *band* cycloheximide at concentrations up to 75% inhibition of growth causes less than a 4-hour change in period length (Sargent, personal communication). Table 1 shows data for calculating the $Q_{10}$ of the period lengths on two different media. On Gray's complete medium the $Q_{10}$ is 1.5 and on Vogel's minimal medium it is 1.8. Table 2 shows the period lengths of *clock* at different levels of inhibition of protein synthesis on these two media. From these data two conclusions are apparent: 1) On both media where the *clock* strain has a significantly higher $Q_{10}$ than either

**Table 1**

Effect of Temperature and Medium on Period Lengths of Circadian Rhythm in *clock* strain of *Neurospora crassa*

| Temperature | Period Length (in Hours) | |
|---|---|---|
| | Gray's Complete Medium | Vogel's Minimal Medium |
| 20° C. | 31.1 | 47 |
| 25° C. | 24.0 | 35 |
| 30° C. | 20.1 | 26 |
| $Q_{10}$ | 1.5 | 1.8 |

**Table 2**

Period Lengthening of Circadian Rhythm of *clock* strain of *Neurospora crassa* by Cycloheximide

| % Inhibition of Growth | %Lengthening of Period (and period length in hours) | |
|---|---|---|
| | Gray's Medium | Minimal Medium |
| 0 | 0% (24 hours) | 0% ( 25 hours) |
| 40 | 14% (27 hours) | 54% ( 54 hours) |
| 50 | 36% (32 hours) | 185% (100 hours) |
| 65 | 69% (40 hours) | 320% (148 hours) |
| 75 | 84% (44 hours) | |
| 80 | 116% (51 hours) | |
| 90 | 292% (94 hours) | 415% (181 hours) |
| $Q_{10}$ of rhythm | 1.5 | 1.8 |
| Equivalent $Q_{10}$ | 1.3 - 1.9 | 1.7 - 2.8 |

*patch* or *band*, cycloheximide has a much greater effect on the period length of *clock* than it does on the period length of *patch* or *band*. 2) Cycloheximide has a greater effect on *clock* on Vogel's medium (with higher $Q_{10}$ value) than on Gray's medium. Both of these observations are qualitatively what is predicted from the model.

Despite the agreement of these results with predictions from the model, so many unanswered questions remain that the model

remains little more than speculation. In addition, these and all the experiments with protein synthesis inhibitors are of such a general nature that it is difficult to know how to pursue this approach further. It is for this reason that we have temporarily turned to a more genetic approach.

## V. Genetic Studies on Clocks

### A. *Background*

Early evidence that endogenous circadian rhythmicity is inherited came from experiments in which plants and animals raised in constant conditions for numerous generations still exhibited rhythms with circadian periodicities (Bünning, 1932; Wahl, 1932; Aschoff, 1955; Hoffman, 1957). In addition organisms exposed to abnormally long or short cycles in most cases revert to their endogenous circadian periodicity when returned to constant conditions (e.g. Hoffman, 1959; Kleinhoonte, 1932). The outstanding exception is the green alga *Hydrodictyon* whose periodicity in constant conditions is determined by the cycle of the previous light-dark regime (Pirson and Doring, 1954).

Several experiments have suggested polygenic control of the period and phase of the rhythm: 1) Hybrids formed by crossing plants with different period lengths showed a continuous distribution of period lengths in the $F_1$ generation (Bünning, 1932). 2) "Early" and "late" emerging *Drosophila* strains, selected over many generations, show altered phase relationships and period differences with respect to each other and to the original parents (Pittendrigh, 1967). 3) Polygenic inheritance has also been suggested as the basis for determination of the critical daylength in the photoperiodic induction of diapause in certain insects (Danilyevski, 1965).

Another class of experiments has identified single genes which control the ability of the organism to express a rhythm. Such genes are now known in *Paramecium* (Barnett, 1966), *Neurospora* (Stadler, 1959; Sussman et al., 1964; Sargent et al., 1966), *Podospora* (Nguyen, 1967), and *Ascobolus* (Chevaugeon and Nguyen, 1969). In all cases, however, these genes probably do not affect the clock mechanism itself but only the particular process controlled by the clock. For example, in *Paramecium* the mutation eliminates rhythmicity in sexual reactivity but the rhythm of cell division persists in these "acyclic" mutants (Barnett, 1969). Again, the distinction between effects on the clock or the "hands" of the clock seem important.

On the other hand mutants affecting either the phase or period of a rhythm have recently been reported in *Chlamydomonas* (Bruce, 1971; Bruce and Bruce, 1972), *Drosophila* (Konopka and Benzer, 1971), and *Neurospora* (Feldman and Hoyle, 1971). Such mutants are much more likely to represent a direct lesion in the clock itself than those described in the paragraph above and should be more helpful in elucidating clock mechanisms.

B. *Experiments with* Neurospora

The unique, and at times startling, properties of circadian oscillations suggests that control processes as yet unelucidated or predicted by modern biochemistry or molecular biology may underlie these oscillations or that currently understood mechanisms are organized in a complex pattern which generate properties unpredicted by our limited knowledge of the component steps.

One of the most powerful tools open to the biologist to unravel complex processes is the genetic and biochemical analysis of *single-gene mutants* in which one step of the process is altered in a clearly measurable way. The best examples of this approach are our understanding of metabolic pathways (e.g., Srb and Horowitz, 1944) and bacterial control mechanisms (Jacob and Monod, 1961), but phenomena at higher levels of organization such as bacteriophage assembly (Wood et al., 1968) and even behavioral responses (Benzer, 1967; Hotta and Benzer, 1970; Ikeda and Kaplan, 1970) are now utilizing similar types of genetic analysis. It should be noted that the use of single-gene mutants to analyze behavioral responses is in contrast to the approach of the population geneticist in this field, whose primary concern is the analysis in natural populations of polygenic systems which control a particular behavioral response. Of the organisms with well-studied circadian rhythms, *Neurospora* is uniquely suited to this biochemical genetic approach because of its vast backlog of techniques and information, and it is for this reason that we have begun a search for biological clock mutants in this organism. It is our ultimate aim to identify specific biochemical lesions in these mutants and correlate such deficiencies with specific alterations in the clock so that we will have specific, rather than general, information about the nature of the biochemical pathways which regulate circadian oscillations.

In our first attempt we isolated three mutants in which the conidiation rhythm in the *patch* strain was affected (Feldman and Waser, 1971). Of these mutants, one had no rhythm, one was less sensitive to light, and one had a slightly shortened period

length at 30°C. These particular mutants, while useful in demonstrating the possibility of isolating clock mutants, for technical reasons have not yielded much additional information. Furthermore, the isolation procedure used in obtaining these mutants was extremely slow and tedious.

Although most previous work on *Neurospora* clocks has been done on the mutants described above, Sussman et al. (1964) showed that one could use certain wild-type strains of *Neurospora* to phenocopy the *clock* mutant — i.e., the wild strains, when grown in the presence of sorbose, expressed a hyphal branching rhythm. By slightly modifying the growth conditions, we have extended our ability to phenocopy the *clock* mutant to a much wider range of strains. In addition, since *Neurospora* grows in tight colonies in the presence of sorbose, it is possible to monitor the hyphal branching rhythm in colonies growing on a petri dish. In this way we can mutagenize a strain and rapidly screen up to 50 colonies/plate while looking for colonies with an abnormal number of bands after several days of growth. By this procedure we have isolated several new mutants (Table 3). Two important results are clear from this table. First, mutants 8 and 9 have a significantly longer period length than the other strains. Second, there is no correlation between growth rate and period length in these strains.

We believe the isolation of these two long-period mutants demonstrates the feasibility of isolating clock mutants in *Neurospora*. Our efforts will be to continue to look for additional mutants, to identify the biochemical lesions in these mutants, and hopefully, then, to identify specific biochemical processes involved in the circadian clock.

**Table 3**

Period Lengths and Growth Rates of New Mutants

0.5% Sorbose, 25° C.

| Strain | Period Length (hours) | Growth Rate (cm/day) |
|---|---|---|
| 1 | 24.05 ± 1.34 (12) | 0.286 ± 0.055 (12) |
| 5 | 26.56 ± 3.63 (11) | 0.255 ± 0.046 (11) |
| 7 | 23.94 ± 4.11 (7) | 0.699 ± 0.164 (7) |
| 8 | 33.26 ± 3.69 (12) | 0.291 ± 0.089 (12) |
| 9 | 35.04 ± 5.66 (12) | 0.304 ± 0.117 (12) |

## VI. Summary and Conclusions

A wide range of behavioral, physiological, and biochemical parameters in most eucaryotic organisms exhibit daily rhythmicity. This rhythmicity is under the control of an endogenous oscillation referred to as a biological, or circadian clock. Although little is known about the cellular basis of this oscillation, experiments with a number of organisms suggest that macromolecular synthesis is necessary for the operation of the clock. A new approach, involving the isolation of single-gene mutants with altered clock properties, has been initiated with the ultimate aim of correlating specific biochemical lesions with specific alterations in the clock. Such correlations should then lead to the identification of cellular or biochemical events which generate these oscillations.

## Acknowledgments

This research was supported in part by grants from the National Science Foundation and the Research Foundation of the State University of New York. The excellent assistance of Mrs. Marian Hoyle is gratefully acknowledged.

## Literature Cited

Aschoff, J. 1955. Tagesperiodik von Maeusetaemmen unter konstanten Umgebungsbedengungen. Pflugers Arch. Ges. Physiol. 262:51-59.

Aschoff, J. [ed.] 1965. Circadian clocks. North-Holland Publ. Co., Amsterdam.

Barnett, A. 1966. A circadian rhythm of mating type reversals in *Paramecium multimicronucleatum,* syngen 2, and its genetic control. J. Cell. Physiol. 67:239-270.

Barnett, A. 1969. Cell division: A second circadian clock system in *Paramecium multimicronucleatum.* Science 164:1417-1419.

Berliner, M., and P. Neurath. 1965. The band forming rhythm of *Neurospora* mutants. J. Cell. Compar. Physiol. 65:183-193.

Bode, V., R. DeSa, and J.W. Hastings. 1963. Daily rhythm of luciferin activity in *Gonyaulax polyedra.* Science 141:913-915.

Brinkmann, K. 1966. Temperatureinflüsse auf die circadiane Rhythmik von *Euglena gracilis* bei Mixotrophie and Autotrophie. Planta 70:344 389.

Bruce, V.G. 1965. Cell division rhythms and the circadian clock, p. 125 to 138. *In* J. Aschoff [ed.] Circadian clocks. North-Holland Publ. Co., Amsterdam.

Bruce, V.G. 1971. Cell division rhythms and the circadian clock. *In* J. Alder [ed.] Behavior of microorganisms. North-Holland Publ. Co., Amsterdam.

Bruce, V.G. 1972. Mutants of the biological clock in *Chlamydomonas reinhardi.* Genetics 70:537-548.

Bruce, V.G., and C.S. Pittendrigh. 1956. Temperature independence in a unicellular "clock." Nat. Acad. Sci., Proc. 42:676-682.

Bruce, V.G., and C.S. Pittendrigh. 1958. Resetting the *Euglena* clock with a single light stimulus. Amer. Natur. 92:294-306.

Bünning, E. 1935. Zuv Kenntnis der erblichen Tagesperiodizitat bei den Primarblattern von *Phaseolus multiflorus.* Jahrb. Wiss. Botan. Ges. 81:411-418.

Bünning, E. 1964. The physiological clock. Springer-Verlag, New York.

Bünning, E., S. Kurras, and V. Vielhaben. 1965. Phasenuerschiebungen der Endogenen Tagesrhythmik durch Reduktion der Atmung. Planta 64:291-300.

Bünsow, R. 1953. Endogene Tagesrhythmik and Photoperiodismus bei Kalanchoë Blossfeldiana. Planta 42:220-252.

Chevaugeon, J., and N.v. Huong. 1969. Internal determinism of hyphal growth rhythms. Brit. Mycol. Soc., Trans. 53:1-14.

Clark-Walker, G.D., and A. Linnane. 1966. *In vivo* differentiation of yeast cytoplasmic and mitochondrial protein synthesis with antibiotics. Biochem. Biophys. Res. Commun. 25:8-13.

Danilevski, A.S. 1965. Photoperiodism and seasonal development of insects. [Transl. by J. Johnston] Oliver and Boyd, Edinburgh.

DeCoursey, P. 1960. Phase control of activity in a rodent. Cold Spr. Harb. Symp. Quant. Biol. 25:49-55.

Edmunds, L.N. 1966. Studies on synchronously dividing cultures of *Euglena gracilis* klebs (strain Z). J. Cell. Physiol. 67:35-43.

Ehret, C.F., and E. Trucco. 1967. Molecular models for the circadian clock. I. The chronon concept. J. Theor. Biol. 15:240-262.

Feldman, J.F. 1967a. Lengthening the period of a biological clock in *Euglena* by cycloheximide, an inhibitor of protein synthesis. Nat. Acad. Sci., Proc. 57:1080-1087.

Feldman, J.F. 1967b. Biochemical and physiological studies on the circadian clock in *Euglena.* Ph.D. Thesis, Princeton Univ.

Feldman, J.F. 1968. Circadian rhythmicity in amino acid incorporation in *Euglena gracilis.* Science 160:1454-1456.

Feldman, J.F., and M.N. Hoyle. in press. Genetic alterations affecting the sorbose-induced hyphal branching rhythm in *Neurospora crassa. In* L. Scheving, F. Halberg, and J. Pauly [eds.] Chronobiology: Proceedings of the International Society for the Study of Biological Rhythms. Igaku Shoin, Tokyo.

Feldman, J.F., and S.B. Stevens. 1973. Protein synthesis and temperature compensation in circadian rhythmicity. *In* A. Perez-Miravepe [ed.] Behaviour of micro-organisms. Plenum Press, London.

Feldman, J.F., and N. Waser. 1971. New mutations affecting circadian

rhythmicity in *Neurospora,* p. 652 to 656. *In* M. Menaker [ed.] Biochronometry. Nat. Acad. Sci., Washington, D.C.

Goodwin, B.C. 1963. Temporal organization in cells. Academic Press, New York.

Guillot, N. 1971. Physiologie bactérienne. Compt. Rend. Acad. Sci. (Paris) 272:2500-2503.

Halberg, F. 1960. Temporal coordination of physiologic function. Cold Spr. Harb. Symp. Quant. Biol. 25:289-310.

Halberg, F., and R.L. Connor. 1961. Circadian organization and microbiology: Variance spectra and a periodogram on behavior of *Escherichia coli* growing in fluid culture. Minn. Acad. Sci., Proc. 29:227-239.

Halberg, F., E. Johnson, W. Brown, and J.J. Bittner. 1960. Susceptibility rhythm to *E. coli* endotoxin and bioassay. Soc. Exp. Biol. Med., Proc. 103:142-144.

Hastings, J.W. 1960. Biochemical aspects of rhythms: Phase shifting by chemicals. Cold Spr. Harb. Sypm. Quant. Biol. 25:131-143.

Hastings, J.W., L. Astrachan, and B.M. Sweeney. 1961. A persistent daily rhythm in photosynthesis. J. Gen. Physiol. 45:69-76.

Hastings, J.W., and B.M. Sweeney. 1958. A persistent diurnal rhythm of luminescence in *Gonyaulax polyedra.* Biol. Bull. 115:440-458.

Helmstetter, C.E. 1969. Regulation of chromosome replication and cell division in *Escherichia coli,* p. 15 to 35. *In* G.M. Padilla, G.L. Whitson, and I.L. Cameron [eds.] The cell cycle. Academic Press, New York.

Hoffman, K. 1957. Angeborene Tagesperiodik bei Eidechsen. Naturwissenschaften 44:359-360.

Hotta, Y., and S. Benzer. 1970. Genetic dissection of the *Drosophila* nervous system by means of mosaics. Nat. Acad. Sci., Proc. 67:1156-1163.

Ikeda, K., and W. Kaplan. 1970. Unilaterally patterned neural activity of gynandromorphs, mosaic for a neurological mutant of *Drosophila melanogaster.* Nat. Acad. Sci., Proc. 67:1480-1487.

Jacob, F., and J. Monod. 1961. Genetic regulatory mechanisms in the synthesis of proteins. J. Mol. Biol. 3:318-356.

Karakashian, M.W., and J.W. Hastings. 1962. The inhibition of a biological clock by actinomycin D. Nat. Acad. Sci., Proc. 48:2130-2137.

Kirk, J.T.O. 1962. Effect of streptomycin on [$^{14}C$] leucine incorporation in *Euglena gracilis.* Biochem. Biophys. Acta 9:476-479.

Klienhoonte, A. 1932. Untersuchungen über die autonomen Bewegungen der Primärblätter von *Canavalia insiformis.* Jahrb. Wiss. Bot. 75:679-725.

Konopka, R., and S. Benzer. 1971. Clock mutants of *Drosophila melanogaster.* Nat. Acad. Sci., Proc. 68:2112-2116.

Linnane, A.W., and P.R. Stewart. 1967. The inhibition of chlorophyll formation in *Euglena* by antibiotics which inhibit bacterial and mitochondrial protein synthesis. Biochem. Biophys. Res. Commun. 27:511-516.

MacDowell, F. 1964. Reversible effects of chemical treatments on the rhythmic exudation of sap by tobacco plants. Canad. J. Bot. 42:115-122.

Menaker, M. [ed.] 1971. Biochronometry. Nat. Acad. Sci., Washington, D.C.

Monod, J., and F. Jacob. 1961. General conclusions: Teleonomic mechanisms in cellular metabolism, growth, and differentiation. Cold Spr. Harb. Symp. Quant. Biol. 26:389-401.

Nguyen, V.H. 1968. Etude de rythmes internes de croissance chez le *Podospora anserina*. Anals Sci. Nat. (Ser. 12) 9:257-360.

Noll, H. 1966. Polysome organization as a control element. M.D. Anderson Tumor Inst., Publ. No. 19, 67-113.

Pirson, A., W.J. Schön, and H. Döring. 1954. Wachstums and Stoffwechselperiodik bei *Hydrodictyon*. Z. Naturforsch. 96:349-353.

Pittendrigh, C.S. 1967. Circadian systems. I. The driving oscillation and its assay in *Drosophila pseudoobscura*. Nat. Acad. Sci., Proc. 58: 1762-1767.

Pittendrigh, C.S., V.G. Bruce, N.S. Rosenzweig, and M.L. Rubin. 1959. A biological clock in *Neurospora*. Nature 184:169-170.

Roberts, S. 1960. Circadian activity rhythms in cockroaches. I. The free-running rhythm in steady state. J. Cell Compar. Physiol. 55:99-110.

Sargent, M.L., W.R. Briggs, and D.O. Woodward. 1966. Circadian nature of a rhythm expressed by an invertaseless strain of *Neurospora crassa*. Plant Physiol. 41:1343-1349.

Schnabel, G. 1968. Der Einfluss von Licht auf die circadiane Rhythmik von *Euglena gracilis* bei Autotrophie und Mixotrophie. Planta 81:49-63.

Schweet, R., R. Arlinghaus, R. Heintz, and J. Shaeffer. 1966. Mechanism of peptide bond formation in protein synthesis. M.D. Anderson Tumor Inst., Publ. No. 19, 47-66.

Srb, A., and N.H. Horowitz. 1944. The ornithine cycle in *Neurospora* and its genetic control. J. Biol. Chem. 154:129-139.

Stadler, D.R. 1959. Genetic control of a cyclic growth pattern in *Neurospora*. *Nature* 184:170-171.

Strumwasser, F. 1965. The demonstration and manipulation of a circadian rhythm in a single neuron. p. 442 to 462. *In* J. Aschoff [ed.] Circadian clocks. North-Holland Publ. Co., Amsterdam.

Sussman, A.L., R.J. Lowry, and T.J. Durkee. 1964. Morphology and genetics of a periodic colonial mutant of *Neurospora crassa*. Amer. J. Bot. 51:243-252.

Sweeney, B.M. 1960. The photosynthetic rhythm in single cells of *Gonyaulax polyedra.* Cold Spr. Harb. Symp. Quant. Biol. 25:145-148.

Sweeney, B.M. 1969. Rhythmic phenomena in plants. Academic Press, New York.

Sweeney, B.M., and J.W. Hastings. 1958. Rhythmic cell division in populations of *Gonyaulax polyedra.* J. Protozool. 5:217-224.

Vanden Driessche, T. 1966. The role of the nucleus in the circadian rhythms of *Acetabularia mediterranea.* Biochim. Biophys. Acta 126:456-470.

Wahl, O. 1932. Neue Untersuchungen über das Zeitgedächtnis der Bienen. Z. Vergl. Physiol. 16:529-589.

Wilkins, M.B. 1959. An endogenous rhythm in the rate of carbon dioxoide output of *Bryophyllum.* I. Some preliminary experiments. J. Exp. Bot. 10:377-390.

Wilkins, M.B. 1960. The effect of light upon plant rhythms. Cold Spr. Harb. Symp. Quant. Biol. 25:115-129.

Wilkins, M.B. 1967. An endogenous rhythm in the rate of carbon dioxide output of *Bryophyllum. V.* The dependence of rhythmicity upon aerobic metabolism. Planta 72:66-77.

Wille, J., and C.F. Ehret. 1968. Light synchronization of an endogenous circadian rhythm of cell division in *Tetrahymena.* J. Protozool. 15:785-789.

Wood, W.B., R.S. Edgar, J. King, I. Lielausis, and M. Henninger. 1968. Bacteriophage assembly. Fed. Proc. 27:1160-1166.

## Discussion

*Dr. Murphey:* Not having any expertise in Dr. Feldman's area, I don't feel confident to make any evaluative statements on it, so what I'm going to do is make one brief comment, ask a couple of questions, and turn it over to you. The comment I would like to make is that I am envious of what you've done. It's a very clever, very sound treatment of a very important topic. It's very nice stuff. It's a good illustration of the kind of taxonomy that I was trying to talk about yesterday. You've broken down an important phenomenon into something that you could work with, into a number of discriminable classes. Probably anyone who learned these techniques could go into the lab and separate those phenomenon the way you did. And now you're in a position to correlate the things you have been talking about with those sub-classes.

*Dr. Feldman:* You mean the one where banding and non-banding segregated?

*Dr. Murphey:* Yes. The banders and non-banders and the fast growers and all that. Well, you know about a lot of variables. You know about a number of variables though that seem to have something to do with this . . . But you also have categories of the ways organisms respond.

*Dr. Feldman:* I'm not sure I understand because I don't see that what we did is very much different than what anybody does with a 2 or 3 point cross in genetics. That is, we just identified several genes and several discrete phenotypes associated with those genes.

*Dr. Murphey:* I'm not saying that it's all that different. I am saying that it's a nice illustration of the kind of taxonomy that can be done. So what you do have is nice discrete phenomena to which you can relate other things, but . . .

*Dr. Feldman:* I just never thought of myself as a taxonomist.

*Dr. Murphey:* Oh, well you're one of the great taxonomists of our time. I have a conceptual problem with biological clocks, that I hope you can straighten me out on. You began your talk by pointing out that a great many variables were under the control of a biological clock. And what occurred to me, and what I've been wondering about for a long time, is a conceptual thing which may be quite trivial but it has to do with whether the clock is a thing which runs the body in some fashion or whether it is some kind of articulation or summation or by-product of a number of ongoing properties. I see this analogous to the James Lange theory of emotion, which says that you don't cry because you are unhappy. You are unhappy because you cry. You see, the feedback mechanism works in such a way that if you are aware that you are unhappy then you start crying. I don't know what difference it makes but I am wondering whether it's a by-product of things that are happening in the body. And your data suggested the latter. For example, you cited a number of studies where one part of the rhythm was knocked out but the others continued. The analogy that sprang to my mind was an automobile where a spark plug was fouled and so then you pour in some gasoline additive and that plug starts to fire again. That's had no effect on the rest of the mechanism and when that fouled plug starts to fire again, you would get a function similar to that which the lady got where she disrupted photosynthetic rhythm and it had no effect on cell division — possibly other rhythms as well.

*Dr. Feldman:* I think you put your finger on a problem that has been raised a number of times and really has no satisfactory answer. I would make a somewhat different analogy than your car. I would say, for example, that you can imagine that you have a clock with two hands. One hand could be torn off and the other hand would still be there and still go around at its normal rate because it's being driven by the clock-work mechanism in the back. This is similar to the example of inhibiting photosynthesis in Gonyaulax. Or if one had Neurospora mutants that did not conidiate, that doesn't mean that you don't have a clock; we just can't see it in the particular way we are looking at it. Now, in terms of a multicellular organism, it's very difficult to say because there are conflicting reports. There is strong evidence that individual cells in a multicellular organism have the capability of expressing independently a circadian oscillation. The parabolic burster in Aplysia is probably the best-studied case but there are a number of others. Also in Aplysia, Jacklet has shown a rhythm in the eye maintained in culture and Wilkins has shown a rhythm in plant tissue culture. On the other hand, the Drosophila mutants of Konopka and Benzer suggest that there are probably two controlling centers in the brain, one on each side. What I favor is the idea that cells in metazoa evolved from micro-organisms which have circadian oscillations and so have the potential to express a rhythm. However there probably is some central nervous system coordination and control over the expression of circadian rhythmicity in a multicellular organism. I'm not sure I've answered your question. One final point, though, is that since there clearly are many reactions that you can knock out and have no effect on the clock mechanism, the clock is not the sum total of all of the chemical, biochemical reactions going on in the organism.

*Dr. Murphey:* Well, bringing up evolution — that takes me to the second question I was going to ask, which is simply to try to get you to speculate on something. Do you think that evolution probably worked on periodicity itself or that periodicity followed the development of other aspects of the organism? For example, if you learned that an animal has highly developed night vision, it's a pretty good bet that he's going to have an activity cycle such that he'll be active at night, but it doesn't work the other way around. By knowing that he is active at night, you can't predict that much about his vision; some don't use vision very much. They develop other means for hunting and escaping

minima come out, and they just come out 28 hours apart or 30 hours apart or 32 hours apart, and that's just counting. I don't see that the slowing up of the clock by inhibitors involves any . . .

*Dr. Best:* What I meant was that the amount of slowing up and the extent to which this is going to affect other things or not affect other things sort of comes out of the kinetics of the transcription process and the translation of the feedback, and that's the problem.

*Dr. Feldman:* I'm not trying to push this.

*Dr. Best:* I know you're not. Two things: The $D_2O$ just incidentally changes the termal stability of virtually every protein in the cell that's put in, so it might be interesting if you were to do any $D_2O$ work to check the dependence of the temperature on the changes in the $D_2O$. Another thing I'm curious about is the relationship of pigments, of cell pigments, to the overall process of light entrainment of Neurospora. Now, there's some effect of light-dark cycle. Is that right? But yet, it's not a photosynthetic organism and not normally heavily pigmented, I would suppose. I was curious as to what the action spectrum of the particular entrainment is, so I could perhaps identify some molecular species of pigment in the particular clock mechanism.

*Dr. Feldman:* Yes. Malcolm Sargent at Illinois has in fact done an action spectrum for light effects on Neurospora and it comes out with a major peak in the blue region. The spectrum is difficult to interpret, but it's almost identical to the spectrum for the light growth response in Phycomyces. It's very difficult to distinguish whether it's a flavin or a carotenoid. Sargent has also looked at the light response of an albino mutant, which has little or no carotenoids. In this mutant, the light response as far as the clock is concerned is completely normal, but one is not sure that there are absolutely no carotenoids present. It's still up in the air but I, for nothing more than intuition, lean towards a flavin of some sort.

*Question:* Do you consider the temperature and protein inhibition compensation mechanism to be independent of the clock or part of the clock mechanism?

*Dr. Feldman:* I think you could have a clock that had no temperature compensation mechanism. You build a clock or you build an oscillation, and then you add parts on to it that make it more adaptive. I would think, for example, on that

period length corresponds to a given rate of protein synthesis. And I don't have any explanation.

*Dr. Peterson:* I'd like to stay with the protein synthesis for a minute, too. To go back to the Euglena, where you don't have this growth-rate problem, is there a periodicity to protein synthesis in general that you know of?

*Dr. Feldman:* Yes. We did some experiments and there is a daily fluctuation in the rate of protein synthesis. We haven't worked out why. I have a feeling from some other experiments that we did, that it's a result of changes in the energy pools in these cells at different times of the day, but that's just a speculation.

*Dr. Best:* There was a report I remember hearing several years ago that you can change these periods with heavy water. I have two questions: First, is that effect still true? Secondly, how do you square that with your view of things?

*Dr. Feldman:* The answer to the first question is 'yes' — heavy water does change the period length of a number of systems. My problem with the heavy water experiments is that heavy water affects so many different things.

*Dr. Best:* Except that there are so many things that you can tamper with that don't affect the period-length.

*Dr. Feldman:* Right, but I mean that heavy water inhibition doesn't pinpoint any particular reaction — maybe $D_2O$ affects the rate of protein synthesis, maybe it affects the membranes. I don't know. I don't have any problem with it because there's obviously some real chemical basis to this clock and $D_2O$ affects that reaction, as well as probably a hundred other things that don't have anything to do with the clocks. This is the trouble with inhibitors (it's even the trouble with our inhibitor experiments), that one is never sure what one is really affecting.

*Dr. Best:* This is one of the parts where you kind of have to take what you say on faith, because you know exactly that the nature of the effect depends sort of on the mathematical analysis, you know, the kinetics of it. You know we don't have to take it on faith, but . . .

*Dr. Feldman:* Well, the calculations on the $Q_{10}$ business, yes. You've got to sit down and go through them, but the changes on the period length that are induced by adding the inhibitor don't depend on any calculations. You just sit and measure where the

the cell division rhythm will shift 6 hours. Now, as I say, this is not absolutely compelling evidence but there is no evidence that you can separate one clock from the other.

*Dr. Schlesinger:* If you consider the model you have on the board, they are at all times the same.

*Dr. Feldman:* One could have lots of these regulatory circuits all operating within a 24-hour sequence and coupled to each other so that they have to maintain fixed phase relationship. That's true. If we could find a mutant, let's say, in a system in which there were several rhythms and we could change the period of one rhythm without changing the period of the other two rhythms, that would be very strong evidence that there were separate clocks controlling these different rhythms.

*Dr. Kates:* In those cases where you have drastically different growth rates, I would expect that you would also have drastically different rates of protein synthesis, is that correct? — or is this just a cell wall extension phenomenon?

*Dr. Feldman:* Yes, this is a problem that bothered us in some of these experiments — that is, we were just looking at changes in growth rate. We haven't measured protein synthesis in Neurospora under the conditions that we assay the rhythm and we have extrapolated everything from the growth rate. The reason is that we do the assay on an agar surface. What I can tell you is that we can show that other kinds of inhibitors that affect growth rate but don't affect protein synthesis such as cyanide . . .

*Dr. Kates:* Let me make my point. If, in fact, these do represent changes in the rate of protein synthesis, I would expect, in a simple minded way, which is obviously not true, that the clock might also change its period.

*Dr. Feldman:* Right.

*Dr. Kates:* But you've got a mutant that changes growth rate but it doesn't change its period, so it seems that the clock is even compensated in a sense for the rate of protein synthesis at any given temperature.

*Dr. Feldman:* Right. In other words, for any given strain, you get a certain rate of the clock with a given rate of protein synthesis, and if you changed that rate of protein synthesis then you change the clock periodicity depending on how well the system is temperature compensated. On the other hand, there is no correlation from one strain to the other that a given clock

predation and so forth. So I'd like you to speculate on that if you don't mind.

*Dr. Feldman:* Sure. You can't test me on that. I think probably what's happened is that there's some original selective value for a biological clock in some biochemical terms, and this is something Pittendrigh has suggested. For example, as I mentioned yesterday, there are many molecules that absorb visible light and in fact, there are many that can be inactivated by visible light. It may be very inefficient for the cell to make those molecules in the daytime rather than in the night time. Pittendrigh's idea is that this may have been the original selective value for a biological clock. Once the clock mechanism has evolved it could then be coupled to any other parameter if it was advantageous for the organism to be rhythmic. For example, in your case, there may have been a circadian clock in this system; the organism developed good night vision and then it was obviously advantageous to the organism to have its activity at night and its rest time during the day. I would be highly skeptical of a scheme in which a clock evolved independently for each observed rhythm.

*Dr. Gamow:* There was an article in the New York Times a few years ago. I think it was by an anthropologist. He made the statement that many, many patients, humans in mental institutions, have very highly developed clocks. Based on this, he said that in early man it was beneficial not to have a clock because when the sun went down if you went to sleep you got eaten and those who could keep the fire alive were selected. And so some of our anxieties and psychoses may have been because of a suppression of a natural clock for survival and then when . . .

*Dr. Schlesinger:* In these varied cellular organisms, you talk as if there's one clock. Is there any compelling evidence that there shouldn't be?

*Dr. Feldman:* No. There's some small evidence, and the kind of evidence is that if you look at an organism that has 3 or 4 rhythms such as Gonyaulax, every rhythm (photosynthesis, luminescence, cell division) has exactly the same $Q_{10}$ (about 0.85) and every rhythm has a fixed phase relationship to every other one. For example, the peak of luminescence is near the minimum of photosynthesis. When these rhythms are phase-shifted by changing the light-dark cycle, the photosynthesis rhythm will shift 6 hours, the luminescence rhythm will shift 6 hours, and

model, that we ought to be able to get mutants which have lost the temperature compensation mechanism but have a perfectly normal clock in all other respects. I don't know what they might represent biochemically. Maybe if I had a good guess at that, we might be able to select them biochemically rather than this horrible screening procedure we have.

# The Snail and the Mouse: Neurophysiological and Genetic Analysis of Behavior

STANLEY B. KATER
and JOSEPH P. HEGMANN
*University of Iowa, Iowa City*

THIS symposium on simple systems, sponsored by a group of behavioral geneticists, is a vivid demonstration of an emerging trend in the study of behavior. This trend might be referred to as the study of behavioral biology. The approach is characterized by freedom from conventional, discipline-bound investigation. The behavior of organisms is recognized as a composite and complex expression of the total physiology of the organism. Furthermore, it is clear that physiological function develops with direction from the genotype and with modification by the environment. In this paper we will discuss the approaches currently being used in our laboratories, and present the rationale for this method.

Our primary goal is to understand individual differences in physiological systems which lead to differences in animal behavior. Perhaps the most powerful tool for the study of individual differences is genetic analysis of specifically defined traits. When those traits are behaviors of animals, it is obvious that physiological systems must be involved. Genetic differences effective in modifying behavior must act by changing those aspects of physiological function that are significant in producing individual

levels of expression of the behavioral output. Environmental sensing and behavioral output by animals depend on neural function. Therefore, gene differences which modify behavior may act by introducing variance in neural function. While we recognize the significance of other physiological systems with respect to their effect on the behavioral output of an organism, we have selected the muscular and nervous systems as foci of our investigations. These systems represent the "final common pathways" for all behavior.

Historically, the study of nervous system function in behavior has been plagued by complexity. Early emphasis was on vertebrate nervous systems, and research centered around the study of large groups of neurons. Even today researchers are forced to interpret the complex electrical output generated by large neuron populations. The development of microelectrodes allowed the investigation of the electrical events occurring in single neurons. The work on the cerebellum attests to the progress resulting from application of this technique to the vertebrate brain. However, only an extremely small proportion of the animal's nerve cells can be investigated at one time, and relocation of cells from one preparation to the next is impossible. Thus, research on the vertebrate brain is presently restricted to either the study of complex output from large numbers of cells, or to discrete electrical events in individual neurons representing an extremely small percentage of the animals' nerve cells.

An obvious alternative to the complexity encountered in vertebrate nervous systems is found in animals with more simple nervous systems. During the past quarter of a century, the number of investigations on invertebrate nervous systems has increased enormously. This increase has had a two-fold basis, a recognition of the value of comparative information, and the idea that such simple systems would facilitate the gathering of information that could be applied to more complex organisms. These two motives are not, of course, mutually exclusive. As has been pointed out by many of the investigators taking this course, the relative paucity of nerve cells in the brains of invertebrates greatly increases the probability of locating and recording from individual neurons involved in specific behavioral acts. Furthermore, recording from a single neuron from an invertebrate may reveal the equivalent information encoded for by a myriad of neurons in a vertebrate. A final point which has greatly influenced the course of neurophysiological and neurobehavioral inves-

tigations on invertebrates is the ability to identify individual neurons of specific function. This has meant that in the investigation of a particular species a given neuron can be recognized from animal to animal.

Identification and relocation of individual neurons within the "simple nervous system" has a counterpart in genetic research. Using discrete phenotypic differences, the geneticist can focus his attention on distinct and relocatable allelic differences at individual loci. This approach has been the *sine qua non* of the rapid increase in our knowledge of the structure and function of the genetic material. During the growth of genetic research, differences with regard to traits were effective in increasing our knowledge of gene differences. However, rarely did this approach increase our knowledge about the trait itself. *Thus, it is important to distinguish between the use of phenotypes to understand genes and the use of genes to understand phenotypes.* This is especially important when the phenotype is a behavior. Rarely do behaviors of interest show the discrete (all or none) differences among individuals which typify the phenotypes effective for the study of genes. Rather, many behaviors display complex (graded) differences. Genetic differences associated with the variance measurable in this class of phenotypes usually involve allelic differences at many loci; and characteristically environmental differences also act to modify phenotypic expression. Too often analyses of these phenotypes resulted only in the determination of whether the character is "genetic or environmental."

Dobzhansky (1964, p. 55) has pointed out that "Asking right questions is, in science, often a large step toward obtaining right answers." Neurobiologists have demonstrated the importance of neurons in the behavior of animals and geneticists have presented convincing evidence that gene differences can modify behavior. We believe that an appropriate question from the standpoint of both neurobehaviorists and geneticists can now be posed.

To what extent are gene differences which affect animal behavior imposing modifications on the structure and function of the nervous system?

A geneticist is not seriously surprised to find genetic variance influencing individual differences in any character. In addition, a biologist is not moved by the information that the nervous system is involved in a particular behavioral sequence. However, knowledge of the functional alteration of the nervous system associated with genetic differences affecting behavior or knowl-

edge of the behavioral differences accompanying genetic alteration of the nervous system would be powerful. This information should allow rather precise physiological dissection of behavior and provide a basis for generalization of physiological behavioral associations across species.

To apply this approach to the study of behavior, we are employing two quite different species. The mouse, *Mus musculus*, is being employed because of the wealth of genetic information already available and despite the difficulties of neurophysiological study of the animal. The pulmonate snail, *Helisoma trivolvis*, is being used because of the facility of neurophysiological investigation and in spite of the paucity of previous genetic research employing this animal.

While the plethora of neurological mutants observed in mice provide clear indication that gene differences at diverse loci can block normal function of the nervous system, observation of these gene differences relies on abnormality of the behavior. Do populations of animals exhibiting non-mutant behaviors display gene differences which result in individual differences in nervous system function? If so, then the genotype of animals may affect not only whether the nervous system functions, but the level of its functioning (degree of gradation). In fact, we have been able to demonstrate this dependence of level of neural function on gene differences in the mouse (Hegmann, 1972).

A salient feature of the nervous system, in contrast to other intraorganismic communication systems, is its rate of information transfer. We have been able to demonstrate that the conduction velocity of action potentials by a nerve trunk is variable among individuals and that this variability is due to complex individual differences associated with gene differences at many loci.

Specifically, a technique was developed for rapid determination of individual differences in speed of conduction in the caudal nerves of *Mus musculus*. The ease of measurement allowed large numbers of behaviorally "normal" animals to be assessed and we chose to measure, in two replicates, 10 males and 10 females of each of 6 highly inbred strains of *Mus*. Since animals within these strains are genetically identical, individual differences within such strains measure variability not associated with genetic differences. However, since genotypes differ between strains, the overall differences displayed (without regard to strain) contain both genetic and non-genetic variability. The average conduction velocities for animals of the 6 strains (displayed in Figure 1

as deviations from the means of the replicates) suggested that mice of different genotypes conduct at significantly different velocities. Comparison of variability within inbred strains to that between strains provided estimates of the heritability of speed of conduction in males (0.14) and females (0.15). The presence of variability and the non-zero heritability suggests that speed of conduction in *Mus* will respond to selection pressure.

An additional 616 animals of known genotype were measured to allow determination of the importance of sex differences, strain differences, and differences between homozygosity and heterozygosity for the determination of speed of conduction in the caudal nerves. Results of this study indicate no sex differeinces, support the initial demonstration of genetic differences, and show that hybrids conduct significantly faster than inbreds. An exhaustive genetic analysis of individual differences in speed of conduction by *Mus* is currently in progress.

While the work discussed provides evidence that gene differences can modify the nervous system within functional limits, the behavioral consequences of that modification remains to be demonstrated. To avoid difficulty in associating functional differences with behavioral effects, we should, really, directly measure the function of individual neurons known to exert direct influence on a behavior. In our work we have begun to exploit the unique characteristics of *Helisoma trivolvis* for such measurements. Several aspects of the behavior of this snail are presently being investigated. Only an exemplary behavior will be discussed at this time.

Feeding behavior is initiated following chemoreception by the tentacles and oral surfaces of this animal. On receipt of the chemoreceptive food stimulus, a snail will typically (showing distinct individual differences) track along a gradient and begin feeding when the food is located. Feeding in *Helisoma* requires coordinated movements of 27 pairs of bilaterally arranged muscles. These muscles position the highly specialized feeding apparatus, the radula, which is contained within the buccal mass (Figure 2A). The radula itself is a thin membranous structure on which are arranged discrete rows of teeth. Feeding is accomplished by the back and forth oscillations of the buccal mass and the successive rasp-like actions of the radula against the substratum. Each successive rasp brings particulate matter into the mouth and buccal cavity from which point it is directed, by ciliary action, down the esophagus.

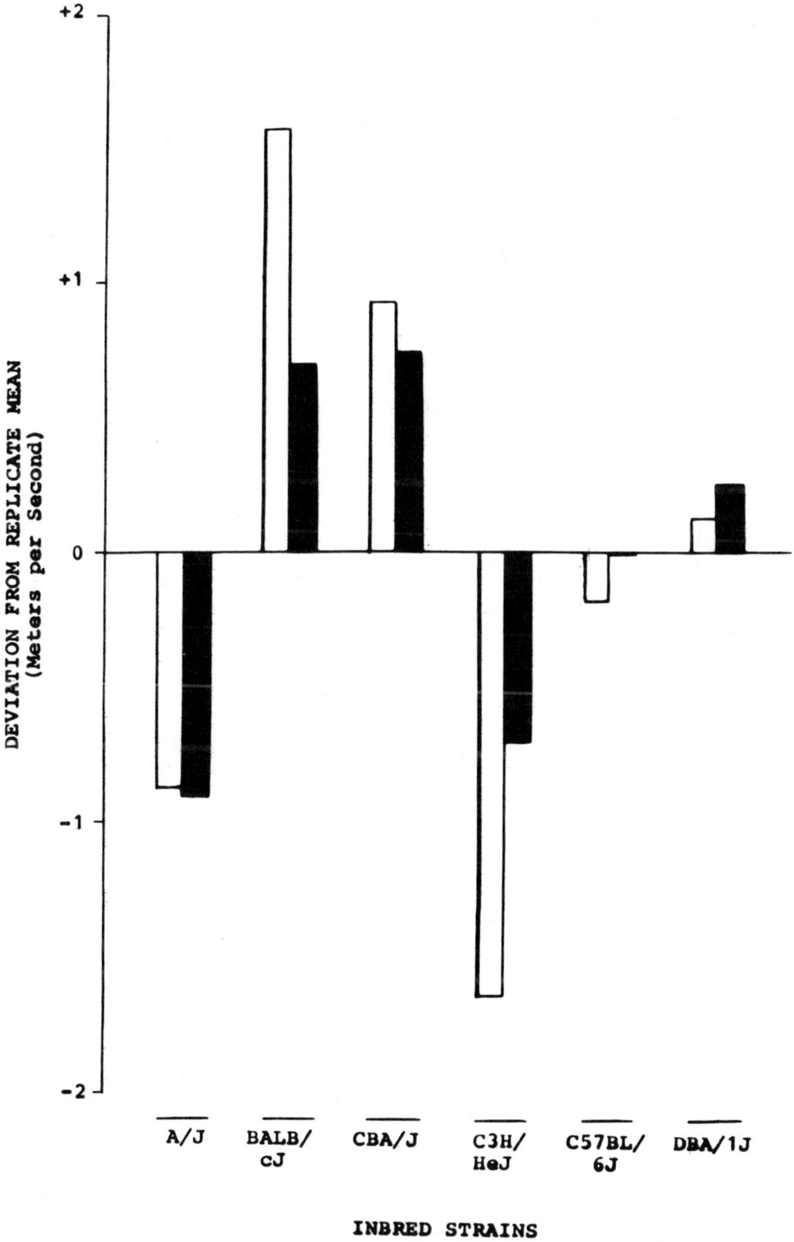


Figure 1. Comparison of caudal nerve conduction velocity for six genotypes of *Mus musculus*. Black and white bars represent replicate determinations from 10 males and 10 females of each genotype.

We have investigated the physiological basis of the feeding behavior of this animal. Without a doubt, the greatest single impediment to physiological investigation on *Helisoma* is one of the most characteristic features of the animal, its shell. However, one need sever only the attachments of a single muscle bundle to obviate this difficulty (Figure 2B). In the absence of predators, animals in this condition live for several weeks, and perform behavioral acts that are qualitatively indistinguishable from those of intact animals. The central nervous system is exposed by a single, three mm longitudinal cut in the dorsal head integument (Figure 2C). The entire central nervous system becomes accessible for investigation when the buccal mass is moved forward. The ganglia of the brain are not "anchored" firmly in the animal. In fact, the ganglia almost float within the body cavity. When a small, precisely fitted "micro-spoon" is employed for stabilization, the ganglia are unaffected by moderate movements of the animal. Under these conditions, we routinely record from as many as four individual central neurons for prolonged periods with no indication of adverse affects from movements of the animal. Neurobehavioral investigations can be performed on snails in this condition for periods in excess of a week.

The architecture of the central ganglia makes them well suited for cellular analyses. The somata of the neurons comprising each ganglion are readily distinguishable. To date, our most extensive observations are on the buccal, visceral, and parietal ganglia. We routinely can identify numerous cells solely on the basis of their morphological characteristics. The yellow to white color of the neural somata is sharply offset by the red of the extraneuronal components of the ganglia. Reproducible differences in intracellular pigment patterns, coupled with the shape and position of the somata allow precise identification of many neurons (see Figure 3).

Feeding can be elicited (by applications of either crayfish or tadpole extracts to the tentacle) in both intact and neurophysiological preparations. Furthermore, specific chemicals such as glutathione and calcium lactate can affect the feeding behavior. The role of the tentacles as chemosensory receptors, specific to these substances, is demonstrated by extracellular recordings from the tentacular nerve (Figure 4). Among the projects being pursued in our laboratories is the assessment of individual differences in chemoreceptive properties of the tentacle.

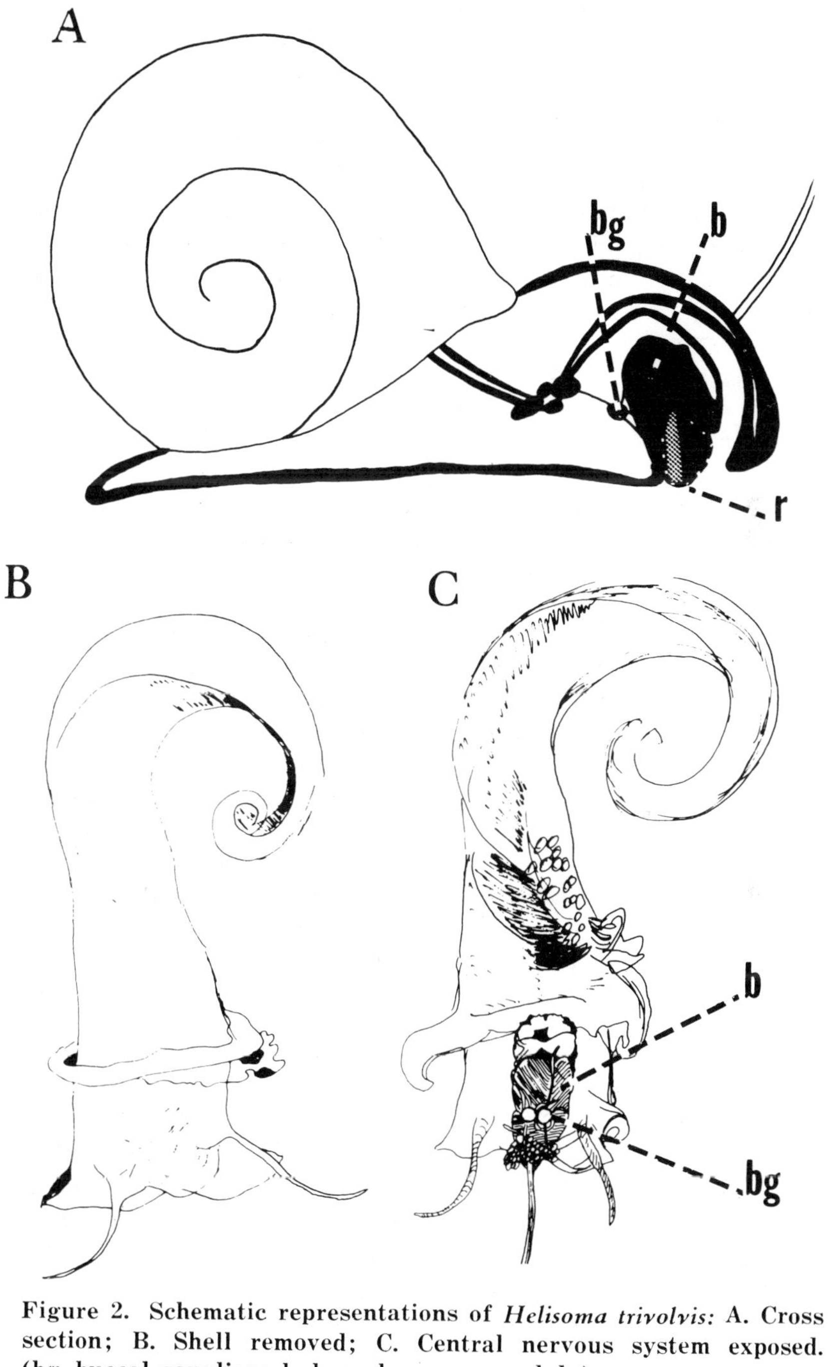


Figure 2. Schematic representations of *Helisoma trivolvis:* A. Cross section; B. Shell removed; C. Central nervous system exposed. (bg, buccal ganglion; b, buccal mass; r. radula).

The feeding behavior elicited by chemostimulation of the tentacles can be quantified using the semi-intact electrophysiological preparation. Employing extracellular myogram recordings from each of the 54 muscles involved in feeding, we have demonstrated the precise phase relationships that exist among these muscles. To accomplish normal feeding each muscle is activated in a precise time relationship with all of the other muscles. This information has allowed us to study the control mechanisms underlying these muscle phase relationships.

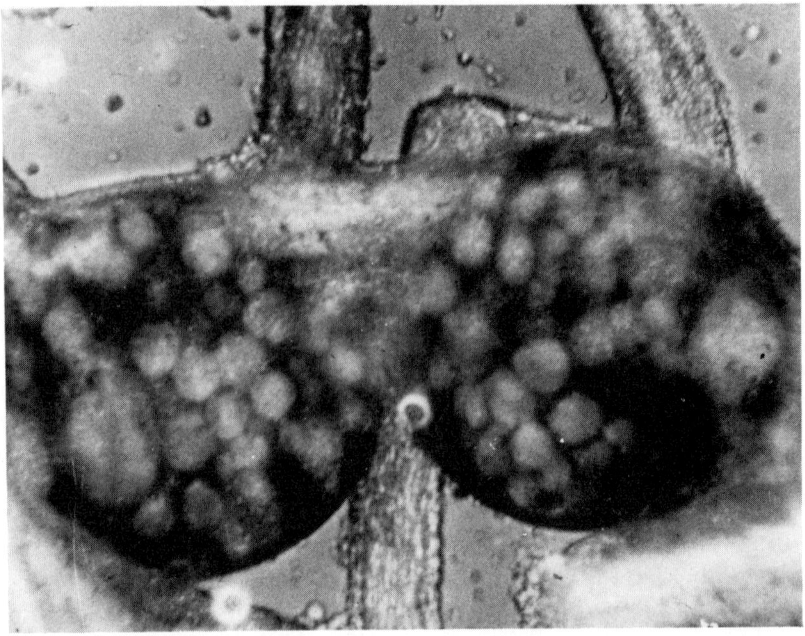

Figure 3. The paired buccal ganglia of *Helisoma trivolvis.* Contrast and clarity of the neural somata are greatly increased in living animals as a result of color differences not shown in black and white reproductions.

Figure 4. An extracellular recording from the tentacular nerve of *Helisoma trivolvis.* The arrow marks the time of application of .002 molar glutathione to the tentacle. (From M. Weis, unpublished). Calibration: 1 sec., 100 micro v.

Each of the muscles involved in feeding appears to be controlled by neurons located in the buccal ganglia. The phase relationships established for the feeding muscles must be generated by discrete interconnections of neural elements in the central nervous system. The next step in the analysis of feeding was the determination of the neuromuscular relationships. Complete understanding required a muscle-by-muscle analysis. A myogram electrode was placed on a particular muscle and successive intracellular penetrations in individual central neurons were made. Each neuron was activated to generate action potentials by passing depolarizing current across the cell membrane. In most cases there was no corresponding activity in the individual muscles being monitored. However, eventually a neuron was found whose action potentials were followed in a one-to-one correspondence by muscle potentials from a single muscle. The activity of two such neurons and their associated feeding muscles is shown in Figure 5. The validity of employing myograms for identification of motoneurons in *Helisoma* has been demonstrated (Kater, Heyer, and Hegmann, 1971). Not only is the myogram an adequate representation of the summated trans-membrane potentials of the individual muscle fibers, but it is also an effective predictor of muscle tension. Knowledge of neuro-muscular relationships is required for precise description of behavior at the physiological level. Our study suggests that the feeding and columellar muscles of *Helisoma* have many characteristics in common with vertebrate muscles. There is no indication of polyneuronal innervation of individual muscle fibers; rather these fibers are arranged in vertebrate-like motor units. We need not be concerned with peripheral integration of motor output such as occurs in the arthropods. The one-to-one transfer function between motoneurons and muscles allows us to focus our attention on the activity of these neurons without loss of information. These facts greatly facilitate the study of integration in *Helisoma*.

What interconnections exist among motoneurons to bring about the phase-locked activity of the muscles associated with the feeding apparatus? While we presently cannot answer this question completely, the clarity of the system has allowed determination of functional interconnections among 16 pairs of motoneurons. These are arranged in a catenulate inhibitory sequence such as that shown in Figure 5. The paired neural recordings in this figure show that action potentials generated in cell 27

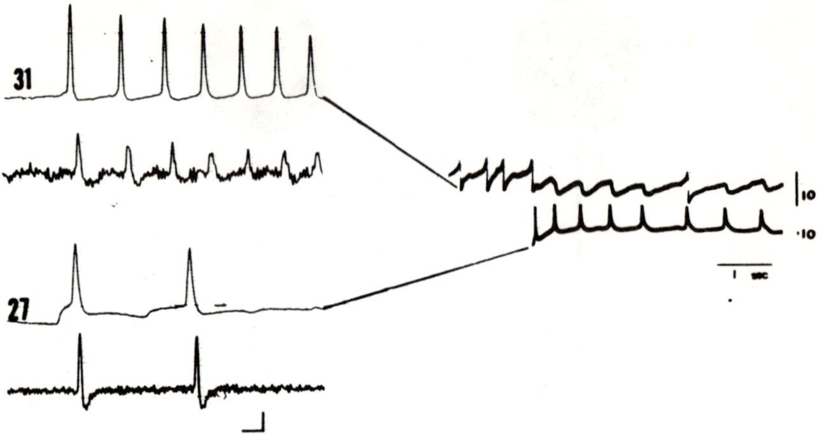


**Figure 5. Electrical recordings from identified nerve cells and muscles in *Helisoma trivolvis*: Left: Intracellular recordings from buccal ganglion motoneurons 31 and 27 with myograms from their associated feeding muscles (calibration: 50 msec, 50 mv for intracellular records). Right: Intracellular records from cells 31 and 27 (tops of action potentials generated by cell 31 are cut off by high gain recording). Cell 27 was inhibited by the injection of hyperpolarizing current. On release from inhibition, action potentials were generated in this cell. These were accompanied by IPSP's in Cell 31.**

(buccal ganglia) result in the generation of inhibitory postsynaptic potentials (IPSPs) in cell 31. The network established by data like that shown in Figure 5 is illustrated in Figure 6. Cell 31 is an excitatory motoneuron to a particular muscle involved in feeding and cell 27 is excitatory to a second such muscle. These muscles are antagonists. The two muscles cannot be active simultaneously because action potentials in motoneuron 27, which yield activity in its follower muscle, also inhibit motoneuron 31 and its follower muscle. The IPSP's observed in cell 31 are the result of a classical chemical inhibitory synapse. This was demonstrated by measurements of specific ionic conductance changes in the cell membrane and determination of reversal potentials. Inhibitory interactions such as those briefly described above play a key role in the organization of the specific phase relationships observed during feeding. Summation of IPSP's during feeding results in the production of sustained hyperpolarizations (Figure 7) in the serially linked motoneurons. These hyperpolariza-

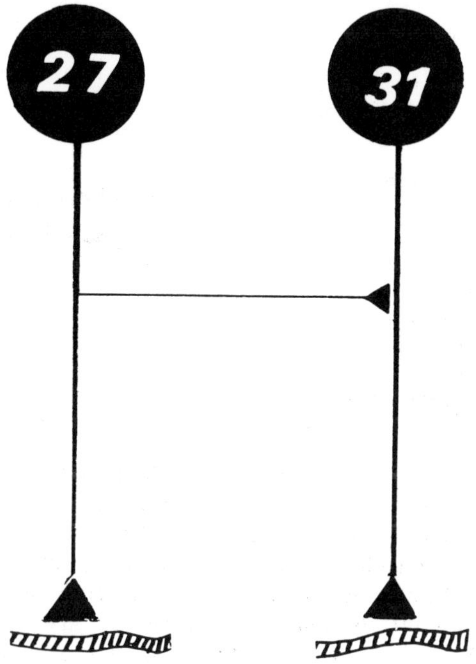


Figure 6. A schematic representation of the functional connections of buccal ganglion cells 27 and 31. Cells 27 and 31 have excitatory chemical synaptic effects on particular feeding muscles. In addition, a collateral of cell 27 forms an inhibitory synaptic contact with cell 31. This contact can be demonstrated by electrical recordings and by intracellular injection of the dye, procion yellow.

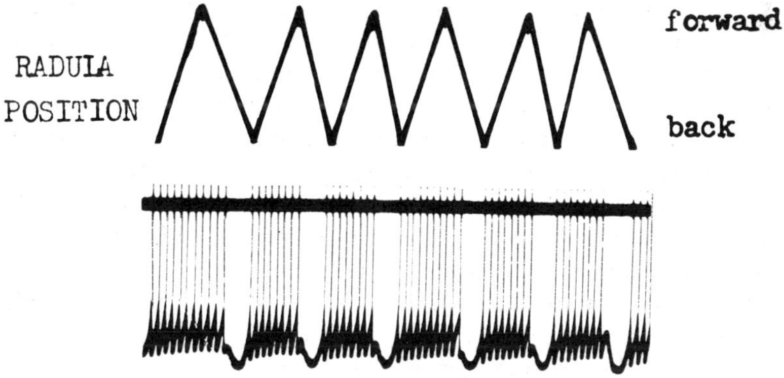


Figure 7. Intracellular record from an identified buccal ganglion motoneuron during feeding (lower trace), zero potential for that cell (middle trace), and a representation of the position of the radula (upper trace). Calibration: 50 mv, 2 sec.

tions are of great significance because action potentials in these motoneurons are generated out of post-inhibitory rebound.

This discussion briefly describes the level at which we can perform neurobehavioral analysis in the snail. Several other systems are currently being investigated by similar methods. The measurements employed are rapid, precise, and at the level of individual cellular components. These are among the major requirements for genetic analysis of fundamental physiological differences in nervous system function. Additional features of *Helisoma* suggest unique advantages of this animal for multidisciplinary investigation.

*Helisoma's* relatively short generation interval allows production of 10 generations per year. Our stocks are of diverse geographic origins and each displays considerable morphological, behavioral, and physiological variability. A most striking feature of this animal for genetic analysis is that it deposits egg masses, allowing offspring of many pairs to be reared together with strict environmental control. Furthermore, we have observed monozygotic twin embryos (Figure 8) in about a quarter of the egg

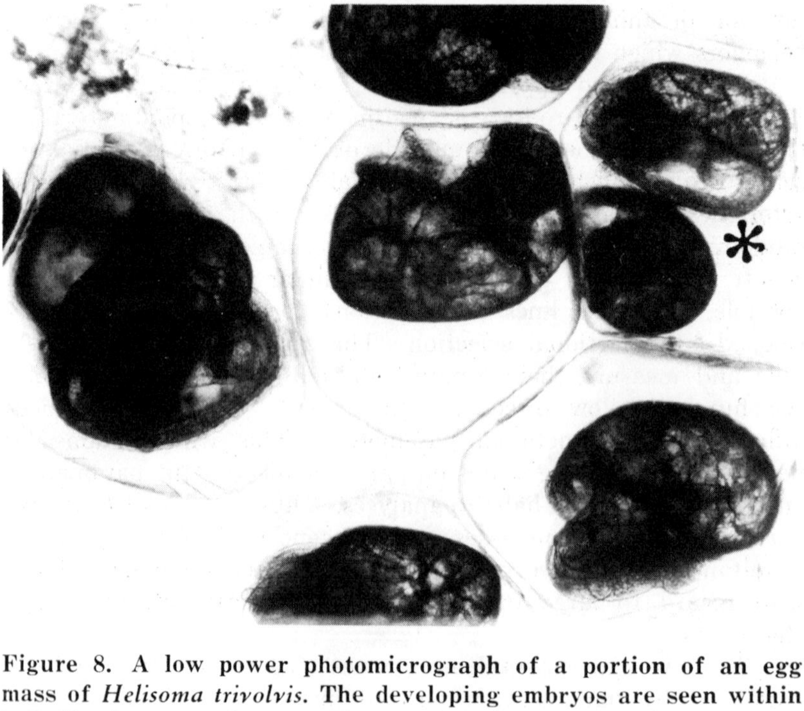

Figure 8. A low power photomicrograph of a portion of an egg mass of *Helisoma trivolvis*. The developing embryos are seen within individual egg membranes. The asterisk shows twin embryos.

masses examined. Since genetic analysis of phenotypes which display complex differences relies on determination of resemblance between relatives of known degrees of relationship, the production by *Helisoma* of neatly packaged groups of full sibs is ideal. The presence of identical twin pairs within full sib groups will facilitate the development of inbred lines and allow additional estimates of environmental variance components contributing to individual differences in measured behavioral and physiological phenotypes. These twin pairs also facilitate genetic analysis of characteristics displaying discrete differences by allowing analysis even when measurement of the trait requires sacrifice of the animal.

Two experimental approaches are being used to determine the relationship between neural, behavioral, and genetic differences in this animal. In the first, behavioral measures are being developed which are appropriate for genetic analysis and which involve neurons from which we can repeatedly obtain functional measures. Discrete electrical events from these neurons will then be compared from animals of known genotype with regard to the behavior. This approach will be employed to compare neural function of animals selected for extreme levels of expression of behaviors which show complex differences.

In the second approach we are measuring multiple parameters of specific neurons (e.g., resting potentials, spike activation threshold and slope and amplitude of postsynaptic potentials of cell 31) to establish the extent of variability of these cellular characteristics in defined groups of snails. Both inbred lines and randomly mating populations will be employed to assess the genetic components of variation in cellular function and, where possible, divergent lines for functional parameters will be developed by directional selection. The ability to relocate single cells and measure their function in individuals of known relationship will allow direct genetic analysis of gene differences influencing the functional parameters. This will be done by regression of offspring scores on parent values for the parameters, and from full-sib and half-sib analyses. Thus, the use of *Helisoma trivolvis* should permit assessment of gene differences that exert simultaneous influence on individual differences among animals with regard to large numbers of physiological and behavioral characters.

The two experimental approaches outlined for the snail are paralleled by our research on the mouse. By simultaneously

directing our attention to these two quite different species, a new level of analysis may be possible; that is, a comparison of the genetic modification of both neural function and behavior between a vertebrate and an invertebrate.

## Literature Cited

Dobzhansky, T. 1964. Heredity and the nature of man. Harcourt, Brace, and World, New York.

Hegmann, J.P. 1972. Physiological function and behavior genetics: I. Genetic variance for peripheral conduction velocity in mice. Behavior Genetics 2 (1):55-67.

Kater, S. B., C. Heyer, and J.P. Hegmann. 1971. Neuromuscular transmission in the gastropod mollusc *Helisoma trivolvis*: Identification of motoneurons. Z. Vergl. Physiol. 74:127-139.

## Discussion

*Dr. Peterson:* I think that having struggled just a little bit with these so-called Simple Systems, the thing perhaps that we all should respect Dr. Kater most for is having had the tenacity and the courage of trying to develop a system like this. It's all very easy to go out and take Mr. Weirsman's crayfish or Mr. Kandel's Aplysia where somebody else has done all the dirty work, but it really is a remarkable job to start with a whole new system and I congratulate you on that. That is a real trick.

*Dr. Kater:* Can I speak to that? It's just one of those little fortuitous things. I was hung up on cockroaches, but I developed an allergy to them. You know you get forced into a lot of good things.

*Dr. Peterson:* One thing that struck me right at the end — one general comment, then I have some other minor ones — the one that struck me was your comment on the mouse work, where you found variations in conduction velocity in this nerve and nothing much different in behavior. From what you've said, I think you would probably agree that the reason for this is that really the conduction velocity changes are probably quite trivial. There may be differences in fiber diameters and mixture of fiber diameters; these are really pretty trivial in the way that the beast is run, and the only reason I bring it up is . . .

*Dr. Kater:* Those differences represent . . . if similar differences, for instance, exist in the sciatic nerve, they would represent a differential of 20% over the reflex time that could occur.

*Dr. Peterson:* No, I'm not knocking . . .

*Dr. Kater:* Yes, I know. We are doing cross sections of these fibers, looking for fiber diameters; we are doing myelin content and this kind of obvious stuff and I don't think, and I don't think Joe really thinks, that's the way to go right now. That did a really big thing for us: it said it could be done. And that nerves will respond to genetic treatment. We expected this but we wanted to see it and we have to see that kind of thing.

*Dr. Peterson:* What it says more importantly, I think, is find an interesting behavior and then look for the genetics rather than finding something you can measure just because it happens to be handy. This is all I'm trying to say. Another specific question I had was: Why do you invoke post-inhibitory rebound when you've got a chronic level of excitation going on in the background?

*Dr. Kater:* That chronic level of excitation is insufficient to bring this about and I will do this very rapidly for those of you who might want to get into this. You do a current-voltage analysis by using the current-clamp technique, and record from many, many cells. We'll make this voltage on this axis and current on this axis. Usual cells look something like this. And at a very particular point which Hodgkin and Huxley have defined as the sodium activation voltage you get action potentials. When you do it to these cells, you can get this if you do it very slowly coming back up. But if you bring it back up with varied slopes from the hyperpolarization (this is negative), you get hysteresis, and get firing down there. I have a very bright student now who is really into this and I think for the first time in 20 years we'll make the first addendum, perhaps, onto the absolutely incredibly good Hodgkin-Huxley equations, and this addendum will be to put a vector into the system. I don't want to talk anymore about this. You not only have to invoke it, you can show it — that's how it does it, that's the only way to explain it.

*Dr. Peterson:* Also where you have these electrical synapses. do you have any reason to believe or reason not to believe that you might simultaneously from the same cells have a chemical synapse also?

*Dr. Kater:* Everywhere you have an electrical synapse, as best I can determine, these synapses are used to tighten things up — they are not drivers — you also have the same kinds of parallel inhibitory interactions from other neurons that go onto these and therefore produce the time at which these fire. You're right,

but within them I see no indication of chemical.

*Dr. Peterson:* I'm trying to think actually what it would look like.

*Dr. Kater:* You cannot fatigue this. You say that there may be an electrical component of that depolarization, and the latency of that depolarization is less than a tenth of a millisecond, which is insufficient to account for any known chemical synapse.

*Dr. Peterson:* What I was actually thinking of was some of the olfactory things where you get a mixed chemical in a synapse that has tight-junctions and synaptic vesicles . . . as though both things are going on.

*Dr. Kater:* Right, I have no indication of this.

*Dr. Peterson:* One other question: When you make those whole-mounts, of those beautiful things, what do you mount them in?

*Dr. Kater:* They're just in mineral oil, and then we take them out after we take the pictures; then we section them, then we make reconstruction models — put the thing together. They're esthetically good — they're not as good as your pretty movies — I really like those.

*Question:* I was curious about the glutathione stimulation of the feeding response. The orientation, I suppose, is initiated by the presence of glutathione. Does the radulla motion start at the same time or does it stay and wait?

*Dr. Kater:* No, it waits until the concentration is high enough and the firing is high enough in the radular nerves.

*Question:* It's all a matter of relative concentrations?

*Dr. Kater:* There are limitations in every bag of tricks you can look at, and I will tell you one of the major limitations here. Look at the locomotion. That foot is made up of a maze-like network of interconnections and crossing muscles, and to attempt to quantify that like we can the reflexive withdrawal in the shell, or mating, is very difficult. So I really haven't gotten into that too terribly much, but what happens apparently is, that . . . you record from neurons that you think and you can show are affecting the muscles in the solar plexus . . . I don't even understand really how the snail moves, to be very honest with you; it's a fairly complicated thing. You can show that at very low concentrations he does these really neat things. I'm going to become a snail. He weaves, then he'll pull one in, and he'll clean it — he'll unadapt it. And all this time he's been homing

in on this thing. He does a really nice job of homing, depending on the gradient and how it's established, of course. He'll start with swings pretty wide, show damped oscillations, and about at this point here, depending on the concentration, of course, he'll start feeding. And so it's not a contact phenomena — it's a true chemoreceptive thing — he feeds before he actually gets to the food.

*Dr. Best:* How big are the eggs and are they fertilized?

*Dr. Kater:* Internal fertilization.

*Dr. Best:* How far is the development by the time they are laid?

*Dr. Kater:* They are still one cell.

*Dr. Best:* How big is the egg? Just roughly?

*Dr. Kater:* 100 microns.

*Dr. Best:* Another one of the things I was wondering was whether one can do nuclear-transplantation on these?

*Dr. Kater:* I don't know. The only thing that we've done is some work on the development of the nervous system. It's development is crystal clear. What would be your direction? I'm just curious.

*Dr. Best:* The point is, you see, that once you have analyzed a particular nervous system and you really know how it is all hooked up, you could recognize one that was a deviant. Suppose you wanted to make an arbitrary number of identical twins of that?

*Dr. Kater:* You can do that by shaking at the two-cell stage. Shake the two cells and they fall apart and you have four of them. They do well that way, so you can produce many twins or quadruplets. Let me caution you for a minute. This is one very small step for snail-kind, and there is really a lot about this system that I don't understand. I spent two years with the animal and that is what I really concentrated on; I mean I related with the snail feeding, and I think I understand a lot of that, but I don't know a lot about satiation. I have some ideas about satiation but there are still some critical experiments to be done, to further nail down the kind of things I talked about. There's a long way to go and I don't think it's Brave New World yet.

*Dr. Best:* Did I understand you to say in our private conversation that . . .

*Dr. Kater:* We're not allowed to talk about private conversations.

*Dr. Best:* All right, I'll make it public. Have you trained them?

*Dr. Kater:* These snails can be classically conditioned. They will display plastic behavior.

*Dr. Best:* Have you tried anything that could be said to be instrumental conditioning?

*Dr. Kater:* No. It's very difficult to do an operant. I would like to try your paradigm.

*Dr. Best:* Do you have a dislike neuron in mind?

*Dr. Kater:* You see, what I want are paradigms that I can do in the dish, when his brain is opened up, because I want to be monitoring some nerve cells during those changes. Certain paradigms are beautiful and very easy to do, and we're concentrating on those.

*Dr. Best:* Then you suspect that there must be some neuron in there that's for disliking things; you know, whatever it may be that he dislikes?

*Dr. Peterson:* Would you make a guess that plastic changes — for example, aversion to eating — might occur in your command cell?

*Dr. Kater:* I don't think we should concentrate too much of our discussion on plastic changes. I haven't demonstrated it to you, but what I can do to add a little variability to this system is to say that either the command cell does this trick at this frequency or it doesn't. And if it doesn't, the animal doesn't feed. There is a very nice cell, which I named — in the terms of Sherrington — a sensory accumulator. It is a final common pathway for every mode of sensory perception that I have been able to define. This cell has input into and is associated with withdrawal. Withdrawal precludes feeding, obviously, and when withdrawal occurs, this cell fires like mad. That firing appears to be inhibitory to the command cell. I really can't talk about a dislike cell right now. The snails have avoidance behavior. You can record from a very particular receptor during avoidance and see that it fires like mad. I don't think that the snail knows he's disliking it. He's going to do the trick, perhaps, because it

is hard-wired in his brain to do so. I would like, someday, to be able to use these techniques with mammals, but I can't now.

*Dr. Kates:* Well, I have a couple of questions. I'll reserve the dumbest question of the symposium and ask this first. It relates to what you said about shaking the cells and getting two embryos. Since these invertebrates, I take it, don't feed while undergoing embryonic development, you should be getting smaller and smaller snails.

*Dr. Kater:* Right. They were smaller and they do get smaller. We're really just getting into the genetics of this thing. I mean, we've done some inbreeding and seven generations are in. We've got all the clean-up work to do, because we are doing electrophoresis patterns after each of those generations. How much of the variability can we show in terms of specific enzymes, just arbitrary enzymes? We seem to be tightening things up a little. For all I know, all these snails will die when they get sufficiently inbred, or they may become infertile. I really don't know yet.

*Dr. Kates:* I think it is a development system; it reminded me of music. In a sense, there's a type of orchestration to all this. Do you like music?

*Dr. Kater:* Somebody else used a music analogy earlier — who was it? My old major professor, in a moment of great philosophical bent, once said to me: "Life is like a Beethoven and there are all these variations around a central theme. What we really like about Beethoven is that recurrent central theme." I think Jay Best is pushing very hard at understanding if there really is this central theme. I will be upset if there isn't, but I won't be surprised if there isn't some sort of a really basic thread — unless that thread, of course, is DNA. We know that is there. But with respect to the system itself, I don't think I'll put any musical bias into it.

# Index

Ikeda (and Kaplan), 10